Evolutionary Processes and Metaphors

Evolutionary Processes and Metaphors

Edited by

Mae-Wan Ho
Department of Biology, the Open University, UK

and

Sidney W. Fox
Institute for Molecular and Cellular Evolution, University of Miami, USA

A Wiley–Interscience Publication

JOHN WILEY & SONS
Chichester · New York · Brisbane · Toronto · Singapore

Library of Congress Cataloging-in-Publication Data:

Evolutionary processes and metaphors.

'A Wiley–Interscience publication.'
1. Evolution. I. Ho, Mae-Wan. II. Fox, Sidney W.
QH371.P76 1988 575 87-25445

ISBN 0 471 91801 6

British Library Cataloguing in Publication Data:

Evolutionary processes and metaphors.
1. Human evolution
I. Ho, Mae-Wan II. Fox, Sidney W.
573.2'01 GN281

ISBN 0 471 91801 6

Typeset by Inforum Ltd, Portsmouth
Printed and bound in Great Britain by the Bath Press, Avon.

Contents

Contributors

PATRICK BATESON FRS Sub-Department of Animal Behaviour, University of Cambridge, High Street, Madingley, Cambridge CB3 8AA

ROBIN CRAW Entomology Division, DSIR, Private Bag, Auckland, New Zealand

CHRIS A. CULLIS Case Western Reserve University, Cleveland, Ohio 44106, USA

WALTER M. FITCH Department of Biological Sciences, University of Southern California, Los Angeles, CA 90089-1481, USA

SIDNEY W. FOX Institute for Molecular and Cellular Evolution, University of Miami, Coral Gables, Florida 33134, USA

BRIAN C. GOODWIN Developmental Dynamics Research Group, Open University, Walton Hall, Milton Keynes, MK7 6AA, UK

DEBORAH M. GORDON Centre for Mathematical Biology, Mathematics Institute, University of Oxford, 24–29 St. Giles', Oxford OX1 3LB, UK

RUSSELL D. GRAY Department of Zoology, University of Auckland, Private Bag, Auckland, New Zealand

JACK P. HAILMAN Department of Zoology, University of Wisconsin-Madison, Birge Hall, Madison, Wisconsin 53706, USA

MAE-WAN HO Developmental Dynamics Research Group, Open University, Walton Hall, Milton Keynes, MK7 6AA, UK

PHILIP KITCHER Department of Philosophy, University of California, San Diego, La Jolla, California 92093, USA

SUSAN OYAMA Dept. of Psychology, John Jay College of Criminal Justice, The City University of New York, 445 West 59th St, New York, NY10019, USA

RODERICK PAGE Department of Zoology, University of Auckland, Private Bag, Auckland, New Zealand

JEFFERY POLLARD MRC Group in Human Genetic Diseases, Department of Biochemistry, King's College, Kensington, London W8 7AH, UK

PETER T. SAUNDERS Mathematics Department, King's College, The Strand, London WC2R 2LS, UK

KIM UPPER Department of Physiological Chemistry, University of Wisconsin-Madison, Madison, Wisconsin 53706, USA

Preface

Human diversity is such that while some of us seem prone to see conceptual revolutions at every turn, others contemplate the ebb and flow of the tide of knowledge with equanimity, secure in their belief that conventional wisdom will endure after all is said and done. But no one should be complacent about recent developments in science and technology, which are in keeping with a new *Zeitgeist* that is making itself felt in all walks of life. To us, it marks the beginning of a new era of biology which follows in the wake of our emancipation from a mechanical world-view.

This world-view had lingered on within our consciousness long after it was displaced by general relativity at the cosmological scale and by quantum theory within the subatomic realm. In his *Science and the Modern World*, Alfred North Whitehead (1925) had already foretold the full implications of our weaning from the Newtonian framework. Today, more than 60 years later, we finally perceive everywhere the blossoming of a deep conceptual revolt against the old order.

The changes within science must be seen in the same context of the many palpable signs of crisis and transition both in western society and in the world at large. The interdependence between our scientific and social perceptions is indisputable; less obvious, perhaps, is the active shaping of sociocultural reality which follows from that. Thus, it is of some importance that as scientists, we should try to articulate the conceptual underpinnings of what may be a very crucial turning point in the history of human civilization; and in so doing, contribute to bringing about a desirable social end.

What are some of the most significant developments in science within the last two decades, and how do they affect our view of reality?

The enormous progress in high energy physics and the extension of quantum theory to explaining the origin of the universe (the new big bang) have indeed pushed back the frontiers of scientific rationality; but the paradox involved in the quantum act of observation remains. It invites us still to enquire whether the universe is not independent of our observation. No 'objective' reality of absolute space and time exists; instead, the human observer may be taking part in constructing physical reality itself.*

The study of cooperative phenomena in physics, chemistry and biology has led to the burgeoning of non-linear mathematics, contrasting fundamentally with the Newtonian world of linear cause and effect. Much attention is now focused on dynamic structures and dynamic stabilities of the sort exhibited by organisms. Form and transformation become intelligible, not as the push and pull of extraneous forces, but as the spontaneous and automatic outcome of physicochemical processes in space-time domains.

In biology itself, applications of recombinant DNA techniques within the past decade have led to the discovery of the 'fluid genome', finally laying to rest the static and mechanical conception of the organism and of heredity epitomized in the 'central dogma' of the 1950s and 1960s. In its place we see the organism in a space–time continuum of 'nature as process', wherein it takes part in shaping its own destiny.

When Darwin delivered biology from vitalism (and from religion) more than a century ago, he sent it straight into the grips of mechanical materialism. In so doing, living process became finally reduced to the endless shuffling of genes by blind 'selective forces'. The current opposition to Neo-Darwinism is above all a move towards the revitalization and humanization of biology. To achieve our goal, we must draw upon new advances, not only within biological and human sciences, but also in physics, chemistry and mathematics. Biological processes must ultimately be explained *as they occur in living organisms and their communities*, and not as poor imitations of mechanical or statistical models.

In keeping with this change of mood is a gradual shift in emphasis in the methods of investigation. Instead of destructive analysis of isolated parts of organisms or of molecules, technological innovations (such as nuclear magnetic resonance imaging, acoustic microscopy and SQUID magnetometry) are now aimed at the non-invasive/non-destructive study of whole organisms in the process of living and developing. This is indeed the new age of the organism. No longer will we have to stop nature dead in order to wrest dubious confessions from her. Instead, we can get to know her in our fellow creatures with sympathy and affection.

As a contribution towards the birth of the new biology, we have collected together a number of articles from evolutionists in all disciplines who have something to offer in the way of spelling out the real processes involved; in the critique of the old paradigm and in the development of new and appropriate methods and metaphors.

A number of the articles (Chapters 2, 3, 4, 7, 10, 11, 14 and 15) are revised versions of papers presented at a Conference on Evolutionary Concepts in

* J.A. Wheeler (1987) The paradoxical universe, lecture delivered at the IBM Conference on Science and Paradox.

Biology and Society, held in Hawk's Cay, Florida Keys, on 18–19 April 1986, which was supported by the Liberty Fund. Chapters 5, 6 and 13 were written afterwards by some of the participants at the Conference; while Chapters 8, 9 and 12 are invited papers which complete the collection. We are very grateful to the other contributors at the Conference who provided excellent comments and ideas without reserve: in particular, Colin Patterson, John Jungck, Martin Barker, Charles Metz, David Price, Neil McLeod and Denis Robinson. Last, but by no means least, we extend a very special thanks to Henry Stanford who chaired the meeting with great finesse and consummate skill. The Conference was a most successful and memorable occasion for all concerned.

MAE-WAN HO (*Milton Keynes*)
1987

Evolutionary Processes and Metaphors
Edited by M.-W. Ho and S.W. Fox

CHAPTER 1

Processes and metaphors in evolution

MAE-WAN HO AND SIDNEY W. FOX

Ten years ago, 'evolutionary processes' would have involved little more than the workings of natural selection under different circumstances, while 'evolutionary metaphors' would not have ranged too far afield of the 'struggle for existence'. Today, descriptions of evolutionary processes are diverse and pluralistic, and include dynamical group properties in social behaviour at one extreme and molecular mechanisms of genomic fluidity at the other. These have in turn inspired a whole new array of metaphors.

The change in attitude is due in no small measure to the vigorous and wide-ranging debate on evolutionary mechanisms that has gone on unabated over the past ten years when alternatives to Neo-Darwinism have been proposed in many disciplines (see, for example, Ho and Saunders, 1984). During the same period, the extension of Neo-Darwinian metaphors to human sociobiology has sparked off strong reactions both among those who ultimately believe in working within Neo-Darwinism (while acknowledging its many shortcomings), and among others who advocate a more radical shift away from the orthodoxy. Increasingly, it has become recognized that almost no scientific theory is a pure logical construct. It both takes root within a particular sociopolitical context *and* feeds back into it. Evolutionary metaphors have a particularly powerful influence on our humanistic perspectives which underpin all social and political action.

In this introductory chapter, we concentrate on discussing a number of themes that recur in many subsequent chapters. They also coincide with the major issues which appear to us to be the most relevant both in the light of contemporary knowledge and for the purpose of advancing the theory of evolution beyond orthodox Neo-Darwinism.

We begin with an enquiry into the origins of the natural selection metaphor, its impacts on our perception of life and on the subsequent development of biological science. The concept of randomness within Neo-Darwinism is contrasted with the determinism inherent in the processes generating form and variation at every level. The nature of heredity comes under re-examination, especially in the light of recent findings from recombinant DNA research. The result is that development and evolution may be seen to be much more intimately connected than previously envisaged. In fact, a predominant theme which runs through nearly all of this book is an emphasis on the integration of the organism with nature. This leads to a view of nature as a multilevel but interconnected process nested in space and time, contrasting with both reductionism and holism. Finally, we ask whether a new paradigm for evolution is emerging and what its distinguishing characteristics may be.

DARWINISM, NEO-DARWINISM AND THE NATURAL SELECTION METAPHOR

Origins

The full title of Darwin's (1859) epoch-making book was *The Origin of Species by Means of Natural Selection or The Preservation of Favoured Races in the Struggle for Life*. Of 'struggle for existence', Darwin wrote:

> I use this term in a large and metaphorical sense including dependence of one being on another, and including (which is more important) not only the life of the individual, but success in leaving progeny. Two canine animals, in a time of dearth, may be truly said to struggle with each other which shall get food and live. But a plant on the edge of a desert is said to struggle for life against the drought, though more properly it should be said to be dependent on the moisture. A plant which annually produces a thousand seeds, of which only one of an average comes to maturity, may be more truly said to struggle with the plants of the same and other kinds which already clothe the ground.
>
> A struggle for existence inevitably follows from the high rate at which all organic beings tend to increase. Hence as more individuals are produced than can possibly survive, there must in every case be a struggle for existence, either one individual with another of the same species or with the individuals of distinct species, or with the physical conditions. It is the doctrine of Malthus applied with manifold force to the whole animal and vegetable kingdoms; for in this case there can be no artificial increase of food, and no prudential restraint from marriage.

These passages make explicit Darwin's reliance on the metaphor of struggle of one against all, and all against nature in the workings of natural selection. They also reveal the origin of the metaphor in the then prevailing socioeconomic view among the ruling classes that suffering was inevitable for the impoverished masses of humanity.

If Darwin liberated the Victorian era from the domination of religion and superstition, he also delivered it to a nature painted 'red in tooth and claw'. Our continuing disharmony with nature derives ultimately from this unedifying image. At the same time, the emphasis on competition between individuals and the implied superiority of the 'favoured races' in the 'struggle for life' were most easily taken to be justification — on the basis of natural law — for the economic exploitation of the masses as for the colonization and oppression of 'inferior' races (Barzun, 1958). Ho (Chapter 7) argues that the metaphor both originates from, and in turn reinforces, a particular perception of life which emphasizes competition and strife at the expense of all other human qualities — a distortion of reality that carries obvious political dangers (Saunders, Chapter 14).

The primacy of teleology — Darwinian impacts in biological science

In the pursuit of biological knowledge itself, the impact of Darwinism was equally considerable. By insisting on explaining everything in terms of selective advantage, the followers of Darwin, far from having done away with teleology as is often claimed, in fact succeeded in reducing nearly all of biology to teleology. Cassirer (1950) wrote:

> Indeed, one can go even further and assert that no earlier biological theory ascribed quite so much significance to the idea of purpose, or advocated it so emphatically, since not only individual but absolutely *all* the phenomena of life are regarded from the standpoint of their survival value. All other questions retreat into the background before this one . . .

The search for natural order inherent in biological forms was the aim of rational taxonomy, a tradition originating with Goethe and elaborated by Cuvier, Geoffroy St Hilaire, von Baer and Richard Owen. Under the influence of Darwinism, it gave way to phylogenetic systematics, which has as its aim the tracing of genealogies guided not so much by a sense of order as by real or imaginary 'adaptive pathways' of transformation. The combination of Darwinism with Mendelian and molecular genetics in the 1950s and 1960s (the Neo-Darwinian synthesis) similarly undermined a great deal of the incentive for the causal analysis of development which Wilhelm Roux began at the turn of the present century. As genes were regarded the ultimate cause of development, the study of development collapsed largely into a search for genes. The late 1970s witnessed the birth of sociobiology which, in its worst excesses, completed the erosion of rational biology (Saunders, Chapter 14 and Kitcher, Chapter 15): the study of animal and human behaviour became little more than an exercise in 'adaptive story-telling' based on hypothetical genes controlling arbitrarily assignated behavioural traits. In the realm of biogeography too, adaptive stories concerning mythical 'centres of origin' were to pose a major stumbling block to a proper understanding of the world-wide distribution patterns of animals and plants (Craw and Page, Chapter 9).

Indeed, biological science as a whole suffered the severing of a longstanding relationship with the so-called 'exact sciences', physics, chemistry and mathematics, which goes back to Harvey, Leibniz, Descartes and Kant, among others. While one grand architect of the Neo-Darwinian synthesis may rejoice in the final emancipation of biology from physics and chemistry (Mayr, 1982), a great deal of current emphasis is placed on integrating biological explanations with contemporary physics, chemistry and mathematics. This is exemplified in the study of protobiotic evolution (Chapter 2), in the problem of how spatial patterns arise in development (Chapters 7 and 8), and in understanding the large-scale distribution of animals and plants (Chapter 9).

The sufficiency of natural selection

The explanatory power of survival value has inspired many Neo-Darwinists to claim that the natural selection of random mutations is necessary and sufficient to account for evolution. A representative list of such assertions is given by Ho, Saunders and Fox (1987). But when the same claim is made by the critics of Neo-Darwinism, it is invariably countered with a firm denial, and the retort that Neo-Darwinists have always admitted the importance of other factors such as 'developmental constraints', 'epistasis' and so on (which would seem to contradict the claim to sufficiency). In practice, however, those other factors are never seriously taken into account. We may quibble about definitions of Neo-Darwinism, but its practitioners do equate *evolutionary* explanations with those based on *natural selection*.

One area where the concept of natural selection has been elevated to that of an *a priori* doctrine, or axiom, is optimality research. Natural selection implies that the behaviour of animals has been selected to be optimally adapted. Hailman (Chapter 6) and Gray (Chapter 11) show that the methodology of optimality research is logically flawed in that it infers cause from effect and can lead to no new predictions. Hailman calls for the development of a rigorous methodology in which 'adaptations' can be recognized independently in terms of measurable variables such as the correlations between character and environment among unrelated species. Gray sees the way ahead in the integration of development with ecology and panbiogeography. Only then would it be possible to overcome the arbitrariness involved in defining hypothetical optima based on the false and unspoken assumption that the organism's behaviour is mechanically fixed in relation to equally static environmental contexts. In reality, an organism's behaviour is strongly influenced by experience; the environment and the organism engage in a continual reciprocation — see Bateson (Chapter 10), Gordon (Chapter 12) and Oyama (Chapter 13) — which can either stabilize an existing regime, or lead to rapid change (see also Ho, Chapter 7).

The use and abuse of natural selection

Natural selection has a very precise meaning within population genetics theory.

It is the differential reproductive success among individuals *because* they possess different alleles. Thus, no selection can be said to occur if individuals do not differ genetically; similarly, deaths incurred which do not depend on the possession of different alleles are non-selective. Kitcher (Chapter 15) points out how difficult it is to make any inference about natural selection in human evolution. In particular, the effects of cultural transmission can mimic those of natural selection. The problem is a deeper one, however.

The relationship between organism and environment forms a seamless whole in development (Ho, Chapter 7): the physiological processes generating variations are themselves set in motion by so-called 'selective forces' in the environment so that it is seldom possible to tell where variations end and selection begins. In particular, it has now been discovered that some of these physiological processes may include directed changes in germline DNA (see Pollard, Chapter 5 and Cullis, Chapter 4). There is thus no rigid separation between a somatic phenotype that varies and a germline genotype that is subject to selection; consequently, the distinction between variation and natural selection lacks any firm ontological basis. In most cases the concept of natural selection is simply misused. For these reasons and because of its ideological nature, Ho proposes to dispense with the concept altogether (also Goodwin, Chapter 8).

Oyama (Chapter 13), writing from a similar perspective of the intimacy and multidimensionality of organism–environment relationship, advocates that the concept of natural selection should remain; but that it should apply to the entire conglomerate of relationships between organism and environment. Thus, physiological and cultural processes would be considered part of natural selection at different levels. In this she also differs from Kitcher (Chapter 15) who retains the conceptual distinction between selection and other factors such as cultural inheritance, psychological states, learning and so on, which affect phenotypic expression of characters controlled by genes subject to selection.

There is another important issue at stake, which is that the assumption of the primacy of natural selection can cloud our thinking and obstruct scientific investigations. A case in point is how it has delayed the general acceptance and application of Croizat's method of tracking the movements of continental landmasses *from* the distributions of animals and plants (Craw and Page, Chapter 9). For decades, biogeographers had attempted to 'explain' faunal and floral distributions by *ad hoc* hypotheses concerning means of dispersal and 'adaptive' radiations.

Saunders (Chapter 14) remarks that Neo-Darwinism fails the test of falsifiability as a scientific theory on two counts: first, the various tests of falsification proposed by Neo-Darwinists are fallacious; second, findings which appear at face value to be falsifications by their own criteria have not led them to abandon the theory. While Saunders does not regard falsifiability as the ultimate criterion of scientific validity, he does question the claim, often made by defenders, that Neo-Darwinism has the same status as theories such as Newtonian mechanics and general relativity, which *do* pass the test of falsifiability.

RANDOMNESS VERSUS DETERMINISM

Randomness, unpredictability and disorder

The emphasis on natural selection within Neo-Darwinism rests largely on the assumption of the 'randomness' of variations on which natural selection is supposed to act. It implies that nothing much could be said about the variations — a sentiment not shared by Darwin (1868) himself who went to great lengths to document the existence of variations in animals and plants under domestication. Bateson (1913), in particular, realized that in order to understand the evolution of biological forms, it was necessary to discover the 'laws' governing the generation of variations. Thus, it was only since the rise of molecular biology in the 1950s and 1960s that the concept of random variations became epitomized in the stochastic and essentially unpredictable nature of genetic mutations involving base changes in nuclear DNA.

The modern concept of randomness derives ultimately from classical equilibrium thermodynamics, which perpetrated the notion that matter left to its own devices will surely degenerate into maximum entropy or disorder. A disorderly universe is a random universe requiring either God or natural selection to put right.[1] Many theorists, therefore, premise their hypotheses concerning the origin of life on a random matrix (see Fox, 1984a), out of which natural selection creates order in effecting a number of 'frozen accidents' (e.g. Crick, 1981).

Molecular selection versus natural selection in protobiotic evolution

In reality, many observations point to the opposite of randomness in what Fox (Chapter 2) refers to as 'molecular selection'. This is evident in the self-ordering which amino acids themselves impose on their own polymerization, by virtue of their stereoelectronic nature. Relics of such a primordial self-ordering process have recently been uncovered by Ivanov and Fortsch (1986): they are preserved in the amino acid sequences of practically all present-day proteins. Moreover, the self-ordered primordial polymers already possess a variety of catalytic and other protocellular functions including electrochemical activities. Non-randomness in the context of protobiotic evolution implies a certain bias within the laws of physics and chemistry which predispose matter towards the evolution of life processes. In other words, there is a general 'fitness' of the environment (Henderson, 1913) for the emergence and evolution of life.[2]

The genetic code and the encoding of vital functions

From a gene-centred perspective, Neo-Darwinists invariably see the problem of the origin of life to be synonymous with the origin of nucleic acids and the genetic coding mechanisms (see Wilkins, 1986, for example). But life entails much more than genetic mechanisms. Indeed, genetic coding mechanisms would be useless were there not vital functions, such as irritability and metabolism, to be encoded. Where Neo-Darwinism sees those functions to be almost exclusively the consequence of the natural selection of fortuitous variations, the new paradigm locates them in the properties of matter and the dynamics of process. The genetic code serves to encode functions that pre-exist in something like thermal proteins and the microspheres which form spontaneously from them.

The question arises as to whether the genetic code itself is arbitrary. Fox answers that it is not; for there is evidence of stereochemical complementarity between amino acids and their respective codons/anticodons as well as between pairs of amino acids. In fact, he suggests that reverse information flow must have occurred in an early phase of protobiotic evolution (see also for example, Root-Bernstein, 1983). Such reverse translation would guarantee the encoding of pre-existing functions, leaving little to the vagaries of chance for the emergence of life. This should be a subject for further experimental investigations.

Fitch (Chapter 3) chooses, however, to start from a state of total ambiguity in which any triplet will code for any amino acid. Evolution of the genetic code is then assumed to involve a progressive reduction of ambiguity by means of natural selection. He reasons that it should be possible to reconstruct the historical sequence from the phylogenies of the transfer RNAs: neighbouring codons, or codons differing by only one base, will have tRNAs which are most homologous, whereas non-neighbouring codons will have more divergent tRNAs. The data so far seem consistent with this hypothesis. However, they are also consistent with the operation of physicochemical complementarity of the sort mentioned above. It may be that Fitch's method will be able to distinguish physical from historical or contingent elements in the genetic code when more data become available on the base sequences of tRNAs and on the stereochemical interactions between codon/anticodons and amino acids.

Determinism and the structure of the probability space

Non-randomness implies a local determinism, but not a mechanical or Laplacean determinism. In other words, it does not lead to finality, or a unique end state. Chance and accident will still be expected to play a role, else there would not have been such a diversity of proteins, organisms, social organizations and cultures. The problem with Neo-Darwinism is not that it places exclusive emphasis on chance or accident in the generation of variations, but that it so often ignores (or does not take into explicit account) any possible *structure* of the *probability space* involved, whether in the emergence of life or in the evolution of organisms and social organizations (see below). Fox claims his experiments show that the probability space for protolife is highly structured. This structure extends from the self-ordering of amino acids to the stereocomplementarity between amino acid and codon triplets, which forms the basis of a reverse translation process in protolife, and finally to the spontaneous generation of protocells when protenoids come into contact with water. Fitch, on the other hand, sees natural selection as the agent which, by reducing codon ambiguity, *constructs* the code itself.

Compared to protobiotic proteins, present-day proteins are, by conventional measures, highly random in amino acid sequence. (The findings of Ivanov and Fortsch, 1986, seem to indicate, however, that much of the non-randomness resulting from the primitive self-ordered polymerization process has in fact been preserved in present-day proteins.) Fox suggests that the 'randomization' introduced by DNA tends to widen the pathway of evolution. In other words, the DNA coding mechanism enables further explorations of the probability space of proteins, *having started from privileged positions within the subset of functional catalytic species*. This could be very important; for the space of total proteins is so large that if polymerization really started at random, it may never arrive at the subset of functional proteins at all. Of course, the crucial question is whether the subset of functional proteins is so very small relative to the total space. It would be extremely interesting to be able to investigate the probability that a randomly generated polypeptide (say, from the translation of random mRNA sequences) has catalytic activity.

Non-randomness of variations

There are two senses of the word 'random' as applied to variations within Neo-Darwinism. The first is the strong sense of the unpredictable. The second is the weak sense in that the particular variations are not correlated with the selective forces in the environment. Recent findings from recombinant DNA research have demonstrated the violation of randomness in both those senses within molecular genetics (Cullis, Chapter 4). These findings present a major challenge to the basic framework of Mendelian and molecular genetics of the 1950s and 1960s (Pollard, Chapter 5; Ho, Chapter 7). Indeed, variations may be generated repeatedly at every level in the organism — including nuclear DNA — directly in response to the environment.

Dynamical structures, forms, transformations and fields

At the morphological level, the same transformations can occur as the result of either genetic mutations or environmental perturbations. This is a clear reflection of a dynamical structure which underlies the generation of form (Goodwin, Chapter 8; also Ho, Chapter 7). Biological forms and their variations should be understood, therefore, in terms of transformations from dynamical structures. Similar arguments may be made for the behaviour of individuals and groups of individuals. Gordon (Chapter 12) presents evidence for the existence of patterns of transformations in the group behaviour of harvester ant colonies.

Goodwin elaborates on the concept of field and its harmonic solutions as the space–time domain of morphological transformations. The concept may be exemplified in a concrete way as a viscoelastic field in which ion fluxes and mechanical deformations give rise to regular patterns and forms. Kortmulder (1983) has recently suggested that ritualized behaviour arises essentially out of play activity. The latter can only be understood in terms of a 'field' extending 'between as well as within the interacting individuals'. Such 'fields' involve the coordinated activities of many individuals. These, too, can be concretized as fields of perception extending between individuals which are in continuity with physiological and psychological processes occurring within individuals (Ho, Chapter 7). The field, as it were, crystallizes out of the activities of many individuals whose numbers and identities may vary within limits without affecting the dynamical characteristics of the whole.

More generally, one might enquire whether dynamical structures underly the evolution of human societies given the processes of cultural and biological transmissions, psychological and physiological developments and the interrelationships among them which make up the rich fabric of human existence. While Kitcher (Chapter 15) takes cognizance of these processes, his chief aim is that they should be included (as parameter?) in a realistic theory of natural selection in order to account for human social and cultural evolution.

But this may be a case of misplaced concreteness, if it should turn out that the dynamics of the processes themselves can account for much of human evolution without the aid of natural selection, or indeed, in spite of natural selection (Ho, Chapter 7). The issue is akin to the lack of a null hypothesis concerning the patterns of observable behaviour against which the hypothesis of optimal behaviour could be tested (Hailman, Chapter 6). The so-called null hypothesis there is none other than *the pattern which would result from the dynamics of processes operating independently of natural selection*. In population genetics, the neutral mutation theory (Kimura, 1983) is the paradigmatic example of such a null hypothesis. It concerns the behaviour of genetic variations in populations in the *absence* of natural selection. Similarly, the articulation of a dynamical theory of sociocultural evolution should start from the *known* processes of cultural, biological transmission, and of psychological, physiological development; without invoking in the first instance, the unknown and unknowable effects of past selection.

HEREDITY, DEVELOPMENT AND EVOLUTION

Weismann's barrier, genotype and phenotype

The basic phenomenon of heredity is that organisms reproduce true to type generation after generation despite variations in detail. The reproduction of organisms naturally includes the process of development. So much so that in 1910, T.H. Morgan considered the problem of heredity to be *identical* to that of development (Allen, 1983). This sentiment was shared by all the leading developmental biologists and evolutionists of the time. Yet, a few years later, as a convert to Mendelian genetics mainly through his own work in chromosomal inheritance, Morgan was to insist on separation between the *transmission* of hereditary 'information' (heredity) contained in the genotype, and the *translation* of this information into phenotype (development). This is a major tenet of Neo-Darwinism which is in turn used to justify the separation of evolution from development (Maynard-Smith and Holliday, 1979).

Actually, the separation of heredity from development began with Weismann's theory of the independence of the germplasm, which split the organism into a vegetative part, the soma, and a reproductive part, the germline. The soma, as the disposable carrier of the germline, was responsible for all vegetative functions and hence subject to variations resulting from environmental exigencies. The germline, on the other hand, was effectively immortal, as it was insulated from the environment, and was passed on unchanged from generation to generation. Functionally, the separation of germline from soma depended on the so-called Weismann's barrier, whose ontological status is ill-defined: embryological studies provided no evidence for the segregation of germ cells from the rest of the body in plants and in well over half of the animal phyla.

In the Watson–Crick era, Weismann's doctrine became restated as the 'central dogma', according to which genetic information flows from DNA to RNA to protein, but never in reverse. Heredity was considered to reside exclusively in nuclear DNA, which is faithfully copied and transmitted from parent to offspring, unchanged except for rare mutations. Thus, it was deemed to account for the general resemblance of offspring to parents as well as for occasional deviations. Associated with the central dogma was a set of basic assumptions of genetics which almost everyone accepted as true. All those assumptions have proved untenable to varying degrees especially in the light of recent findings from recombinant DNA research (Chapters 4, 5 and 7). Genomic DNA, once thought to be a static and unchanging 'information store', turns out to be extremely fluid. Amplifications, deletions, rearrangements and mutations occur frequently during development and in response to environmental stimuli. In addition, large-scale 'reverse information flow' is effected by the reverse transcription of processed RNA back to germline

DNA. It seems that present-day organisms too, may conform to the principle of protobiotic evolution enunciated by Shapiro (1986): 'protein begat RNA, and then both begat DNA'.

The reformulation of heredity as process

Ho (Chapter 7) suggests that heredity should be reformulated as a process involving multilevel feedback interrelationship between organism and environment. Heredity thus depends on the dynamic interconnections between genotype and phenotype (or soma and germline); and not on their rigid separations as previously conceived. Moreover, it depends on the fluidity of process and not on the constancy of nuclear DNA (see also Ho, 1987).

In a sense, a case for reformulating heredity has always existed (see also Oyama, Chapter 13). There is nothing in the replication of DNA or the transcription/translation mechanisms which tells us how organisms are reproduced. But so long as the strict separation between soma and germline, and between phenotype and genotype, held sway, development was considered apart from genetics, and it seemed a simple article of faith that once we understand how genes control morphogenesis, we shall see how organisms are put together. This is the predominant guiding principle behind much current work in 'developmental genetics' aimed at unravelling the 'genetic programme' of development. The successes within the past few years in locating specific gene products during early embryogenesis have considerably enriched the observations of development, but by themselves do not offer any explanation of how the spatial organization of the gene products can come about. Goodwin (Chapter 8) demonstrates the insufficiency of genes as determinants of morphogenesis by analysing their detailed transcription patterns in *Drosophila* embryos.

The equivalent action of genes and environment in the generation of pattern, as mentioned earlier, reflects the dynamical properties of a developmental system of which genes and environment are part. It is those dynamical properties that are responsible for heredity: the stable generation and reproduction, not only of patterns and forms, but of particular life histories.

The organism is an integrated whole, genotype with phenotype, and soma with germline. This whole is in turn part of a nexus of ecological and sociocultural relationships which extends from the past and into future generations. Development and evolution are therefore intimately connected by processes nested in space and time; they are not separate as conceived within Neo-Darwinism.

LEVELS AND PROCESSES

The present collection of articles encompass evolution from protobiotic molecules to human societies. To varying degrees, all authors are positively committed to a view which we shall refer to as the 'unity of nature' (see Ho, Chapter 7): nature is continuous from the inorganic to the organic and cultural domains.

This unity or continuity neither legitimizes rampant reduction of all phenomena to the 'lowest level' of the molecules, nor does it entail a nebulous holism where everything is connected to everything else. Levels should be put together intelligently in the sense of working out what their interconnections are.

The strongest objection to reductionism is that there is no direct rule of translation from one level to the next 'down' or 'up'. Thus, changes in genes do not parallel changes in morphological and life histories characteristics of the organisms (see Ho and Saunders, 1984; also Fitch, 1985). At the genic level changes have been predominantly neutral, that is, they are neither selected for nor against; and occur at approximately constant rates in the course of evolution (Kimura, 1983). Morphological evolution, on the other hand, occurs at a much more uneven tempo; long periods of stasis are punctuated by relatively rapid bursts of major changes (Eldredge and Gould, 1972). Recently, it appears that saltatory changes in copy numbers in non-genic DNA sequences can occur in response to environmental perturbations, and that these may be correlated with morphological effects. Such saltatory events could be associated with rapid speciation (Cullis, Chapter 4; Pollard, Chapter 5). None the less, there is no obligatory relationship between genomic and morphological change. This is not to say that genetic mutations cannot result in morphological or physiological change, but that the mutations *per se* are not sufficient explanations of the change. Instead they must be considered in the context of the developmental or epigenetic system as a whole.

Just as organisms cannot be reduced to a sum of genes, societies are not reducible to a sum of individuals (Kitcher, Chapter 15). This applies equally to animal societies; for example, ant colonies exhibit repeatable patterns of group behaviour which are not apparent at the individual level (Gordon, Chapter 12). Thus, there are level-specific regularities that should be investigated in their own right and, more importantly, these regularities constitute parameters which can determine the behaviour of the constituent parts. The global fragmentation and fusion of continental landmasses in the evolution of the earth's crust determined the large-scale patterns of animal and plant diversity (Craw and Page, Chapter 9). So much so that the latter can be used to infer the movements of continental landmasses themselves in geological time. Gray (Chapter 11) sees the biogeographical context as a main factor which must be considered in animal behaviour. Ho (Chapter 7) and Cullis (Chapter 4) complete the line of determinative influence from the environment to behaviour, physiology and biochemistry, right 'down' to the genes or nuclear DNA.

It seems clear that the directions of causation are both 'upwards' from the genes to the environment and also in reverse: from the environment to genomic DNA. But even this is an oversimplification. First and foremost, it leaves out the organism as an active agent (see Chapters 7, 10 and 13), not to mention the psychology of the organisms involved (Chapter 15). Second, it is far from the case that levels are ordered as neatly and separately as conceptualized. In reality, there is one continuous process nested in space and time. It may be legitimate to ask separate level-specific questions, or to ask separate questions about proximate causation, development, function and evolutionary history (Bateson, Chapter 10). But those aspects are not ontologically separate. And if we wish to understand evolution, we ignore the other aspects at our own peril (Ho, Chapter 7; Gray, Chapter 11).

IS A NEW PARADIGM EMERGING AND IS IT NECESSARY?

There are divergent views on this matter. A number of authors to varying extent acknowledge the existence of 'uncomfortable gaps' in the old Neo-Darwinian paradigm, but believe that they can be suitably amended to give an evolved or transformed Neo-Darwinism (e.g. Kitcher, Bateson, Oyama, Gray). Others (Ho, Goodwin, Fox, Saunders) see the real emergence of a coherent framework of ideas for informing a research programme; in other words, a new paradigm, according to Kuhn's (1962) own definition.

The fundamental assumption of the new paradigm is that form and variation are neither random nor arbitrary, and that much of evolution can be understood, not in terms of the maximization of fitness, but through the processes which generate form and variation at every level before natural selection could be said to act (see above). This is most clearly exemplified in protobiotic evolution (Chapter 2), molecular genetics and development (Chapter 4, 5, 7 and 8), where the new approach leads to ideas and practices very different from those based purely on naturally selection.

Another characteristic of the new paradigm — also enthusiastically endorsed by 'transformed' Neo-Darwinists — is an emphasis on integration, which transcends all discipline boundaries; especially that between biology on the one hand, and physics, chemistry and mathematics on the other. In protobiotic evolution, physics and chemistry are responsible for molecular selection (Chapter 2) and for the fitness of the environment for life. In the generation of biological forms, physics and chemistry provide organizational principles which can often be expressed precisely in mathematical terms (Chapters 7 and 8; see also Saunders, 1984). These organizational principles coordinate the detailed biological mechanisms which will include viscoelastic changes of the cytoskeleton, reaction and diffusion of molecules, as well as the expression of different genes.

Those who see a new paradigm emerging also regard it as necessary. To them, 'the way forward is to recast the problem of evolution in a new light and to discover (and rediscover) the relevance of both new and neglected lines of research. Only then would we be able to see nature as she really is and not through a neo-Darwinian glass darkly' (Ho, Saunders and Fox, 1987).

SOME NEW METAPHORS

If a new paradigm is emerging, what are some of its new metaphors? Fox's 'self-organization' and 'constructionism' (Chapter 2) in the emergence of life are directly opposed to Neo-Darwinian natural selection and the *re*construction of evolutionary history from present-day genetic coding mechanisms. In a similar vein, Gray (Chapter 11) cites with approval, Oyama's 'reciprocally constrained construction' (Chapter 13) in the development of behaviour in relation to the animal's ecological and sociocultural environment.

In harmony with the cosmology of the day, Darwin's guiding metaphor for biogeography was absolute space and time, while that of Croizat is *relative* space–time (Craw and Page, Chapter 9). The result is an emphasis on the integration of the evolution of life with that of the earth.

Goodwin's field concept (Chapter 8) in development is reminiscent of various physical field theories but with emphasis on dynamic transformation. In contrast with the particulate conception of the organism which the gene theory inspires, it offers an understanding of form as organized space–time domains. It would be of interest to explore the implications of a generalized field concept applied to the study of social behaviour. The latter would then be understood in terms of group dynamics (Gordon, Chapter 12) rather than selective advantage or cost and benefit to individuals.

Ho (Chapter 7) advocates a Whiteheadian 'process' view in which organisms are seen not as the consequence of natural selection of past random mutations but as dynamic structures which are immanent and simultaneous with process. The organism, as well as the human observer as organism, are firmly located within nature where they are enpowered to shape their own evolution and destiny (see also Bateson, Chapter 10 and Oyama, Chapter 13). It may be noted that in quantum physics, the observer is also considered to be within the theory in a very fundamental way. The observer, in the act of observing, determines the outcome of particular phenomena; and may therefore be said to take part in the construction of physical reality itself (Wheeler, 1987). Just as the constructed physical reality and quantum theory are interdependent; so too, the constructed sociobiological reality and evolutionary theory are mutually reinforcing. This is the real basis of the creative power inherent in process, which has hitherto remained untapped within a conceptual framework that completely negates it.

It is nearly 70 years since Whitehead (1920) wrote, 'There is no holding nature still and looking at it'. It was very apt then, and all the more so today. Living organization is dynamic and fluid right down to the organism's genome. Surely, there is a lesson here for us: unless we too are intellectually supple, we shall never really come to grips with nature.

All nature dances with process and creativity. We can do no more than dance along with her.

NOTES

1. The assumption of randomness is fallacious in another, more subtle respect. An initial symmetry-breaking to a non-equilibrium situation was involved in the origin of the universe. It is on account of this that life could have evolved as dissipative structures (Prigogine, 1980). Thus, Prigogine's own slogan of 'order out of chaos' is, strictly speaking, quite inaccurate.
2. Actually, the universe itself is biased in the non-conservation of parity. This may in turn account for biomolecular handedness: the fact that living organisms contain almost exclusively L-amino acids, whereas chemically synthesized amino acids are invariably mixtures of D- and L- forms (see Mason, 1984). Indeed, quantum physicists now regard the very origin of the universe to be governed by the laws of physics (Davies, 1987). Evolution therefore, proceeds as 'order out of order' (Fox, 1984a), and not as 'order out of chaos'. Some may even go so far as to claim that the origin of mind precedes, or is simultaneous with, the origin of living systems (see Wald, 1984; Löwdin, 1984; Fox, 1984b).

REFERENCES

Allen, G.E. (1983). T.H. Morgan and the influence of mechanistic materialism on the development of the gene concept 1910–1940, *Amer.Zool.*, **23**, 829–43.
Barzun, J. (1958). *Darwin, Marx, Wagner*, Doubleday Anchor, New York.
Bateson, W. (1913). *Problems of Genetics*, Yale University Press, New Haven.
Cassirer, E. (1950). *The Problem of Knowledge*, Yale University Press, New Haven.
Crick, F. (1981). *Life Itself: Its Origin and Nature*, Simon and Schuster, New York.
Darwin, C. (1859). *Origin of Species*, John Murray, London.
Darwin, C. (1868). *Variations of Animals and Plants Under Domestication*, John Murray, London.
Davies, P. (1987). *The Cosmic Blueprint*, Simon and Schuster, New York.
Eldredge, N., and Gould. S.J. (1972). Punctuated equilibria: an alternative to phylogenetic gradualism. In T.J.M. Schopf (ed.), *Models in Paleobiology*, Freeman, San Francisco, pp. 82–115.
Fitch, W.M. (1985). Rapid evolution in inbred strains of mice. In *Third Int. Conf. Sys. Evol. Biol. Abs.*, Brighton.
Fox, S.W. (1984a). Proteinoid experiments and evolutionary theory. In M.W. Ho and P.T. Saunders (eds), *Beyond Neo-Darwinism*, Academic Press, London, pp. 15–60.
Fox, S.W. (1984b). Molecular selection in the roots of evolved life and mind, *Int.J. Quant. Chem.: Quantum Biology Symposium*, **11**, 17–29.
Henderson, L.J. (1913). *The Fitness of the Environment*, MacMillan, New York.

Ho, M.W. (1987). Genetic fitness and natural selection: myth or metaphor? In *Proc. 3rd T.C. Schnierla Conferences*, Lawrence Erlbaum Assoc., NJ.

Ho, M.W., and Saunders, P.T. (eds) (1984). *Beyond Neo-Darwinism: Introduction to the New Evolutionary Paradigm*, Academic Press, London.

Ho, M.W., Saunders, P.T., and Fox, S.W. (1986). A new paradigm for evolution, *New Scientist* (27 Feb.), 4–43.

Ho, M.W., Saunders, P.T., and Fox, S.W. (1987). Through a neo-Darwinian glass darkly, *BioEssays*, **6**, 3–4.

Ivanov, O.C., and Fortsch, B. (1986). Universal regularities in protein primary structure: Preference in bonding and periodicity, *Origins Life*, **17**, 35–49.

Kimura, M. (1983). The neutral theory of molecular evolution. In M. Nei and R.K. Koehn (eds), *Evolution of Genes and Proteins*, Sinauer Associates, Mass. pp.213–33.

Kortmulder, K. (1983). Play-like behaviour: an essay in speculative ethology, *Acta Biotheoretica*, **32**, 145–66.

Kuhn, T.S. (1962). *The Structure of Scientific Revolutions*, 2nd edn. University of Chicago Press, Chicago.

Löwdin, P.O. (1984). Some aspects of the relationship between natural sciences and the human mind: an introduction to a panel discussion on the 'Origin of Life and Mind', *Int. J. Quant Chem.: Quantum Biology Symposium*, **11**, 31–43.

Mason, S.F. (1984). Origins of biomolecular handedness, *Nature*, **311**, 19–23.

Maynard-Smith, J., and Holliday, R. (1979). Preface. In J. Maynard-Smith and R. Holliday (eds), *The Evolution of Adaptation by Natural Selections*, The Royal Society, London, pp.v-vii.

Mayr, E. (1982). *The Growth of Biological Thought*, Harvard University Press, Cambridge.

Prigogine, I. (1980). *From Being to Becoming*, Freeman, San Francisco.

Root-Bernstein, R.S. (1983). Protein replication by amino acid pairing, *J. Theor. Biol.*, **110**, 99–106.

Saunders, P.T. (1984). Development and evolution. In M.W. Ho and P.T. Saunders (eds), *Beyond Neo-Darwinism*, Academic Press, London, pp.243–63.

Shapiro, R. (1986). *Origins*, Summit Books, New York.

Wald, G. (1984). Life and mind in the Universe, *Int. J. Quant Chem.: Quantum Biology Symposium*, **11**, 1–15.

Wheeler, J.A. (1987). The paradoxical universe, lecture delivered at IBM Conference on Science and Paradox, London.

Whitehead, A.N. (1920). *Concept of Nature*, Cambridge University Press, Cambridge.

Wilkins, A. (1986). Is there really a new evolutionary paradigm or just an uncomfortable gap in the old one? *BioEssays*, **5**, 195–6.

Evolutionary Processes and Metaphors
Edited by M.-W. Ho and S.W. Fox

CHAPTER 2

Evolution outward and forward

SIDNEY W. FOX

ABSTRACT

An holistic view has been attained by experimenting with chemical systems that yield cells and that trace organic evolution from its beginnings forward. The product of these steps is a kind of protein remarkably like protein produced by organisms, although the prebiotic polymer would have arisen in the geological realm in the absence of cells.

The special enabling chemistry, largely completed before 1960, is reviewed. The tendency of such protein (thermal protein, proteinoid) to organize itself into cells is reviewed. The numerous integratively associated biomimetic properties of such microstructures are reviewed elsewhere, as in Ho and Saunders: *Beyond Neo-Darwinism: An Introduction to the New Evolutionary Paradigm* (1984).

Since Darwinian theory has in the past been constructed from external observation of organisms, examination of evolution outward and forward permits a new look at the evolutionary sequence. Principles derived from such chemical study may be compared with processes in Darwinian evolution, in which natural selection has hitherto played the major role. With laboratory protocells and ancient microfossils exhibiting virtually identical morphology, and with protocells as a doorway between protobiotic and biotic protein, origin of life and subsequent evolution are now conceptually united.

The central enabling process at all stages is non-random polymerization of amino acids in the formation of prebiotic, and of biotic, protein. This emphasis is discussed.

INTRODUCTION

The tremendous accumulation of knowledge that constitutes modern bioscience has developed almost exclusively from analysing living things that are here now. Analysis, of course, proceeded by the logical steps of taking whole organisms apart. But we want to know how these organisms can come into existence and we have more recently wanted to learn how the first ones did so in the evolutionary past. For the latter objective the main approach has been to back-extrapolate. What understanding we have is based on what is here now, on what we can see from the outside of organisms, from what we infer from older and older fossils, and from what we can perceive by chemical analysis of the components obtained by reducing true organisms to their components.

Thus, we have extrapolated our knowledge from the outside in and, for the evolutionary questions, from the present to the past. The analyses have unquestionably yielded a glorious catalogue of knowledge plus some theoretical understanding. *The direction of synthesis of polyamino acid, its assembly and evolution is, however, the reverse of analysis and back-extrapolation.* Evolution has occurred through synthetic assembly of precursor components of organisms and through stepwise modification in a forward direction rather than by back-extrapolation. We can attribute this direction to the fact that, as Wald (1954) said, 'Given the right molecules, one does not have to do everything for them; they do a great deal for themselves'. Since Wald made that statement, experiments have shown that molecules and their assemblies can do virtually everything for themselves in their internal reactions and in their reactions with the environment. When retracement of evolution is managed by a human, the process is 'construction' (Fox, 1977). The forces of self-organization (Fox, 1986), which are central to the transition from precursor to cell, have only begun to receive attention (Cherkin and Flood, 1986).

CELL CONSTRUCTION

Cell construction is a term that signifies a working direction opposite to that of reduction. In our connotations, constructionism is *not* anti-reductionism. Rather, the two are complementary, somewhat like night and day. On this planet, which is dependent upon time, solar system and the universe, night and day come in succession. Reduction, as a mode of study, had indeed to precede construction. For example, until we knew that cells are composed primarily of water and protein, we could not test an hypothesis that a special kind of protein and water would react to yield a cell.

Cell construction, as defined, occurs in two steps: (a) synthesis of large molecules (protein) from small ones (amino acids); and (b) self-assembly, or self-organization, of the large molecules to supramolecular systems. These two steps can be followed and individually studied when laboratory protocells are made.

Can we ever learn to retrace the actual route and mechanisms of constructionistic evolution? Some believe that, since the processes occurred out of sight in the distant past, we cannot do so. I believe that history has already shown

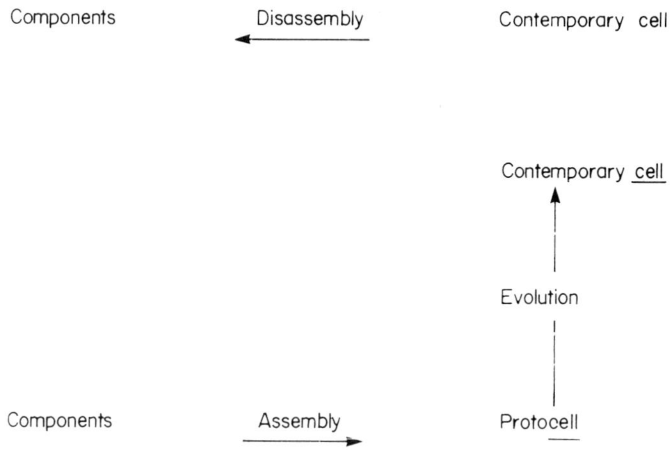

Figure 1 Current reductionistic research requires disassembly. Constructionistic research is in the direction of assembly and evolution

such an evaluation to be unduly pessimistic. We are all aware of accomplishments based on the atomic theory and on the theory of the chromosomal gene; these are two examples of what has already been attained by indirect reasoning by investigators at a distance from their subjects. At the outset, no one could see an atom, or a gene. This inaccessibility to handling in space, comparable to the inaccessibility of handling in time, has not prevented the development of the atomic theory nor of a civilization built on atoms and new molecules. Inability to see genes has not prevented indirect understanding in medical practice dealing with mongolism, sickle-cell anaemia and a growing number of other maladies.

Despite the logical impossibility of back-extrapolating to understand the emergence of the first organism (Figure 1), back-extrapolation has been tested as an approach biochemically by Lipmann (1965), Eigen (1971), and others. Much can be learned in this fashion about organisms, but none of it has yet included the primordial precursor-to-organism transformation, which required assembly. The anachronistic inadequacy of back-extrapolation for understanding the first assembly step forward is, however, beginning to be recognized (Young, 1984; Schuster, 1986). The constructionistic direction of approach is meaningfully expressed by a 1985 film title, *Back to the Future*, but preceded by a scientific paper titled 'Looking forward to the present' (Fox, 1975).

The only method so far developed to understand initial assembly within evolution requires the retracement of steps in evolution. For this method, hints obtained from reduction have been crucial, e.g. the 'unity of biochemistry' and the geological occurrence of amino acids.

If we are to understand how the chemical processes led to the gamut of biological phenomena, we need to use whatever kind of research investigation that will advance our understanding in both detail and generality, directly or indirectly.

THE PRIMORDIAL CHEMISTRY

Amino acids have been known for a long time to undergo tar formation, deamination and decarboxylation when heated (Katchalski, 1951). This is true whether they are heated dry or in water; in the case of aqueous solution, even mild heating vaporizes the water and the treatment becomes heating of residual solid amino acids. The destruction is largely limited to α-amino acids. Heating non- α-amino acids like ε-aminocaproic acid in fact gave the nylon family of polymers (Flory, 1953), which are polypeptides of a kind. The use of ω-amino acids is a way to avoid the thermal destruction of α-amino acids, but it does not give a biological type of peptide or protein. Nor are they found in nature as are the α-amino acids (Fox, Harada and Hare, 1981).

However, another way to minimize thermal destruction without avoiding α-amino acids was derived from the early knowledge that one α-amino acid could be heated non-destructively, i.e. aspartic acid. This appeared to yield small peptides, although the products were for a long time of doubtful constitution (Katchalski, 1951).

The enabling finding of our laboratory was that aspartic acid could be heated with other α-amino acids to give *co*poly-α-amino acids (Fox and Middlebrook, 1954). This could be extended to all α-amino acids simultaneously to yield a polymer with composition capable of simulating that of modern proteins quite exactly (Fox and Harada, 1958; Fox and Waehneldt, 1968). Along the way we learned that glutamic acid also could be thermally copolymerized with other α-amino acids (Harada and Fox, 1958), which when heated alone would decompose. In the case of glutamic acid alone, the product was pyroglutamic acid (α-pyrrolidone carboxylic acid) instead of peptides. When heated with other α-amino acids, however, glutamic acid yielded peptides (Harada and Fox, 1958).

Since the initial examination revealed these heteropolymers to have many of the properties of modern proteins, they were called *proteinoids*. Upon continued examination, they were found to have nearly all such properties, except antigenicity. *Chemical Abstracts* coined for these polymers the term 'proteins, thermal' and has indexed them as such since 1972.

Eigen (1971) has discussed self-organization of RNA in connection with the hypercycle (Eigen and Schuster, 1979). RNA, however, does not directly yield a cell, whereas thermal protein does so to a remarkable degree (Fox, 1980). Eigen *et al.* (1981) conceptually defer the cell to much later in evolution and do

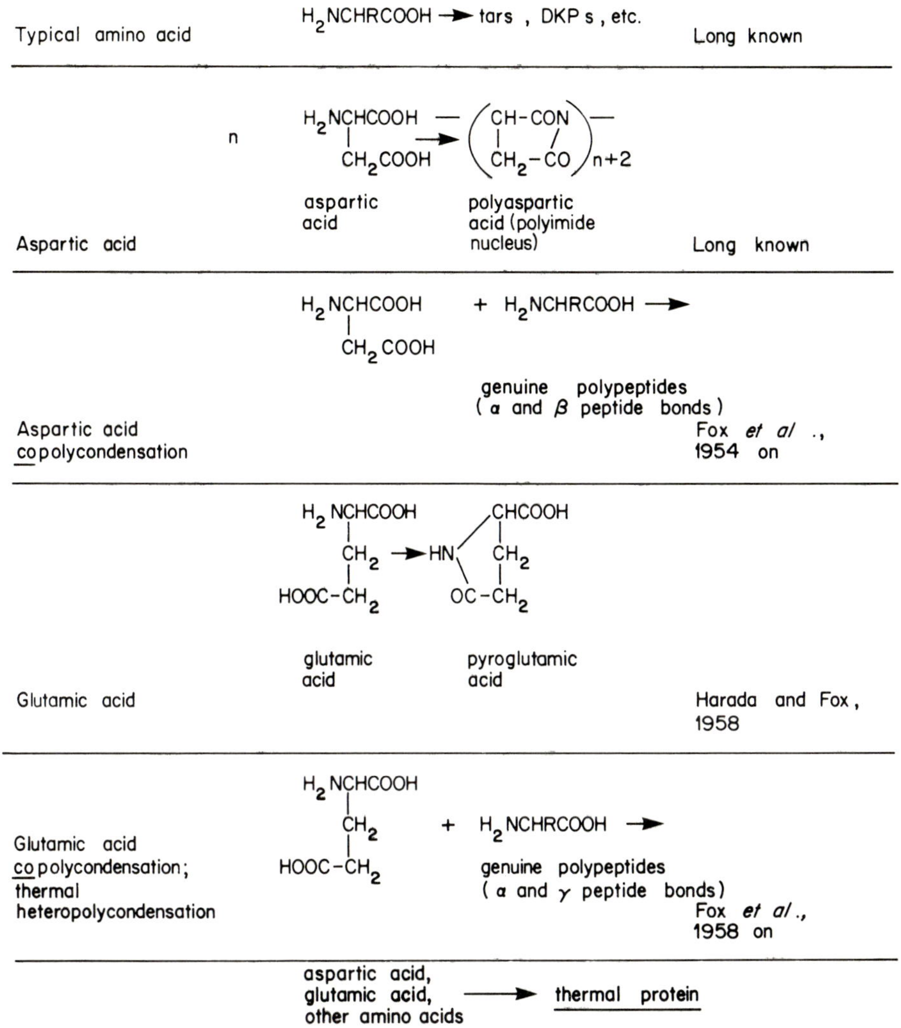

Figure 2 Aspects of the special amino acid chemistry involving thermal *co*polycondensation (Fox and Dose, 1977; Matsuno, 1984)

not even hint at the mode of its emergence. An early thermal protein cell, however, would have been lipoidal and was already close to a modern cell in its properties. Cells-early is now seen as a crucial circumstance for evolution (Fox, 1985).

While the first proteins were prepared by heat, according to these inferences, modern cells use ATP as energy source. The constitution of ATP-energized proteins is remarkably similar to, yet discernibly different from, thermal proteins. The thermal proteins were probably throwaway substances

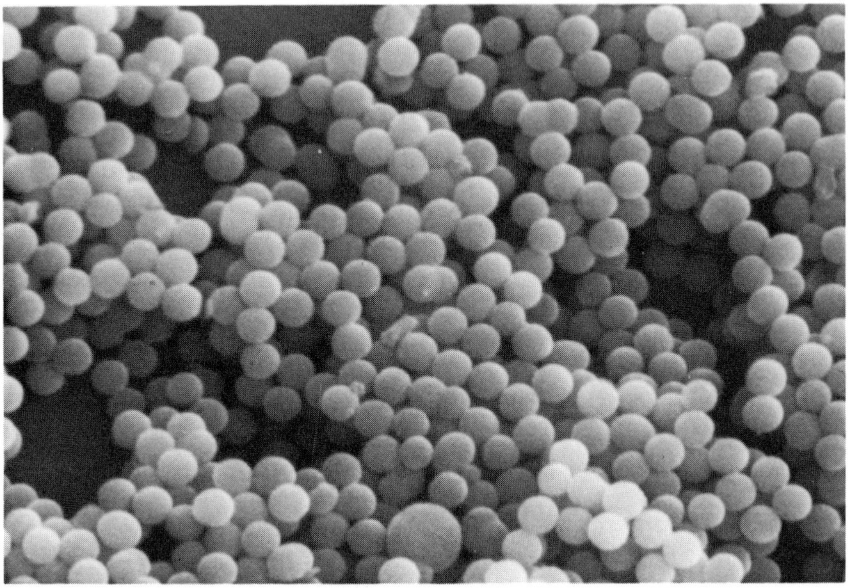

Figure 3 Scanning electron micrograph of proteinoid microspheres. Uniform size from non-random macromolecules. One gram of thermal protein typically yields 10^{10} microspheres. (Courtesy of Steven Brooke)

in evolution. As proteins, they were a doorway to the evolvable cell, but they were replaced by ATP-synthesized polyamino acids within the evolving cell.

The chemistry that enabled the production of thermal proteins is summarized in Figure 2.

PROTO SELF-ORGANIZATION

For rationalizing early evolution, the most ostensible germinal property of thermal proteins is their high propensity to yield huge numbers of cell-like units, proteinoid microspheres (Figure 3). This requires only contact with water; it is a process of utmost simplicity, maximal rapidity and high efficiency. The properties of the cellular structures formed are those of the constituent proteins (Fox, 1980a) *plus* special properties rooted in membranes, such as selective diffusion of molecules through those membranes. One should note that the thermal proteins, like modern proteins, have hydrocarbon side chains (leucine, proline, etc.) that confer on the thermal protein lipid quality. This is the reason microspheres are a kind of lipoidal unit at the same time that they are peptidic.

The generation of peptides by heating of amino acids jointly, whereas most individual amino acids decompose, is related to, and overshadowed by, the fact

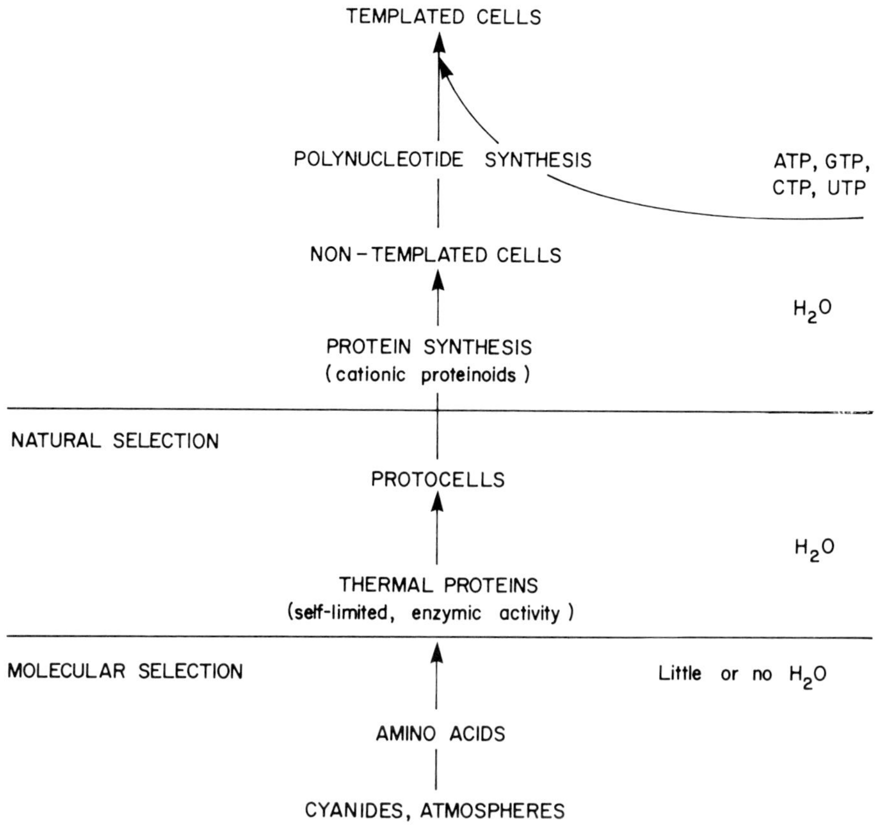

Figure 4 Flowsheet of molecular evolution to and through the first cells. Molecular selection evolves to natural selection; organisms, however, continue to experience internal molecular selection. The flowsheet also emphasizes stepwiseness

that the amino acids order, or sequence, themselves. The model of the emergence of protein within a larger model of emergence of living microunits stresses the endogenicity of the evolutionary sequence (Fox, 1986b). Almost none of the US programme on the origin of life has applied construction from amino acids to the problem of life's beginning (Young, 1984). The self-ordering of amino acids when they are heated may be characterized as a 'biomacromolecular big bang' (Fox, 1986b), since it serves as an original source of information.

The perspective that evolution needs to be understood from the inside out, as well as from the outside in, is brought into focus by the observation that the matrical reactions are endogenously self-ordered and by the laboratory formation of lipoprotein protocells (Figure 4).

TABLE 1 Principles of (molecular) evolution

1. Stepwiseness of processes
2. Self-ordering of amino acids (subsequently replaced or supplemented by genetic coding mechanism)
3. Self-organization of proteins to cells
Derived principles
4. (Informational) proteins first
5. (Proto) cells-early as stepwise evolutionary precursors to modern cells

EVIDENCE FOR SELF-ORDERING

The change in modes of thinking in moving from analysis (from the outside) to synthesis (from the inside) is highlighted in the subject of amino acid sequence (Fox, 1945; Sanger, 1953; Rosmus and Deyl, 1972) in proteins. What was needed was to identify principles and phenomena in the constructionistic processes of evolution, but analysis necessarily preceded synthesis.

The consequences of joining amino acids together under geological conditions proved to be several. Study in a forward direction became *molecular evolution* (Fox, 1953). The most significant principle to be uncovered (Table 1) was that of self-ordering of amino acids (Fox, 1981). As an initial source of biological information, this process appears to have been a *sine qua non*.

The fact that heated amino acids sequence themselves in chains in highly precise and reproducible non-random order has been supported by numerous laboratories; the extensive evidence for the phenomenon has been reviewed a number of times (e.g. Fox, 1980b; Dose, 1984; Melius, 1982). When looked at as a chemical phenomenon, some chemists see such ordering as normal, whereas others do not (Hartman, Lawless and Morrison, 1985). When looked at in an evolutionary context, many other chemists, like Eigen (1971), have assumed that the resultant proteins must be random and their synthesis not reproducible. The formation of non-random thermal proteins has, however, been rigorously shown to be highly reproducible (Fox and Windsor, 1984), and to be determined by the precursors.

EVOLUTION FROM WITHIN

The degree to which evolution has been analysed from outside the individual organism is arresting. Very little of the mechanism of evolution from inside the organism has been attempted, and those attempts have been largely unfocused (e.g. Whyte, 1965). Since attention to the organism has been directed from the environment, it is understandable that the environment could easily be over-emphasized — as in Darwinian evolution. While the view of natural selection is exercised in the environment in which, of course, the evolutionist works, what has needed understanding are events in the interior of the organism and their variations. Darwin (undated) foresaw that need when he said, 'A grand and almost untrodden field of inquiry will be opened, on the causes and laws of variation . . . '.

However, the Neo-Darwinists abandoned, or failed to carry on, the search. Crick (1981), for example, holds a perception of variation that differs from that of Darwin. In speaking of 'accidental' production of proteinoids before nucleic acids, Crick said, 'It is this latter process which was needed if natural selection was to have operated freely'. For Crick, the primal production of nucleic acids and the process of a dominant natural selection are mutually reinforcing assumptions. Neither assumption, to my knowledge, is otherwise or independently supported.

EVOLUTION FROM EARLIER TIMES

In our own earliest defences of the technique of retracement of steps in molecular evolution, the most frequent challenges were those in which the critic did not separate his perception of what existed at the earliest stage of evolution from what exists now. When a property in the protein model protocell failed to be evident in as full measure, or in the same form, as that of a modern cell the experiment was considered to be unsuccessful. What was expected by many critics was that each cellular function would be present in full measure and exactly in modern form in this model of protolife produced in the laboratory. Even though it is not immediately perceived as such, such an expection is tantamount to demanding special creation. What is needed, instead, is recognition of the evolutionary guideline that cell construction proceeds by steps.

In actuality, the steps following the original assembly brought the protocell to its modern abilities. Not only were the steps needed; the steps were irreversible. In other words, one cannot analyse assembly steps by dismantling an organism, nor can one describe the steps of evolutionary assembly by proceeding in reverse. Failure to recognize an inability to proceed in reverse at each step appears to be the reason that so often the perception of evolutionary direction is opposite to that which is viewed as modern cellular direction. An example of this is found in the direction of flow of cellular information. On the modern scene, the flow is from DNA to RNA to protein. The only experimentally defended retracement of evolutionary experiments, however, has the flow of information from the protein to nucleic acids. So far we do not know how to find evidence of a protein → nucleic acid informational flow in modern organisms, although information flow in that direction has been identified for RNA → DNA (Baltimore, 1977).

One other example of the mind-setting nature of back-extrapolation is found in the fact that such an approach leads easily to the view that life began with a unique event. Mayr (1984), for example, has stated that, 'Even Darwin's daring speculation that all life on earth might have had a single origin, was finally proven'. This is indeed a logical inference by back-extrapolation from the genetic code in plants and animals. The only origin of the genetic coding mechanism that has been proposed from experiments, however, is that in which (a) the initial bioinformation arose in thermal proteins by self-ordering of amino acids and (b) this information was transferred to the first nucleic acids, themselves produced by (informed) proteins.

On the basis of those experiments, one learns that almost any time a set of amino acids arose, it would have become thermal protein, which in turn would self-assemble into protocellular units. The nature of the syntheses indicates that such generation(s) would not be irregular and that highly constrained, non-random, variation of protein (Fox and Nakashima, 1984) would occur within each generation. We cannot anticipate that all viewers of the scene will quickly agree to this inference, but it is the only one based on experiments, and there is no way to understand how its evolutionary occurrence might have been recognized by back-extrapolation.

DARWINIAN EVOLUTION

Correlates of the process of self-sequencing of amino acids include non-random products, order out of order, and a non-random matrix for Darwinian selection (Fox, 1986a, b). Mutation is also more recognizable as a non-random process (Ho and Saunders, 1984; Li, Wu and Luo, 1984). With non-random variation built into the process and non-randomness recognizable from the 'big bang' onward by extrapolating *forward* to its later consequences (Barrow and Silk, 1983; Weinberg, 1977) and from the chemical precursors to life (Fox, 1980b), an interdigitated picture of stellar, molecular and biological evolution (Ho and Saunders, 1984; Vrba, 1985) emerges. Non-random molecular beginnings thus yield constrained evolution through life, mind, behaviour and society (Löwdin, 1984; Fox, 1986b).

The presumed function of DNA/RNA to convert true chaos to deterministic inheritance becomes irrelevant. The nucleic acids serve in this newer paradigm the crucial function of coding for inheritance and probably also for a randomization that has speeded variation and thereby widened the pathway of evolution.

This newer picture differs basically from Neo-Darwinism, but it does not deny Darwinian selection. It focuses on the variational steps in evolution, which are rooted in the shapes of molecules, i.e. as morphomolecular evolution (Fox, 1984a). This analysis of variation is, we believe, the kind that Darwin asked for.

Whether the larger picture permits us to say that a mechanism of *variation*, strictly speaking, has been uncovered is doubtful. Rather, in our present state of incomplete understanding, what we may have learned is that the 'big bang' set initial limiting *boundaries* for evolution, especially molecular evolution. All of evolution has occurred within those boundaries by hierarchial construction, by self-organization, by recombination, by extinction, etc., but not by any process that we can truly call creative.

Differences between natural selection and the newer term molecular selection have been spelled out (Fox, 1986c). It may be said further that, while the idea of natural selection is historically derived from artificial selection and breeding practices, molecular selection is an internal process based on morphomolecular specificities. Molecular selection first occurs in the variation step of evolution prior to natural selection which operates at a higher hierarchical level.

The problem that has existed was expressed by Szent-Györgyi (1972):

> We know that nature can build complex organic molecules without the intervention of life.
>
> Another paradox still awaits solution. Mutation is a random process. How could a random process lead to such highly ordered structures as a multicellular living organism? The usual answer to this question is that there was plenty of time to try everything. I could never accept this answer. Random shuffling of bricks will never build a castle or a Greek temple, however long the available period. A random process can build meaningful structures only if there is some kind of selection between meaningful and non-sense mutations. A certain selection was already supposed by Darwin, who made the struggle for life responsible for selection. The selection, however, in my opinion has to take place much earlier, not only with the final product.

EVIDENCE FROM MICROPALAEONTOLOGY

Some evidence for these processes in nature is available from micropalaeontology. The initial step of assembly was spanned inferentially when spheroidal microstructures were found in ancient sediments (Barghoorn and Tyler, 1963) *after* morphologically similar microstructures (Figure 5) were assembled in the laboratory under geophysical conditions (Fox, Harada and Kendrick, 1959). The putative relationship was strengthened when what looked like 'Archean microfossils showing cell division' (Knoll and Barghoorn, 1977) imitated similar structures in the laboratory (Fox and Yuyama, 1964). The parallelism was further strengthened when Francis, Margulis and Barghoorn (1979) applied a method of artificial silicification to both algae and proteinoid microspheres and found that artificial fossils of the two types were essentially indistinguishable. This work showed, in addition, that protein can undergo lithofossilization, a significant point since protein molecules are Wald's 'right molecules'.

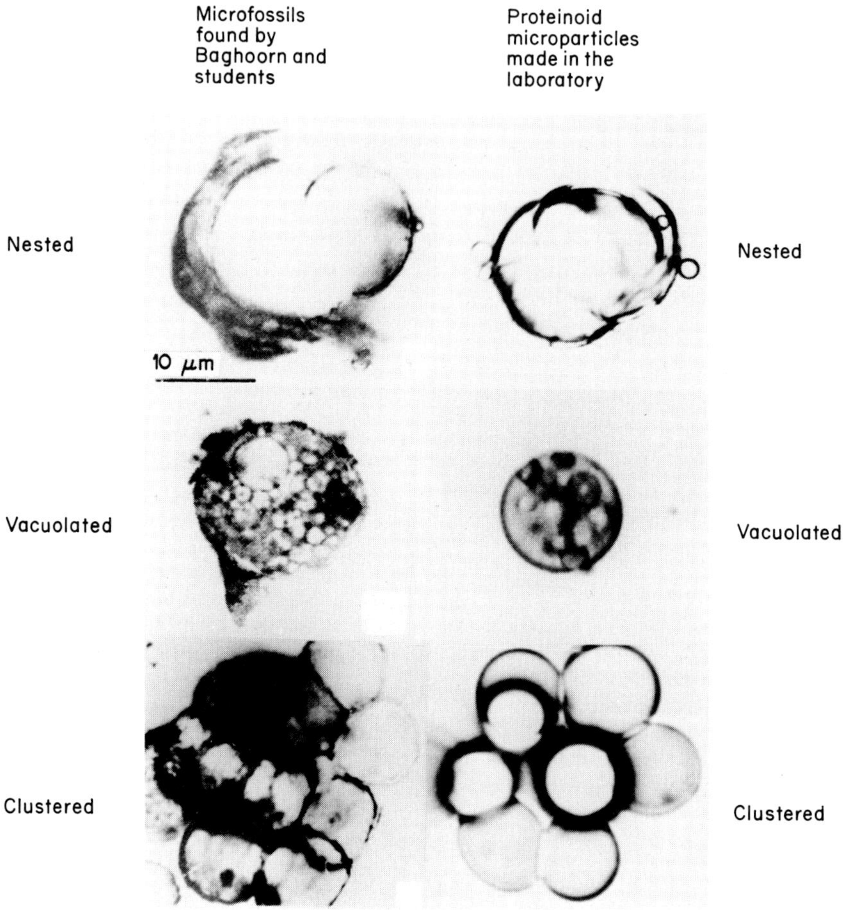

Figure 5 Microfossils from ancient strata in left-hand column. Proteinoid microparticles in right-hand column

Information not otherwise obtainable is, thus, derived from *assembly* experiments. These are the first stage of attempts at retracing steps in early evolution, especially those of molecular evolution prior to organisms (Fox, 1956; Rohlfing and Oparin, 1972).

We may also note that a laboratory protocell was assembled artificially more than a quarter of a century before a modern cell was assembled in the laboratory. The latter has in fact not occurred yet; the attempts were discussed by Danielli (1966) and they may deserve to be reviewed again. The power of the constructionistic method, which includes macromolecular synthesis, has been demonstrated by progress in the attack on the problem of 'origins'. Also demonstrated is the inexorable strength of the evolutionary processes themselves.

The constructionistic approach has yielded principles of self-assembly (Fox, 1985); some of them are listed in Table 1. The first of these, self-ordering or self-sequencing, of amino acids is the most fundamental. The second, self-organization, is the most dramatic. The manifestation of self-sequencing is seen, for example, in the fact that three amino acids, glutamic acid, glycine and tyrosine, yield profoundly non-random peptides when these free amino acids are heated together (Nakashima *et al.*, 1977). The non-random nature of the product would not have been learned if we had not, in fact, first become aware that genuinely specific peptides can be made by what used to be thought of as a brutally destructive process, that of heating amino acids (Katchalski, 1951).

EPILOGUE

This chapter has discussed some of the aspects of evolution that are looked at from the inside out and the past forward. Some students would like to expand the consideration to 'outward, forward, and upward'. Upward begins to connote value judgements more than does forward, and much more than does outward. With the development of bioethics, and of sociobiology that includes ethical considerations (Ruse and Wilson, 1985), we may be overcoming an urge to apologize for the leap from molecular to everyday human activity (Fox, 1985b).

We can begin to explore through the evolutionary sequence the source of what meets our value judgements of human liberty, especially because of, and not in spite of, the fact that they have come out of a constrained evolution. We may even place before high school students the possibility that following the 'straight and narrow' rather than the permissive path (randomness) is a manifestation of the human tendency to 'go with the flow' (Fox, 1986b). When we are prepared to take all this seriously, we are truly looking at evolution from the inside out, the past forward, and the bottom upward.

ACKNOWLEDGEMENTS

The research that yielded the perspectives in this chapter has been supported by Grant No. NGR 10-007-008 of the National Aeronautics and Space Administration and by a contract with the National Foundation for Cancer Research.

REFERENCES

Baltimore, D. (1977). Viruses, polymerases and cancer. In *Nobel Lectures in Molecular Biology 1933–1975*, Elsevier, New York, pp. 495–506.

Barghoorn, E.S., and Tyler, S.A. (1963). Fossil organisms from Precambrian sediments, *Ann. NY Acad. Sci.*, **108**, 451–2.

Barrow, J., and Silk, J. (1983). *The Left Hand of Creation*, Basic Books, New York.

Cherkin, A., and Flood, J.E. (1986). Modulatory interaction: an intrinsic attribute of selforganization. In S.W. Fox (ed), *Selforganization*, Adenine Press, Guilderland, New York.

Crick, F. (1981). *Life Itself: Its Origin and Nature*, Simon and Schuster, New York.

Danielli, F. (1966). Remarks on de novo cell synthesis. In G.J. Jacobs (ed), *Conference on Theoretical Biology*, NASA SP-104, Washington, DC pp. 111–12.

Darwin, C. (undated). *The Origin of Species by Means of Natural Selection*, The Modern Library, New York, p. 372.

Dose, K. (1984). The evolution of individuality at the molecular and protocellular levels. In S.W. Fox (ed), *Individuality and Determinism*, Plenum Press, New York, pp. 33–50.

Eigen, M. (1971). Selforganization of matter and the evolution of biological macromolecules, *Naturwissenschaften*, **58**, 465–523.

Eigen, M., and Schuster, P. (1979). *The Hypercycle*, Springer-Verlag, Berlin.

Eigen, M., Gardiner, W., Schuster, P., and Winkler-Oswatitach, R., (1981). *Scientific American*, **244** (4), 88–118.

Flory, P.J. (1953). *Principles of Polymer Chemistry*, Cornell University Press, Ithaca, New York.

Fox, S.W. (1945). Terminal amino acids in peptides and proteins, *Adv. Protein Chem.*, **2**, 156–75.

Fox, S.J. (1953). A correlation of observations suggesting a familial mode of molecular evolution as a concomitant of biological evolution. *Amer. Natur.* **876**, 253–6.

Fox, S.W. (1956). Evolution of protein molecules and thermal synthesis of biochemical substances, *Amer. Scientist*, **44**, 347–59.

Fox, S.W. (1969). In the beginning — life assembled itself, *New Scientist*, **4**, 450–2.

Fox, S.W. (1975). Looking forward to the present, *BioSystems*, **6**, 165–75.

Fox, S.W. (1977). Bioorganic chemistry and the emergence of the first cell. In E.E. van Temalen (ed), *Bioorganic Chemistry III. Macro- and Multimolecular Systems*, Academic Press, New York, p. 21.

Fox, S.W. (1980a). Metabolic microspheres, *Naturwissenschaften*, **67**, 378–83.

Fox, S.W. (1980b). Life from an orderly Cosmos, *Naturwissenschaften*, **67**, 576–81.

Fox, S.W. (1981). Copolyamino acid fractionation and protobiochemistry, *J. Chromatography*, **215**, 115–20.

Fox, S.W. (1984a). Morphogenesis within evolution: morphomolecular evolution. In J. Miklovsky and V.J.A. Novak (eds), *Evolution and Morphogenesis*, Academia, Prague, pp. 263–78.

Fox, S.W. (1984b). Creationism and evolutionary protobiogenesis. In A. Montagu (ed), *Science and Creationism*, Oxford University Press, New York, pp. 194–239.

Fox, S.W. (1985). Protobiological selforganization. In E. Clementi, G. Corongiu, M.H. Sarma, and R.H. Sarma (eds), *Structure and Motion: Membranes, Nucleic Acids and Proteins*, Adenine Press, Guilderland, New York, p. 101.

Fox, S.W. (1986a). Molecular selection and natural selection, *Quart. Rev. Biol.* **61**, 375–386.

Fox, S.W. (1986b). The evolutionary sequence: origin and emergences, *Amer. Biol. Teacher*, **48**(3) 140–9, 169.

Fox, S.W. (1986c). *Selforganization*, Adenine Press, Guilderland, New York.

Fox, S.W., and Harada, K. (1958). Thermal copolymerization to a product resembling protein, *Science*, **128**, 1214.

Fox, S.W., Harada, K., and Hare, P.E. (1981). Amino acids from the Moon: notes on meteorites, *Subcellular Biochem.*, **8**, 357–73.

Fox, S.W., Harada, K., and Kendrick, J. (1959). Production of spherules from synthetic proteinoids and hot water, *Science*, **119**, 1221–3.

Fox, S.W., and Middlebrook, M. (1954). Anhydropolymerization of amino acids under the influence of hypothetically primitive terrestrial conditions, *Federation Proc.*, **13**, 211.

Fox, S.W., and Nakashima, T. (1984). Endogenously determined variants as precursors of substrates for natural selection. In S.W. Fox (ed), *Individuality and Determinism*, Plenum Press, New York, pp. 185–201.

Fox, S.W., and Waehneldt, T.V. (1968). The thermal synthesis of neutral and basic proteinoids, *Biochim. Biophys. Acta*, **160**, 246–9.

Fox, S.W., and Windsor, C.R. (1984). Reproducibility of amino acid compositions in repeated copolymerization of amino acids, *Int. J. Quant. Chem.: Quantum Biology Symposium*, **11**, 103–8.

Fox, S.W., and Yuyama, S. (1964). Dynamic phenomena in microspheres from thermal proteinoid, *Comp. Biochem. Physiol.*, **11**, 317–21.

Francis, S., Margulis, L., and Barghoorn, E.S. (1978). On the experimental silicification of microorganisms. II. On the time of appearance of eukaryotic organisms in the fossil record, *Precambrian Res.*, **6**, 65–100.

Harada, K., and Fox, S.W. (1958). The thermal condensation of glutamic acid and glycine to linear peptides, *J. Amer. Chem. Soc.*, **80**, 2694–7.

Hartman, H., Lawless, J.G., and Morrison, P. (1985). *Search for the Universal Ancestors*, NASA, Washington, DC.

Ho, M.-W., and Saunders, P.T. (1984). *Beyond Neo-Darwinism: An Introduction to the New Evolutionary Paradigm*, Academic Press, London.

Katchalski, E. (1951). Poly-α-amino acids, *Adv. Protein Chem.*, **6**, 126–85.

Knoll, A.H., and Barghoorn, E.S. (1977). Archean microfossils showing cell division from the Swaziland system of South Africa, *Science*, **198**, 396–8.

Li, W.-H., Wu, C.-I., and Luo, C.-C. (1984). Nonrandomness of point mutation as reflected in nucleotide substitutions in pseudogenes and its evolutionary implications, *J. Mol. Evol.*, **21**, 58–71.

Lipmann, F. (1965). Projecting backwards from the present stage of evolution of biosynthesis. In S.W. Fox (ed), *The Origins of Prebiological Systems*, Academic Press, NY, pp. 259–73.

Löwdin, P.-O. (1984). Some aspects on the relation between the natural sciences and the human mind; an introduction to a panel discussion on the 'origin of life and mind', *Int. J. Quant. Chem.: Quantum Biology Symposium*, **11**, 31–43.

Mayr, E. (1984). *Times Literary Supplement*.

Melius, P. (1982). Structure of thermal polymers of amino acids, *BioSystems*, **15**, 275–80.

Nakashima, T., Jungck, J.R., Fox, S.W., Lederer, E., and Das, B.C. (1977). A test for randomness in peptides isolated from a thermal polyamino acid, *Int. J. Quant. Chem.: Quantum Biology Symposium*, **4**, 65–72.

Rohlfing, D.L., and Oparin, A.I. (1972). *Molecular Evolution: Prebiological and Biological*, Plenum Press, New York.

Rosmus, J., and Deyl, Z. (1972). Chromatographic methods in the analysis of protein structure. The methods for identification of N-terminal amino acids in peptides and proteins. Part B, *J. Chromatography*, **70**, 221–339.

Ruse, M., and Wilson, E.O. (1985). The evolution of ethics, *New Scientist* (Oct. 1985), 50–2.

Sanger, F. (1953). The arrangement of amino acids in protein, *Adv. Protein Chem.*, **7**, 1–67.

Schuster, P. (1986). Mechanisms of molecular evolution. In S.W. Fox (ed), *Self-organization*, Adenine Press, Guilderland, New York, pp. 57–91.

Szent-Györgyi, A. (1972). The evolutionary paradox and biological stability. In D.L.

Rohlfing and A.I. Oparin (eds), *Molecular Evolution: Prebiological and Biological*, Plenum Press, New York, pp. 111–12.
Vrba, E. (1985). Comment in article on evolution by revolution, *Science 85*, **6** (9), 78.
Wald, G. (1984). The origin of life, *Sci. American*, **191** (2), 44–53.
Weinberg, S. (1977). *The First Three Minutes*, Basic Books, New York.
Whyte, L.L. (1965). *Internal Factors in Evolution*, George Braziller, New York.
Young, R.S. (1984). Prebiological evolution: the constructionist approach to the origin of life. In K. Matsuno, K. Dose, K. Harada, and D.L. Rohlfing (eds), *Molecular Evolution and Protobiology*, Plenum Press, New York, p. 45.

ADDENDUM

A number of relevant papers have appeared since the Hawk's Cay meeting. They relate to three main areas.

1. *Roles of individual dicarboxylic amino acids in thermal polycondensation* The individual contributions of aspartic acid and glutamic acid to the production of thermal proteins have been further clarified by Luque-Romero, deMedina and Blanco (1986). Facile copolycondensation is aided primarily by aspartic acid. Endogenous sequence control is aided primarily by glutamic acid. The ease with which glutamic acid is internally condensed to pyroglutamic acid converts it quickly from a trifunctional amino acid to a monocarboxylic acid. In this form it is eminently suitable as an N → C polymerization initiator (Fox, 1980), which is the basis for directional synthesis and self-ordering.
2. *Evolutionary evidence for primordial self-ordering (self-sequencing) of amino acids* If the initial evolutionary fountain of biological information was the self-ordering of amino acids, evolutionary relics of that process should be discoverable (Calvin, 1969). A thorough study reported in 1986 (Ivanov and Förtsch) now claims to have identified such relics, and to have found that the relic condition is more evident in such proteins as histones and ferredoxins than in others such as ribonuclease.

   With the corresponding realignment in thinking, one may now re-examine earlier evidence, such as that of Lipmann (1974). Lipmann reported the endogenous control of polypeptide antibiotic synthesis using only reactive amino acids and enzymes in the sequentialization experiments as a 'possible predecessor of the ribosomal mechanism'.
3. *The genetic coding mechanism as the source of biological information* The assumption that biological information began with nucleic acids is widespread (this volume, Chapter 3). However, the origins of the nucleic acids and the information in nucleic acids has not been explained without prior informed proteins. Nor has the transfer of information from the assumed original DNA → RNA and from RNA → protein been explained without the agency of prior protein, also of the informed variety.

The determinism that is evident in the evolved world is often assumed to depend solely on nucleic-acid-centred genetics (see Chapter 3) or on divine decisions. These two assumptions, while reinforcing each other (Fox, 1987), ignore a third explanation: that the thermodynamic order of living systems emerges from even more order in the universe. The determinism delineated by the order of life having arisen out of even more orderly thermal protein precursors (Fox, 1980) is supported by the newly recognized order emerging from the 'big bang' (Barrow and Silk, 1983). A problem related to this, which is partly but incompletely solved (Fox, 1986) is the relationship between galactic order and molecular order (Seiden, 1986).

The data that have been accumulated for self-ordering are obtained constructionistically; it depends upon experimenting in a forward direction from appropriate primordial conditions. As was remarked at the conference, 'You can't get there from here'.

Related to the above in Chapter 3 by Fitch and Upper are some statements which are potentially confusing because they state for 'Fox's proteinoids' (p. 37) assumptions that differ markedly with the factual experimental observations.

First, the thermal proteins are not random, quite the opposite (Fox, 1980). Fitch's related assumption that bioactive proteinoid molecules are therefore rare is much like that of Eigen, who simply asserted randomness in proteinoids. This assumption would, if correct, prevent repeated biofunctionality as Eigen claims, but the concept of randomness has been shown to be in disagreement with the facts (Fox and Windsor, 1984).

These several assumptions are paradigmatically related (Fox, 1984) to a dominant view of order out of chaos, which has been interpreted as untenable not only by the proteinoid experiments, but by the inferences from the 'big bang' (Barrow and Silk, 1983).

REFERENCES TO ADDENDUM

Barrow, J.D., and Silk, J. (1983). *The Left Hand of Creation*, Basic Books, New York.

Calvin, M. (1969). *Chemical Evolution*, Oxford University Press, Oxford.

Fox, S.W. (1980). Life from an orderly Cosmos, *Naturwissenschaften*, **67**, 576–81.

Fox, S.W. (1981). Creationism, the random hypothesis, and experiments, *Science*, **213**, 290.

Fox, S.W. (1984). Proteinoid experiments and evolutionary theory. In M.-W. Ho and P.T. Saunders, *Beyond New-Darwinism, an Introduction to the New Evolutionary Paradigm*, Academic Press, London, p. 42.

Fox, S.W. (1986). Molecular selection in a unified evolutionary sequence, *Int. J. Quantum Chem.: Quantum Biology Symposium*, **13**, 223–35.

Fox, S.W., and Windsor, C.R. (1984). Reproducibility of amino acid compositions in repeated copolymerizations of amino acids, *Int. J. Quantum Chem.: Quantum Biology Symposium*, **11**, 103–8.

Ivanov, O.C., and Förtsch, B. (1986). Universal regularities in protein primary structure: preference in bonding and periodicity, *Origins Life*, **17**, 35–49.

Lipmann, F. (1974). The search for remnants of early evolution in present-day metabolism. In K. Dose, S.W. Fox, G.A. Deborin and T.E. Pavlovskaya (eds), *The Origin of Life and Evolutionary Biochemistry*, Plenum Press, New York, pp. 321–30.

Luque-Romero, M.M., de Medina, L.S., and Blanco, J.M. (1986). Fractionation and amino acid composition of an aspartic acid-containing thermal proteinoid population, *BioSystems*, **19**, 267–72.

Seiden, P. (1986). The organization of galaxies. In S.W. Fox (ed), *Selforganization*, Adenine Press, Guilderland, New York, pp. 1–21.

Evolutionary Processes and Metaphors
Edited by M.-W. Ho and S.W. Fox

CHAPTER 3

The evolution of life — an overview of general problems and a specific study of the origin of the genetic code

WALTER M. FITCH AND KIM UPPER

ABSTRACT

Physics deals with repeatable material causes in nature whereas history is a sequence of unrepeatable events. The origin of life is a problem both of physics and history. Starting from some general principles, we present a theory of the origin of the genetic code in which an initially nonspecific genetic codon assignment becomes differentiated by ambiguity reduction. This predicts that amino acid transfer RNAs that are most closely related in evolution should possess anticodons that differ by a single nucleotide. A test of the hypothesis is initiated.

INTRODUCTION

The world in which we live is partly material (made of matter) and partly non-material. We wish to know about both kinds of entities. The ways of knowing include instruction, inspiration and revelation. Testing the truth of our knowledge can be difficult. In matters of theology, morality or aesthetics we can ask for consistency if, in fact, consistency is required for truth (must one like, or dislike, oranges on Monday as well as Saturdays?). For the material world, however, we have a method, called scientific, which is our way of knowing that world and testing the truth of that knowledge.

In science we may ask 'Do rocks fall to earth?' but not 'Does God forgive?' In sciences we ask how? but not why? Science is about 'is-ness', not 'ought-ness'.

Do people kill? Ought they? To seek a scientific answer means to seek a materialistic answer. If there is no material answer, it is not a scientific question. The great success of science as a way of knowing about material things in no way makes science or materialism superior to inspiration or morality. Knowing right from wrong is more difficult than knowing plants from animals and more important.

PROBLEMS IN THE EVOLUTION OF LIFE

The material in this section draws heavily on an excellent but, unfortunately, insufficiently known and understood paper by Smith and Morowitz (1982).

Thermodynamics teaches us that the universe is running down, becoming more random, less complex, less ordered, more chaotic, increasing in entropy; but life is very complex, very ordered, non-random, neg-entropic, which seemingly poses a problem. How can one create order in a universe that moves inexorably towards chaos?

We first note that the increasing entropy is the overall result in a closed system. It does not mean that local ordering cannot be a by-product of a global disordering. Indeed, local ordering is highly probable. Consider a beaker of water in equilibrium with the environment. The water molecules move in all directions, randomly, chaotically. Now place a small flame under the beaker. The molecules at the bottom warm up and start moving towards the top where they cool off and move back down to replace other molecules moving up. We have set up a convection cell that is highly ordered compared to the original beaker, a cell in which molecules cycle among states.

It has been demonstrated that cycles (birth and death is a cycle) require an energy flow; at a minimum, an energy source, an intermediate system where the order may be induced and a sink to which the energy is dumped; the flame, the beaker full of water and the room; for earthly life, these are the sun, the earth and space.

If the energy is sufficient to cause exciting electronic transitions of molecules, there will be material cycles within the system:

$$H_2 \rightarrow 2H\cdot,\ H_2O \rightarrow H\cdot + \cdot OH,\ CH_4 \rightarrow H\cdot + \cdot CH_3, NH_3 \rightarrow H\cdot + \cdot NH_2,\ CH_3\cdot + \cdot NH_2 \rightarrow CH_3NH_2,\ CH_3\cdot + \cdot OH \rightarrow CH_3OH,\ 2CH_3\cdot \rightarrow CH_3CH_3,\ CH_3CH_3 \rightarrow CH_3CH_2\cdot + \cdot H, \text{ etc.}$$

C_2H_6 and CH_3OH are more complex than CH_4 and H_2O and they show the principle that local order in an open system is not a violation of the second law of thermodynamics (Smith and Morowitz, 1982). The increased order in this example is a miniscule step towards the order in a human being or in *Escherichia coli*, but it is already sufficient to exemplify that the scientific creationists' charge that life, as an ordered entity, violates that law. Their repeated assertions to the contrary, despite countless refutations made to them, prove they are not scientific and suggests they are not even honest.

At this point we need to consider that in our knowledge on the origin of life we sometimes forget that there are two very different types of processes involved — physics and history (Smith and Morowitz, 1982). Physics is the study of events that have a probability of occurrence so close to one that it does not make sense to ask about exceptions. (Will the chair move when the vast majority of the atoms by chance have a component in the same direction? Will the Cl^- fail to precipitate the Ag^+ this time?) It is the phenomenon of the law of large numbers, the case where the ensemble average taken over all the molecules is precisely knowable even if not a single molecule has that precise value (temperature for instance).

History is the study of events that happen only once, or thereabouts. What is the probability that a random collection of amino acids would by chance assemble some of themselves into the sequence of a molecule of human cytochrome c, a molecule of 104 amino acids? If any one of 20 amino acids could be present at each of the 100 positions, there are 20^{104} possible sequences $= 2 \times 10^{135}$. Assuming the assembly process was one per second on every planet on the universe for 10 000 million years, the probability is essentially zero of getting the human sequence and cytochrome is only one, and a particularly simple one at that, of thousands of molecules that are present in the human. Therefore, the creationists claim, life required divine intervention.

Human cytochrome c did not arise by the accidental coming together of 104 amino acids as the computation assumes. It involved a process of evolution on genes subject to selection, on a replicative mechanism of informational material. Of course, we still have the problem of providing a materialistic, scientific explanation of the course of events that did lead to human cytochrome c and other proteins and structures and ontogenies. However, it is helpful to turn that computation around. What if we do construct some random proteins 104 amino acids long? Suppose only one in a million is water soluble; suppose only one in a million of the water-soluble molecules has catalytic activity of any degree and suppose only one in a million of those is, for unseen reasons, usable in a biological context, then even one gram of random protein would contain half a dozen useful molecules and Fox's experiments with the proteinoids (Fox, this volume), to which I shall return later, demonstrate the pertinence of that calculation. Thus there is a multitude of ways of solving the same catalytic or structural problem so that the finding of a particular solution may be infinitesimal (as the creationists calculate) while the probability of finding at least one solution may be close to one. There are hundreds of human haemoglobin variants without any known deleterious effects. There are millions of organisms each with different haemoglobins. Thus there must be billions of different ways of transporting oxygen by haemoglobin alone to say nothing about the non-haemoglobin methods of transport. The probability of a solution, even a poor solution since selection may improve it, is the origin problem and may be close to one (a physical process) even though the probability of a particular solution is essentially zero (historical).

Notice that, following Smith and Morowitz, we have defined two types of processes, physical and historical, involving the repeatable and the non-repeatable. Before there were genes and a genetic code, the process must have been physical and the events repeatable or else we are in the realm of requiring the one lucky event, a special creation, a postulate not significantly different from divine intervention. Of course, it might really have been a lucky event unlikely to occur again anywhere else in the universe at any time, but, if so, it falls outside the realm of science and materialistic explanation.

Once there are genes and a genetic mechanism, there is a mechanism for selecting for a rare but fortuitous opportunity. A million fruit flies will produce in each generation one variant of each gene. Each variant is only one in a million and in a physical process is simply noise that would rarely be seen (e.g. an amino acid sequence of the pool of these genes' product would never detect the 0.0001 per cent contamination this variant represents). But if that variant is advantageous, selection will amplify that noise so that, many generations later, that rare variant will have spread through the population, replacing the old gene to become the current wild type. If one were able to repeat that experiment, the result would very likely be different. Some other advantageous mutant would arise before this specific variant arose again (even, possibly, in a different gene whose change no longer made the original variant advantageous). Therefore, we are in the realm of history. The evolution of life on earth, like the history of our civilizations, can never be repeated, but it can be scientifically studied and understood.

In the material to follow, I shall assert some principles that I have found useful for viewing problems on the origin of life. These principles are not original but stem from the paper by Smith and Morowitz (1982) and the book by Kenyon and Steinman (1969, Ch. 1). Following this, I shall present a brief outline of the major theories on the origin of the genetic code. Finally, I shall present a scientific study designed to test one of those theories.

PRINCIPLES FOR EXAMINING EVOLUTIONARY QUESTIONS ABOUT THE ORIGIN OF LIFE

1. The laws of nature are unchanging. This is clearly a metaphysical construct in which all scientists believe. Without that belief there is no reason to perform experiments since today's rules need not apply tomorrow and we could not demand that experiments be repeatable. We, as scientists, accept this with as great a faith as anyone accepts beliefs in other areas of thought.
2. Evolution proceeds by creating local order out of chaos. The second law of

thermodynamics is irrelevant because the earth is an open system, receiving energy from the sun and dissipating it into space, and the second law applies only to closed systems.

3. Evolution proceeds from the simple to the complex. Examples: atoms → molecules (monomers) → polymers; cells → tissues → organs → organisms; no cells → unicellular → multicellular (this last example answers that old conundrum, which came first, the chicken or the egg?).
4. Evolution proceeds from the general to the specific. Examples: all cells do everything → some cells perform special tasks some of the time → all cells perform special tasks all the time; hydrolases → peptidases → trypsin-like → thrombin (fibrinogen specific).

   Principles 2, 3 and 4 are not absolute although they are surely correct in general. However, the processes are not irreversible. Indeed, the second law of thermodynamics tells us that the equilibrium lies to the left in the exemplary sequences. But while movement from the complex to the simple or from the specific to the general is probable, the complexity and specificity must have first been created before the process could be reversed. Thus the value of these three principles lies primarily in recognizing the weakness of hypotheses that presume a pre-existing complexity whose origin has itself not first been explained.
5. Reaction systems are progressively separated. There may have been some separation from the start, as between those that took place in the atmosphere and those in the oceans, tide pools, or wherever. With time, other phase separations occurred as lipids (or perhaps proteinoids?) created alternative environments in which some reactions occurred more readily. Ultimately, this would separate the biotic and abiotic worlds. Later came compartmentation via organelles such as nuclei, mitochondria, chloroplasts, Golgi apparatuses and specialized vacuoles.
6. Elements (molecules and structures) of living organisms found universally were already incorporated into the scheme by the time of the most recent ancestor of all things now a part of the living world. Thus we can be confident that, by that time, there were 20 amino acids, five nucleotide bases, glucose, fructose, ribose, deoxyribose, lipids, various coenzymes (NAD, FAD, etc.), membranes, ribosomes, replicases, the genetic code and a number of metabolic pathways. The utility of this principle is in assisting in the evaluation of competing evolutionary hypotheses.

   These elements do not have to be completely universal. Clearly if we found a compound present in all organisms except potatoes, monarch butterflies and brook trout, we would not be likely to err in assuming that the absence of that compound was the result of these three taxa having lost the need for this compound rather than retaining the primitive condition in these three taxa and in all their ancestors. The latter would require that the others gained the compound by massive occurrences of parallelism.

THEORIES ON THE ORIGIN OF THE GENETIC CODE

We must now ask whether the origin of the genetic code is physics or history. It has been suggested that there is a physicochemical relationship between the amino acids and the triplets that encode them that determined the code assignments. This is the *instructive theory* of the origin of the genetic code and makes the origin a matter of physics. It was first proposed by Woese (1966) and is also known as the *direct interaction theory*. The evidence for this is unconvincing to me but it is a very viable alternative to other theories. Indeed, the various theories are not all mutually incompatible and physical interaction properties between amino acids and nucleotides were probably of some influence.

Another theory is the *frozen accident theory* (Crick, 1968). It says the code could have been any which way and the current code just happens to be the way it turned out. This is clearly an historical theory. On this basis there is little point in asking about its origin. But in its extreme form it is certainly wrong. Under this theory, four-fold degeneracy would not be restricted to differences in the third codon position.

A third theory is the *vocabulary expansion theory* in which an originally limited set of amino acids, say 15, was, as time went on, expanded to include additional members, perhaps tryptophan and methionine. This was initially proposed in papers by Jukes (1965) and Crick (1967). A particularly provocative version that connects expansion of the code with the expansion of amino acid synthetic pathways has been proposed by Wong (1975).

A fourth theory is the *ambiguity reduction theory* in which I proposed that originally many amino acids, probably more than 20, were involved in a relation to oligonucleotides to form polypeptides and/or polynucleotides (Fitch, 1966). With the passage of time, the amino acids were reduced to 20, the oligonucleotide was, so far as coding was concerned, restricted to a trinucleotide and the nature of the first two nucleotides gradually came to differentiate among the various amino acids. This is largely an historical process built upon an assumption that amino acids (and hence proteins) and nucleotides (and hence nucleic acids) were associated in their work from a very early stage. This avoids the problem of which came first, the protein or the nucleic acid. Another conundrum best avoided is which came first, function or structure or information?

A fifth theory is the *translation error theory* (Woese, 1965). It assumes that there were amino acids assigned to codons but the error was initially great. With time and selection, the translation errors decreased until the underlying code as we see it today stood revealed. Presumably the initial assignment was by his instructive theory. Thus this is not a theory of how the amino acids came to be assigned to their codons. If there were no pre-assignment, this theory would be equivalent to the ambiguity reduction theory.

A sixth theory is the *lethal mutation theory* proposed by Sonneborn (1965). This theory suggests that the code evolved under pressure to possess the property that the effect of point mutations (those that substituted one nucleotide for another) would, in so far as possible, have either no phenotypic effect (the same old amino acid was still encoded by the new codon) or had a minimal phenotypic effect (the new amino acid encoded was very similar in its properties to the old amino acid). This theory appears to suffer from a teleological fallacy; the process is goal directed.

A seventh theory is the *coding expansion theory* of Lesk (1970). This theory suggests there was a time when the number of nucleotides was less than four so that, in order to accommodate all the amino acids by trinucleotides, it was necessary to increase the number of nucleotides from two or three to four. Like the translation error theory, this is not a theory about how the amino acids came to be assigned to their codons.

ON THE ORIGIN OF THE GENETIC CODE

It was the principles in the preceding that suggested to me (Fitch, 1966) the ambiguity reduction hypothesis whereby the genetic code developed from a system where amino acids and nucleotides were intimately involved with each other in the synthesis of proteins from the very beginning, first with no specificity (any trinucleotide, NNN, specifies all 20 or more amino acids), then through stages of increasing specificity. This is illustrated in Figure 1. There are many possible alternative routes from the initial condition to the present assignments, but they are a very small subset of all the ways the assignment process might have occurred. Moreover, there is the definitive prediction that the amino acid transfer RNAs (tRNAs) that are most closely (recently) related evolutionarily should possess anticodons that differ by a single nucleotide.

This is a testable hypothesis. One first reconstructs the ancestral sequences of various tRNAs and then constructs a phylogenetic tree of those ancestors. That tree of ancestral tRNAs will represent the history of gene duplications and should provide a pattern consistent with the ambiguity reduction hypothesis if the hypothesis is correct. It should not make met-tRNA and tyr-tRNA nearest neighbours nor lys-tRNA and phe-tRNA; it might be expected to put asp-tRNA and glu-tRNA together. We have initiated such a test.

The first task was to select tRNAs for amino acids for which there were many known sequences spanning the range of phylogenetic possibilities (i.e. covering the *five major groups*: eubacteria, archaebacteria, eukaryotes, chloroplasts and mitochondria). The amino acid tRNAs selected were met_i-RNA (initiator, AUG : CAU, 45), met_m-tRNA (elongator, AUG : CAU, 16), ile-tRNA (AUH : GAU, 22), val-tRNA (GUR : UAC, 24), val-tRNA

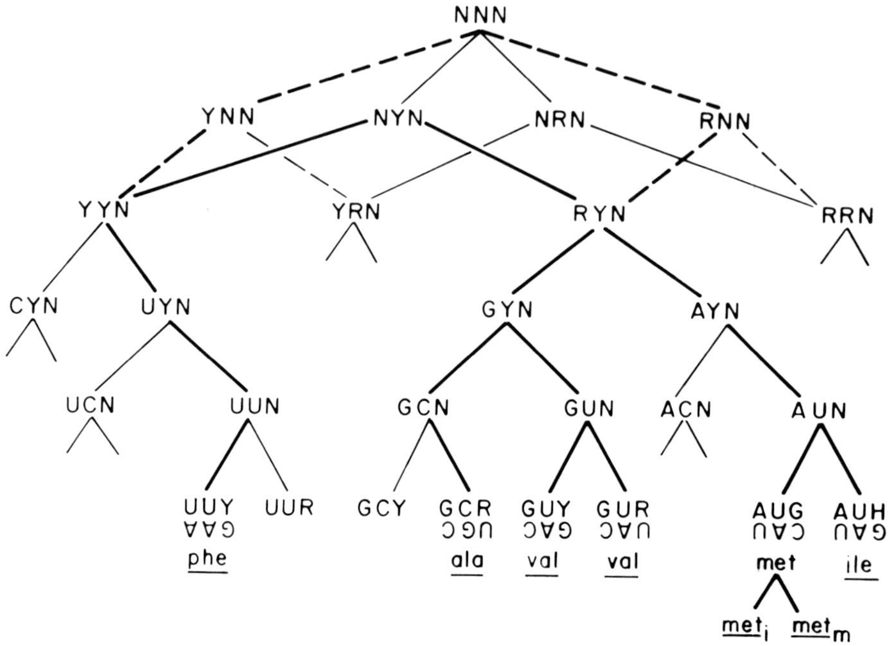

Figure 1 Hypothetical example of ambiguity reduction in the origin of the genetic code. The example shows a totally ambiguous trinucleotide, NNN, at the top, implying that originally any amino acid might be charged to any tRNA so that coding was effectively non-existent. Following the solid lines, we see that the charging mechanism differentiated so that some amino acids are preferentially charged to tRNAs that recognize a pyrimidine in the first position of the codon while others are preferentially charged to tRNAs that recognize a purine in the first position. Further subdivision gives rise to the triplets YYN (preferentially recognized by tRNAs charged with pro, leu, ser, phe), YRN (gln, his, arg, tyr, cys, trp), RYN (thr, ala, met, ile, val) and RRN (lys, asn, asp, glu, gly, ser, arg). Successive reductions, not all shown, would eventually lead to each triplet being recognized by tRNAs chargeable with only one amino acid. The order in which different reductions in ambiguity occur cannot be specified and the dashed lines for the upper tiers show another alternative leading to the same four disjoint sets of 16 ambiguous trinucleotides. At the bottom are shown the anticodon sequences of the tRNAs we used that pair with those codons; also shown is the amino acid specified in today's code. Those amino acids underlined indicate tRNAs specifically studied here. The thicker lines refer to matters discussed in the legend to Figure 4. The IUB single letter code for nucleotides is used in which H = A, C, or U (not G), R = purine, Y = pyrimidine, N = A, C, G or U (any)

(GUY : GAC, 10), phe-tRNA (UUY : GAA, 29) and ala-tRNA (GCR : UGC, 34). The values inside the parentheses are the codons : anticodons and the number of tRNAs we found and used to construct phylogenies.

The phylogenies, obtained by the maximum parsimony method, were

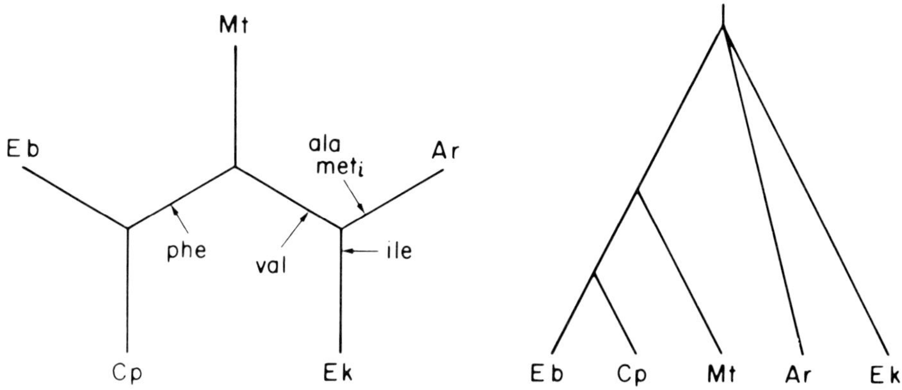

Figure 2 Rooted and unrooted trees for five major groups of tRNAs. The abbreviations are Ar = archaebacteria, Cp = chloroplast, EB = eubacteria, Ek = eukaryotes, Mt = mitochondria. The tree on the left shows the unrooted arrangement of tRNAs shown by four of the six tRNAs examined and for a single sequence composed of all six sequences in tandem. No root (the location on the tree that would be ancestral to all five groups) is shown because parsimony cannot give evidence to the root's location. The shape is designed to prohibit the consideration of the tree in phylogenetic terms. The tree on the right shows the same topology rearranged to show a slightly ambiguously rooted tree obtained by a method described in Figure 3. The trifurcation arises from an inability of the data being analysed to consistently resolve the order of divergence of the three lineages descending from the root; it does not imply that all three diverged simultaneously. The length of a line carries no implication about the amount of change in that interval

compared for consistency since each phylogeny for any one tRNA should be in accordance with the phylogeny of the organisms whence they came if all tRNAs with the same anticodon are orthologous. There was not strict agreement between any two tRNA phylogenies but it was clear that the five major groups generally segregated out cleanly. Moreover, most of the time (four out of six tRNAs; GUR-val-tRNA was not represented by all five groups), the five major groups clustered in a pattern such as that shown in Figure 2, left.

When concatenated sequences were used (all six tRNAs contiguous in a single sequence), the result was that shown in Figure 2, left. Thus we believe this to be the best representation of the true relationships of the five major groups. It disagrees with the trees from Woese's laboratory only in that the mitochondria do not arise from within the eubacterial group. This may be a result of too restricted a range of eubacteria in this study but this study does include *Escherichia coli* (Gram negative), *Bacillus subtilis* (Gram positive) and *Anacystis nidulans* (blue-green). *A. nidulans* always clustered with the chloroplasts, never with the eubacteria. The tree is unrooted because the parsimony procedure cannot specify the location of the root, the minimum number of nucleotide substitutions being the same irrespective of the location of the root.

Not knowing where the root is means we do not know the order of divergence and hence cannot reconstruct the ancestral tRNA.

The relatively clear distinction among the five major groups suggested the next step, namely to reconstruct the ancestors of the groups and construct a phylogeny for all of the tRNAs combined, not to find the answer to our initial question, but rather to establish the root of trees for each tRNA. The principle will be explained below. In this case, however, we did not use ancestors obtained from the most parsimonious tree for each tRNA but imposed a tree that was most reflective of biological opinion. In this fashion, each tRNA tree is constrained to reflect the same evolutionary relationships. This costs additional nucleotide substitutions, makes the ancestors slightly more ambiguous and may well be introducing some noise, but it seemed preferable to the alternatives. To have accepted the most parsimonious trees as correct would have required assuming that all the phylogenetic inconsistencies were the results of paralogy. There are indeed many gene duplications known from these data, but they generally give the appearance of being recent and we felt that the inconsistencies were better considered to result from homoplasy (parallel, reverse and multiple nucleotide substitutions).

For this purpose the five major groups were reduced to four. Since the eubacteria and chloroplasts always grouped together, we chose to replace their respective ancestors with the ancestor common to both of them. Thus we analysed the ancestral sequences for each of the now four major groups for each of the six tRNAs and looked for the most parsimonious tree for all 24 ancestral sequences. What we discovered was that, except for met_m-tRNA, the tRNAs grouped together according to the amino acid they encoded. This is important since it should be so if, as postulated, the gene duplications creating the various amino-acid-specific tRNAs all occurred prior to the most recent ancestor common to all present-day organisms.

From this same tree, with many ancestral group tRNAs for several amino acid specificities, one may hope to discover how the tree in Figure 2, left, should be rooted. The principle is shown in Figure 3 for just three major groups (one could imagine them to be the ancestral Eb, Ar and Ek) and three tRNAs. We see them in the upper portion of the figure unconnected.

With the attempt to find a tree for all nine sequences we might get them connected in a way that did or did not group the tRNAs according to the amino acid they encode. This did not happen. It is important that this did not happen as it would mean that they have diverged so much as to render their evolutionary history obscure or their evolutionary relationships are not those postulated by the theory of ambiguity reduction.

Alternatively, the tRNAs might group according to the amino acid, but as shown on the left, middle. This tree can be rearranged to give a tree (lower left) topologically identical to it but which now shows three pendant (hanging) trees (one for each aa-tRNA) that, unfortunately, show different subtrees for each tRNA; that is, each tRNA subtree has a different root.

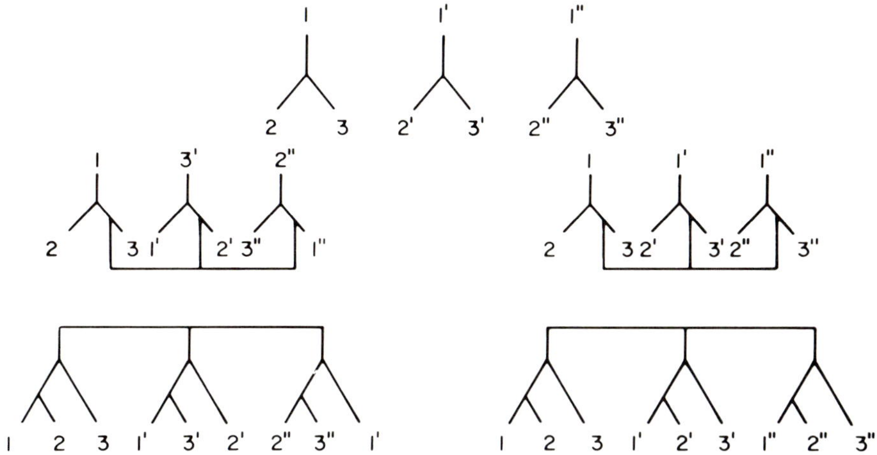

Figure 3 Method for rooting trees when paralogous genes are present. In the upper row are three unrooted trees, each for a different hypothetical tRNA which is known for three hypothetical taxa, 1, 2 and 3. Given that the three tRNAs are paralogous (arose by gene duplications that here preceded species divergence), then one can, instead of building three separate trees, build one large tree comprising all nine sequences. The resultant tree might mix up all the tRNAs but in the middle row are shown two of the many possible results that leave, for any one tRNA, the relationship of the three taxa unaltered. The tree on the left shows the connections between the three subtrees to be two different branches (3, 2′, 1″) of those subtrees, whereas the tree on the right shows the connections between the three subtrees to be all to the same branch (3, 3′, 3″). The bottom two trees are topologically identical to the trees above them so that the gene duplications that led to three tRNAs are implicitly separated from the species divergence that is depicted in lower portions. The tree on the lower left shows a *different* species divergence for each tRNA and one would be forced to conclude that these tRNAs were not able to tell one where the root on the tree belongs. On the right is a similar depiction of an alternative result that leads to a different conclusion. All three trees show the same species divergence and hence we may conclude that the root belongs on the branch to species 3 leaving species 1 and 2 as sister groups, i.e. more recently diverged from each other than either is from group 3. When this method was applied to the five groups shown in Figure 2 using six tRNAs, five of them gave the same result for the groups excluding archaebacteria (the met_m tRNAs behaved aberrantly in not giving an isolated subtree). The root for those five tRNAs, however, could be on the branch to the archaebacteria or on the branch on either side of where the archaebacteria joined the subtree formed by the other four groups. Thus the location of the root could be at any one of the three points shown by arrows in Figure 2, left. It is this ambiguity that leads to the trifurcated root shown in Figure 2, right. The trifurcation arises from an inability of the data being analysed to consistently resolve the order of divergence of the three lineages; it does not imply that all three diverged simultaneously

Finally, they might group as shown on the right, middle. This can be similarly rearranged to give a tree (lower right) topologically identical to it, again with three pendant trees. This time, however, the three tRNA subtrees all show the same rooting of the three subtrees. Their agreement, if this were not a

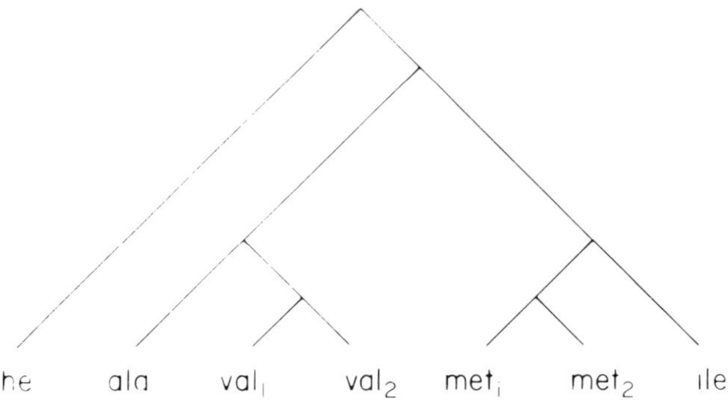

Figure 4 Evolutionary tree for seven tRNAs. The phylogenetic tree shown in Figure 2, right, was assumed to be correct and, for each of six tRNAs, the ancestral tRNA corresponding to the root of that tree was reconstructed by the parsimony process (Fitch, 1971). Those six ancestral sequences, one for each amino-acid-specific tRNA, were then examined in a programme that looked at all 105 possible unrooted tree arrangements. There were three equally most parsimonious results. One of those is shown in this figure. To one of those trees, a seventh tRNA, possessing the other valine anticodon, was added as the closest relative to the other valine tRNA on the basis of its very great similarity to the other valine tRNA. It was not included in the original set because there were not representatives of it in all five major groups. The tree shown here, topologically, corresponds exactly to either of the two trees shown in Figure 1 by the heavy lines. The length of a line carries no implication about the amount of change in that interval

hypothetical case, would demonstrate where to locate the root on the unrooted tree for these taxa.

When this procedure was done for the 24 major group ancestral tRNAs (6 aa-tRNAs x 4 groups), the result was ambiguous. The root was, however, always located at one of the positions shown by the arrows on the left-hand tree of Figure 2. This leads to the result shown in Figure 2 right, where there is an unresolved trichotomy at the root of the tree which means the data do not reveal which two of the three lineages are most closely related genealogically.

Knowing the phylogenetic relationship of the major taxonomic groups and this much about where the root is, permitted us to reconstruct the ancestral aa-tRNA sequence for each of the six tRNAs, again constrained to reflect consistency of phylogeny. These six sequences were then examined in all 105 possible unrooted tree relationships. There proved to be three equally parsimonious trees which are represented in Figure 4 in the consensus form with another unresolved tetrachotomy. The three ways of resolving this tetrachotomy give the three unrooted most parsimonious trees. The root of the tree is placed arbitrarily so as to show a maximum correspondence to the hypothetical tree structure in Figure 1. Indeed, the heavy lines in Figure 1 are a phylogeny congruent to one of the three most parsimonious trees for these tRNAs.

If one assumes that the first reductions in ambiguity were in the first or second position and differentiated purines from pyrimidines and that these two reductions preceded any differentiation between A and G or between C and U, then there are only two possible unrooted trees that are consistent with the current gene code assignments. Given two acceptable tree solutions and three equally parsimonious trees, the probability of obtaining an acceptable tree is $2 \times 3/105 = 0.06$. That misses the 95 per cent confidence level but is close enough to be encouraging. Nevertheless, there are many other tRNAs to be fitted into the scheme and they may not fit well, and even if they do, one should remember that other views on the origin of the genetic code are not totally incompatible.

If we ask, for instance, what was the basis for the choices made in reducing codon ambiguity, it may well have had a physical basis. While there may not have been a direct recognition of amino acids by trinucleotides, perhaps hydrophobic amino acids do tend to associate preferentially with trinucleotides possessing a central pyrimidine, especially a uracil. Thus a theory of the origin of the genetic code that focuses on the evolution of the code, as the ambiguity reduction theory does, may be *a priori* compatible with any theory that focuses on the reasons why a particular choice may be made during that evolution. It remains to be seen if this method of reconstructing evolutionary history is sufficiently powerful to improve our knowledge of the code's origin, but the beginning is most encouraging.

REFERENCES

Crick, F.H.C. (1967). Origin of the genetic code, *Nature*, **213**, 119–121.

Crick, F.H.C. (1968). Origin of the genetic code, *J. Mol. Biol.*, **38**, 367–79.

Fitch, W.M. (1966). Relation between frequencies of amino acids and ordered trinucleotides, *J. Mol. Biol.*, **16**, 1–8.

Fitch, W.M. (1971). Toward defining the course of evolution: minimum change for a specific tree topology, *Syst. Zool.*, **20**, 406–16.

Jukes, T.H. (1965). The genetic code II, *Amer. Scientist*, **53**, 477–87.

Kenyon, D.H., and Steinman, G. (1969). *Biochemical Predestination*, McGraw-Hill, New York, 301 pp.

Lesk, A.M. (1970). On the possibility of a state in the evolution of the genetic code in which replication was imprecise, *Biochem. Biophys. Res. Comm.*, **40**, 235.

Smith, T.F., and Morowitz, H.J. (1982). Between physics and history, *J. Mol. Evol.*, **18**, 265–82.

Sonneborn, T.M. (1965). Degeneracy of the genetic code: Extent, nature and genetic implications. In V. Bryson and H. Vogel (eds), *Evolving Genes and Proteins*, Academic Press, New York, p. 377.

Woese, C. (1966). On the fundamental nature and evolution of the genetic code, *Cold Spring Harbor Symp. Quant. Biol.*, **31**, 723.

Woese, C. (1965). On the origin of the genetic code, *Proc. Nat. Acad, Sci.*, **54**, 1546–52.

Wong, J.T. (1975). A co-evolution theory of the genetic code, *Proc. Nat. Acad. Sci.*, **72**, 1909–12.

Evolutionary Processes and Metaphors
Edited by M.-W. Ho and S.W. Fox

CHAPTER 4

Control of variation in higher plants

C.A. CULLIS

ABSTRACT

In many organisms, including all higher plants, the life cycle is such that somatic 'mutations' can be inherited by subsequent generations. The rate at which such mutations arise in higher plants may be affected by a number of factors including the external growing environment and the internal nuclear composition. One well-documented case in which this occurs, namely, the environmental induction of heritable changes in flax, is considered in detail. In this system phenotypic and genomic changes can be brought about by the growing environment and these alterations can be subsequently inherited. The genomic changes appear confined to a specific subset of the genome although what distinguishes this subset from the rest is not known. A role for these types of changes in generating bursts of variation in response to stress in higher plants is considered.

INTRODUCTION

One of the major considerations of the Darwinian theory of evolution is that it is equally applicable to all organisms. This notion dismisses any differences that might exist between organisms, especially in life histories, which may influence the path of evolution and the generation of variation. At the same time a fundamental tenet of modern genetics is that the germline and the soma have distinct destinies, a distinction that was made by Weismann (1883). The somatic cells form the body of the organism and support and nourish the reproductive cells. Thus any changes that occur in the somatic cells cannot be transmitted to the subsequent generations. The separation of these two types of cells means that so-called Lamarckian evolution, in which useful acquired characteristics are transmitted via the germline to the offspring, is all but impossible. Weismann was a zoologist and applied his principle only to animals, and especially insects, where there is an early separation of the germ cells from the somatic cells. Subsequently, the principle has been extended to all organisms (Mayr, 1978). The clear separation of germline and soma is not a ubiquitous phenomenon and it does not apply to higher plants, colonial animals and protists and is unsupported in nineteen phyla of multicellular animals (Buss, 1983). Thus, in groups where the early sequestration of the gametes does not occur, the germplasm in sexual organisms is formed from the somatic cells and so any somatic mutation could be inherited. A consideration of these life forms may shed a different light on the processes of evolution.

The genetic behaviour observed in plants can only be appropriately considered in the context of the organization of their life cycle. Plants have no formal germline (Walbot, 1985) and their life cycle has two distinct phases: the haploid gametophyte which produces the gametes; and the diploid sporophyte which contains cells capable of undergoing meiosis. After fertilization, in higher plants, the embryonic development establishes the basic body plan of the root–shoot axis and the first adult tissue and organ systems soon differentiate. In contrast to the animal embryo the plant embryo only contains a fraction of the components of the final adult body and contains neither reproductive cells nor a germline. The plant grows, after germination, by division and expansion of the pre-existing cells and subsequently initiates organ proliferation by setting up a series of stem cell populations, the meristems. These meristems allow the plant to produce more organs and also to increase the number of meristems. The shoot apical meristems produce vegetative tissue until an environmental signal triggers the switch to floral development. Individual apices become reprogrammed to form the organs of the flower including the gametes. In many species the gametes are drawn from only a small subset of the original apical cells, but in all species the floral structures are derived from a set of cells which have previously been involved in the generation and organization of the vegetative body of the plant. In annual plants most, if not all, of the active apices are converted to floral structures, while in perennial plants only a subset of the apices are converted from vegetative to floral development, leaving some to form vegetative buds to support the following year's growth.

The plant can be considered to be an assemblage of multiple repeating units (the meristems) each of which is capable of reproduction and which compete with one another. Control over which of the units currently dominates the developmental form is exercised through apical dominance. However, the release of alternative units usually follows biological or physical damage to the dominant meristem. In addition, in long-lived plants the growth of new sectors each year allows the propagation of new somatic variants. These variants then have the opportunity to differentiate into reproductive structures and so become incorporated into the gene pool.

The presence of somatic variants, known as bud sports, has been utilized in the horticultural industry. The world's entire crop of pink grapefruit owes its existence to a sport branch on an ordinary grapefruit tree. Buds from that branch have been propagated and the mutation successfully maintained. In a similar fashion seedless grapes, navel oranges and so on stem from sport branches. The mere appearance of sport branches is not a problem for evolution provided that the rate at which they appear is not greatly in excess of the prevailing mutation rate. The possibility to be considered here is that the rate at which variants appear in higher plants can vary, and that some of the major contributing factors in the appearance of the variation are the 'stresses' or 'shocks' experienced by the plant over evolutionary time. This would give the environment a dual role. On the one hand it would be the agent causing variants to arise and on the other it would act to select among the variants (Cullis, 1977).

Investigations into such interactions require amenable systems of study where the changes occur over short time scales and so can be characterized. Ideally, such changes should occur within a single generation and need to be characterized both at the phenotypic level as well as at the genomic level to determine the relationship between genomic variation and the appearance of new phenotypic characters.

The current complex arrangement of the higher eukaryotic genome is a legacy of the changes which have occurred over the course of evolution. Comparisons of the DNAs between species, and in rare cases, between members of the same species, given an indication of the rate of change and of the type of change which has occurred (Flavell, 1980). It is clear that there is a considerable *fluidity* of the genome and that there have been amplifications, deletions and transpositions occurring frequently over an evolutionary time scale. However, there is much less evidence for the occurrence of these processes from generation to generation, or within plants during a single generation, simply because the experiments have not been performed. There are indications that chimera are common in some species of fern (Klekowski, 1984) while in the herbaceous perennial spring beauty, *Claytonia virginica*, 68 per cent of the population show differences in chromosome number within a plant (Lewis, Oliver and Luikart, 1971). However, we do not know much about the extent of variation, at the DNA level, between individuals within inbred lines, how this may be dependent on the genotype of a particular inbred line and what effect the external environment may have on the variation generating process.

Among the higher eukaryotes, the plant genomes may be especially flexible. In the main they are larger than the animal genomes and they have a higher proportion of repeated DNA (Flavell, 1980; Bennett and Smith, 1976). Genomic diversity in plants has been recognized by a number of different criteria: variation in genome size, ploidy, karyotype and in specific nucleotide sequences. The organization of the higher plant genome with its high level of repetitive DNA and many interspersed sequences is suggestive of a large amount of amplification, deletion and transposition events having occurred. This has been characterized in detail in the molecular analysis of the cereal genomes (Flavell, 1982). However, when the DNAs from different genera are compared their organizations can differ dramatically. Thus, the flax genome contains many tandemly repeated families but very few intermediately repeated dispersed sequences (Cullis, 1981b), while the pea genome contains many dispersed sequences and very few tandemly repeated sequences. The maize genome appears somewhat intermediate between these two, with about one-third of its repeated sequences in tandem arrays while the remainder are dispersed (Hake and Walbot, 1980). Are these examples simply the result of random fluctuations and the fact that we are sampling at the present time, or do they provide evidence for an underlying difference in the relative rates of the amplifications/deletions and transpositions in the various species? If there can be, and has been, a difference in the rate of these processes then the organization of the genomes of the three species could have been derived as follows. Firstly, if the rate of tandem duplications/deletions in flax was very much greater than the rate of transposition then a long period interspersion pattern would have resulted. Secondly, the pea pattern would have to be the result of much more frequent transposition events than tandem duplications. Finally, the maize pattern of DNA arrangements would have been built up by the two types of process having about the same frequency of occurrence.

There are a number of questions to be addressed. Firstly, do changes occur at random within the genome? Secondly, can the frequency with which they occur be influenced by environmental components? Thirdly, do any of the changes seen as a result of the environmental stress have any adaptive value?

In order to characterize increases in variation due to environmental components, it is first necessary to define the stability of the genome under what may be considered 'normal' conditions. Once this has been done it is possible to perturb the system and measure the resulting changes. The results obtained with two inbreeding plant species, namely flax and pea, will be considered in some detail. Inbreeding species were taken so that the effects of residual heterozygosity would be minimized. Flax has been extensively characterized since it has been demonstrated that the genome can change rapidly and in response to external environmental stimuli (Durrant, 1962; Cullis, 1981b).

STABILITY OF THE PEA GENOME

A comparison has been made between individuals of four inbred pea lines using a number of specific DNA sequences as probes. The sequences involved include the genes for the large ribosomal RNAs, the 5S RNA and the two major groups of the storage proteins. In addition to these coding sequences a number of undefined, dispersed repetitive sequences were also used as probes. The results showed that for three of the pea lines there was no significant variation for any of the probes between individuals within a line, although there were significant differences between the lines (Cullis, unpublished). However, the fourth line was somewhat exceptional. Here it was found that there were significant differences between the individuals for all except one of the probes (the invariant probe was that which proved to be most stable when all the lines were compared with each other). This variation also appeared when the individuals compared were derived from a single inbred plant rather than from a bulk, selfed sample. Thus, it would appear that this one line is throwing up continuous variation. What is not yet known is if this type of variation generating system is dominant in its effects. This can be tested in crosses between the apparently stable lines and the variable line. However, it is clear from the study thus far that the amount of variation observed is genotype dependent in peas.

In an earlier study it was found that the variation in rDNA amount between pea lines, and even *Pisum* species, was considerably less than that found in other groups of species (Cullis and Davies, 1975). This is not unexpected if the major form of genomic variation in *Pisum* is by transposition rather than amplification/deletion events. Also supporting this notion is the evidence that in the pea genome the only two sequence families which appear to be present as long tandem arrays are the rDNA and the 5S DNA.

ENVIRONMENTALLY INDUCED HERITABLE CHANGES IN FLAX

Heritable changes can be found in some flax varieties after they have been grown in different, characterized environments for a single generation (Durrant, 1962; Cullis, 1981b). The stable lines produced, termed genotrophs, differ from one another, and the original line from which they were derived, in a number of characters. These include plant weight and height, total nuclear DNA amount (Evans, Durrant and Rees, 1966), the number of hairs on the seed capsule septa (Durrant and Nicholas, 1970), the mobilities of the isozyme bands for peroxidase and acid phosphatase (Cullis and Kolodynska, 1975), the number of genes coding for the 18S, 25S and 5S RNAs and for most of the other highly repetitive sequence families (Goldsbrough and Cullis, 1981; Cullis and Cleary, 1986b).

There is no obvious relationship between the defined aspects of the environmental conditions and the variations in the characters mentioned above. For this reason the environmental conditions have been termed 'inducing conditions' in the absence of any clear molecular mechanisms. However, since only a small number of phenotypic characters have been studied it is possible that only a minor portion of the variation is of adaptive significance with respect to the inducing environments. The remainder may be neutral. In a particular inducing environment adaptive changes may appear insignificant among the wide range of variation observed. Certainly, in the selection of mammalian cell lines with amplified gene numbers for a particular gene, e.g. that coding for the enzyme dihydrofolate reductase, the structural gene only accounts for a small proportion of the total amplified DNA (Schimke, 1980).

Induction of changes

The changes in the rDNA and the total nuclear DNA have been followed while plants have been grown under inducing conditions (Evans, Durrant and Rees, 1966; Culis and Charlton, 1981). It is clear from these experiments that the changes actually occur during the vegetative growth of these plants under inducing conditions. By the time that the plants grown under inducing conditions flower, the changes observed in the apical cells are at the level seen in the following generation. Thus the changes do not appear to be generated during meiosis although this stage may play a part in the stabilization and subsequent heritability of the induced changes.

Observations of the plants grown under inducing conditions suggest that there may be some element of selection by the inducing conditions. The plants grow sub-optimally under the most effective inducing conditions although all the individuals usually survive and contribute to the subsequent generations. However, after some time of growth under inducing conditions the growth rate can increase, either in the main meristem and/or in one of the branches (Figure 1). This portion of the plant grows as if it were now under more favourable conditions. One explanation for this is that there has been a change in the composition of the cells in either the apical and/or adventitious meristems which are now composed of cells more fitted to grow under the prevailing conditions. Since all the progeny tested breed true the change would have to be in all the cells in the meristem otherwise a chimeric plant with variable progeny would be expected.

Changes in the flax genome

Flax has a diploid chromosome complement of 30 chromosomes and contains 1.5 pg DNA/2C nucleus. The DNA is organized in a long period interspersion pattern, which means that the vast majority of the highly repeated sequences are arranged in long tandem arrays (Cullis, 1981a). Representatives of all the highly repeated sequence families have been cloned and their presence and organization in a number of genotrophs determined (Cullis and Cleary, 1986b).

Quantitative variation in all but one of the repeated sequence families has

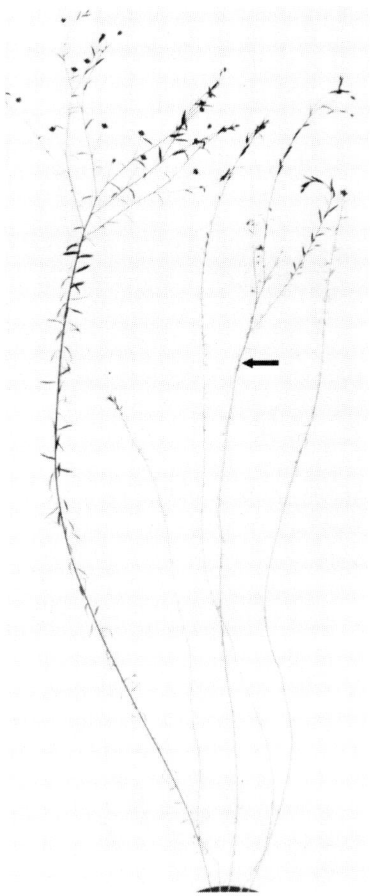

Figure 1 A flax plant grown under inducing conditions (in inert support, a mixture of perlite and grit, and fed with one-tenth Hoaglands solution once a week) (Cullis and Charlton, 1981). Note the one strong side shoot which was the only one to give fertile flowers on this plant. This was the last of the shoots to grow out on this plant. The main stem of the plant is arrowed and its branched top is due to the failure of the originally dominant meristem

been found. The extent of the variation appears to be family specific. The range in the quantitative variation is from more than three-fold for the ribosomal RNA genes to less than 30 per cent for one of the other highly repeated families (Cullis and Cleary, 1986b). The one invariant highly repeated sequence family was the most abundant sequence in the flax genome. In three of the cases the variation was shown to occur in particular subfractions of the repeated sequence family. One of the families for which this was especially obvious was the 5S ribosomal RNA gene family (Goldsbrough, Cullis and Ellis, 1981).

When comparisons were made between the flax and linseed varieties, and the wild linum species (*Linum bienne*), which is one of the putative progenitors of flax (Ockendon and Walters, 1968), then the quantitative differences between the genotrophs were as large as those between various *Linum usitatissimum* cultivars and the *Linum bienne* samples. Thus the environmental induction process appeared to be reproducing the range of variation which had accumulated over a much more extended time. This correspondence between the changes in the genotrophs and between species only holds for the quantitative nature of the changes. It was noted that the particular subfractions of any family were not identifiable as labile when the species comparisons were made, particularly with regard to the 5S gene family (Cullis and Cleary, 1986b). The accumulation of changes over a time scale much longer than a single generation can obviously occur in a much larger fraction of the genome than that varying in the generation of rapid changes.

If flax is taken through a cycle of tissue culture and regeneration, rather than through inducing conditions, then again genomic changes are found in the resulting plants (Cullis and Cleary, 1986a). As in the case of the induced changes the same subfractions of the genome appear to be preferentially varied. There did appear to be a significant difference between the changes induced by tissue culture and those induced by the environment in that the frequency of change in specific families varied. Thus, when changes were induced by growing whole plants under inducing conditions, one of the most common changes observed was in the 5S RNA genes. However, this was a very rare change observed in the plants regenerated from tissue culture. In the same context, when genotrophs were induced as a result of altering the growth temperature rather than the nutrient supply, then although similar morphological forms were obtained the genomic changes found in the two sets of genotrophs were quite different (Durrant, 1971). Thus, it would appear that the range of the changes induced may well depend on the physiological status of the cell at the time that the inducing stimulus is applied.

DISCUSSION

It appears that the maintenance of a pool of variability within a population on which selection can act would be advantageous to most organisms. In the higher plants this can be done in outcrossing populations by the maintenance of heterozygosity, perhaps through hybrid vigour. However, the situation in inbreeding species is not as clear-cut. The maintenance of heterozygosity would necessitate the discarding of a significant portion of the gametes and/or zygotes at each generation. Secondly, a homozygote which was particularly fitted to that niche would be vulnerable to extinction on a change in the prevailing conditions. Thus, in inbreeding species, the capacity to produce bursts of variation at times of change may be necessary. A comparison of outbreeding versus inbreeding species for induced variation has not been made to test whether or not the inbreeding species are more prone to this type of variation. However, such a comparison would be difficult to organize as homogeneous starting material is required for the induction experiments and this would not be available for the outbreeding species. One possibility to circumvent this problem would be to use clonally propagated material.

The strategy proposed here is that inbreeding species have the capacity to generate variation anew in each generation. This generation is controlled by a genetic system in so far as the extent and timing of the variation. Thus, in the pea example described above, the line in which the progeny from a single selfed individual differed from one another would have had a fully competent variation generating system. However, in stable conditions, it would be counter-productive to change an adapted genotype. Thus, the system must be responsive to the strains placed upon it by the environment. In accordance with the ideas published by McClintock (1984) on the biological significance of responses of the genome to stress, the variation generating system could be activated by such a stress.

In the same way as it would be inefficient to be continually generating variation when there is no necessity for it, it would also be inefficient to have the genome rearranged at random as all of those changes which disrupted essential functions would be lethal. Thus, within an evolutionary framework, the labile fractions of the genome would have evolved such that genomic changes would cause non-lethal variation in the phenotype. A consequence of this containment is that the rapid changes observed in response to stress should be localized in an identifiable subfraction of the genome. There should not be any such clear distinction between stable and labile sets of sequences in the comparisons between more distantly related material. This is the situation which has been observed with the comparisons between the flax genotrophs and the related wild species (Cullis and Cleary, 1986b), in so far as there do appear to be specific, labile subsets of the flax genome which are not readily identifiable in *L. bienne*. Whether or not *L. bienne* is, itself, able to respond to 'environmental stresses' by undergoing genomic changes is currently under investigation. If induced variation is observed the expectation is that it should be limited to a subset of the genome and that this subset will be analogous to the one defined in flax. However, if the response to the environment is mediated by position effects then there would not be any sequence specificity to the labile sets. The important parameters would then be sequence organization and position and not the primary sequence.

The fractions of the genome which appear to be most susceptible to change are the highly repeated sequences and transposable elements. The transposable elements in maize appear to be normal components of the genome which can be activated under certain conditions, especially following the disruption of chromosomes. Their transposition is responsible for numerous spontaneous unstable mutations as well as the restructuring of chromosomes which can contribute to heritably altered offspring. This type of instability may play a significant role in genome evolution and may cause accelerated rates of diversification in plant populations (Dellaporta and Chomet, 1985). The highly repeated sequence families in flax are also normal components of the genome and appear to be destabilized under specific conditions. A comparison between the organization of the genomes of the two species, flax and maize, and the labile sequences may throw light on the evolutionary process. In flax, the majority of the changes occur in the highly repeated tandem families with little evidence for transposition events, while the reverse is true in maize. Thus the present organization of the higher plant genome may be considered to be the result of the relative rates of amplifications and deletions compared to transposition events occurring, perhaps in response to environmental challenges, over evolutionary time.

Systematic studies have not been done to determine how general is the response to environmental challenges. However, changes of varying degrees of stability have been demonstrated in a number of species, although few of these examples have been investigated at the nucleic acid level. The species responding are tabulated in Table 1. Included in this table is the phenomenon of somaclonal variation in which genetic variation is associated with the passage through tissue culture although a relationship between this phenomenon and other changes has not been demonstrated.

How are the changes in DNA in the genome related to the phenotypic effects observed in subsequent generations? In the case of the changes in the highly repeated tandemly arrayed sequences in flax one possible mechanism by which these phenotypic changes could be mediated would be a position effect, that is, the change in the DNA at a specific site altering the expression of genes located near that site. Examples of this type of effect have already been described in *Drosophila* (Hilliker and Appels, 1982). In the case of transposable elements there are a range of options by which they can alter gene expression depending on the exact location of their insertion (Dellaporta and Chomet, 1985).

The genetic changes which occurred could be a change from a dominant to a recessive condition in a single generation. This type of alteration has been observed for two of the characters compared between the genotrophs, namely the band pattern for peroxidase (Cullis and Kolodynska, 1975) and the hair number on the false septa of the seed capsules (Durrant and Nicholas, 1970). In the case of the peroxidase isozyme band pattern the change was from the dominant pattern in the original line to the recessive pattern in the small genotroph. The change in the hair number was from hairy (approximately 60 hairs per septum) in the original line to hairless in the large genotroph with

TABLE 1 Species in which changes have been observed in response to the environmental conditions or 'genomic shocks'

Species	Treatment	Reference
Flax	Nutrients	Durrant, 1962; Cullis, 1981b
	Temperature	Durrant, 1971
	Tissue culture	Cullis and Cleary, 1986a
	Specific crosses	Cullis, 1979
Maize	Wide crosses	McClintock, 1984
	Brome mosaic virus infection	Dellaporta and Chomet, 1985
Microseris	Inter- and intra-specific crosses	Price *et al.*, 1983
Pea	Temperature	Highkin, 1958
Phaseolus vulgaris	Temperature	Moss and Mullett, 1982
Many species	Tissue culture	Larkin and Scowcroft, 1983

hairy dominant to hairless. In each case the recessive character was expressed in the next generation so it had to be homozygous and there was no evidence for heterozygosity in the original line for either of the two characters. Thus, a similar, or identical, change must have occurred on both the homologous chromosomes within the inducing generation.

The rapid changes appear to be confined to specific subsets of the genome but we do not know anything about the way in which these subsets are recognized in the genome. One possibility is that the susceptibility of a site to rearrangement could also be dependent on the chromatin structure of that site. Thus the spectrum of DNA sequences affected by a stress would be dependent on the physiological status of the cell to which the stress was being applied since the chromatin structure of various regions of the genome alters with the state of the cell. This is consistent with the observation that different stresses do not affect all the labile sequences equally (Cullis and Cleary, 1986a).

The highly repeated sequences and transposable elements are the most susceptible to rapid variation. These two classes appear to have the potential for inciting phenotypic variation and so it would not be surprising to find them particularly labile. The range of stresses, or 'genomic shocks', which can induce changes are many. In flax it has been shown that changes in the nutrients supplied, the growth temperature or the passage through tissue culture all induced DNA variations (Cullis and Charlton, 1981; Durrant, 1971). In addition, outcrossing between the inbred genotrophs produced a destabilization of the genome, a situation which has been previously reported in maize (McClintock, 1942). In a number of species, including maize and *Microseris* species (Price *et al.*, 1983) it has been shown that the progeny from specific outcrosses are not simply composed of the sum of the parental contributions, some form of interaction has occurred after the formation of the zygote so that the progeny are now a new set of combinations. These types of interactions are dependent on the genotypes involved in the cross so that there is a genetic basis for the interaction. If the crosses are continually repeated then the interaction recedes, as if the initial burst of variation is subsequently brought under control.

In summary, then, it apears that higher plants have a genetically controlled variation generating system. This can be activated by a number of 'shocks' to the organism and it generates variation in a specific subset of the genome. This genomic variation can be manifest as phenotypic variation from which better adapted lines can be selected. In the absence of any 'shocks' the variation system is not active, or active at a very low level, and it is the exposure of the organism to the 'shock' which causes the variation. The limitation of the variation to a subset of the genome, which is controlled by the physiological state of the cell, gives the variation a Lamarckian dimension in that repeated exposures to the same shocks generate the same range of variants. However, this is not caused by a pre-knowledge of what would be advantageous but simply by the generation of a subset of variants which are subject to selection by the same environment each time the experiment is repeated.

REFERENCES

Bennett, M.D., and Smith, J.D. (1976). Nuclear DNA amounts in angiosperms, *Proc. Roy. Soc. Lond. Ser.B*, **274**, 227–74.

Buss, L.W. (1983). Evolution, development and the units of selection, *Proc. Nat. Acad. Sci. USA*, **80**, 1387–91.

Cullis, C.A. (1977). Molecular aspects of the environmental induction of heritable changes in flax, *Heredity*, **36**, 129–54.

Cullis, C.A. (1979). Quantitiative variation of ribosomal RNA genes in flax genotrophs, *Heredity*, **42**, 237–46.

Cullis, C.A. (1981a). DNA sequence organisation in flax, *Biochim. Biophys. Acta*, **625**, 1–15.

Cullis, C.A. (1981b). Environmental induction of heritable changes in flax: defined environments inducing changes in rDNA and isozyme band pattern, *Heredity*, **47**, 87–94.

Cullis, C.A., and Charlton, L. (1981). The induction of ribosomal DNA changes in flax, *Plant Sci. Lett.*, **20**, 213–17.

Cullis, C.A., and Cleary, W. (1986a). DNA variation in flax tissue culture, *Can. J. Genet. Cytol.*, **28**, 247–52.

Cullis, C.A., and Cleary, W. (1986b). Rapidly varying DNA sequences in flax, *Can. J. Genet. Cytol.*, **28**, 252–9.

Cullis, C.A., and Davies, D.R. (1975). Ribosomal DNA amounts in *Pisum sativum, Genetics*, **81**, 485–92.

Cullis, C.A., and Kolodynska, K. (1975). Variation in the isozymes of flax (*Linum usitatissimum*) genotrophs. *Biochem. Genet.*, **13**, 687–97.

Dellaporta, S.L., and Chomet, P.S. (1985). The activation of maize controlling elements, *Pl. Gene Res.*, **2**, 169–216.

Durrant, A. (1962). The environmental induction of heritable changes in *Linum, Heredity*, **17**, 27–61.

Durrant, A. (1971). Induction and growth of flax genotrophs, *Heredity*, **27**, 277–98.

Durrant, A., and Nicholas, D.B. (1970). An unstable gene in flax, *Heredity*, **25**, 513–37.

Evans, G.M., Durrant, A., and Rees. H. (1966). Associated nuclear changes in the induction of flax genotrophs, *Nature*, **212**, 697–9.

Flavell, R.B. (1980). The molecular characterisation and organisation of plant chromosomal DNA sequences, *Ann. Rev. Pl. Physiol.*, **31**, 569–96.

Flavell, R.E. (1982). Sequence amplification, deletion and rearrangement: major sources of variation during species divergence, In R.B. Flavell and G.A. Dover (eds), *Genome Evolution*, Academic Press, New York, pp. 301–23.

Goldsbrough, P.B., and Cullis, C.A. (1981). Characterisation of the genes for ribosomal RNA in flax, *Nucleic Acids Res.*, **9**, 1301–9.

Goldsbrough, P.B., Ellis, T.H.N., and Cullis, C.A. (1981). Organisation of the 5S RNA genes in flax, *Nucleic Acids Res.*, **9**, 5895–904.

Hake, S., and Walbot, V. (1980). The genome of *Zea mays*, its organisation and homology to related grasses, *Chromosoma*, **79**, 251–70.

Highkin, H.R. (1958). Temperature induced variability in peas, *Amer. J. Bot.*, **45**, 626–31.

Hilliker, A.J., and Appels, R. (1982). Pleitropic effects associated with the deletion of heterochromatin surrounding rDNA on the X chromosome of *Drosophila, Chromosoma (Berl.)*, **86**, 469–90.

Klekowski, E.J. (Jr) (1984). Mutational load on clonal plants: A study of two fern species, *Evolution*, **38**, 417–26.

Larkin, P.J., and Scowcroft, W.R. (1983). Somaclonal variation and crop improvement. In T. Kosuge, C.P. Meredith and A. Hollander (eds), *Genetic Engineering in Plants*, Plenum Press, New York, pp. 289–314.

Lewis, W.H., Oliver, R.L., and Luikart, T.K. (1971). Multiple genotypes in individuals of *Claytonia virginica, Science*, **172**, 564.

Mayr, E. (1978). Evolution, *Sci. American*, **239** (3), 38–47.

McClintock, B. (1942). The fusion of broken ends of chromosomes following nuclear fusion. *Proc. Nat. Acad. Sci., USA*, **28**, 458–63.

McClintock, B. (1984). The significance of responses of the genome to challenge, *Science*, **226**, 792–801.

Moss, G.I., and Mullett, J.H. (1982). Potassium release and seed vigour in germinating bean (*Phaseolus vulgaris* L.) seed as influenced by temperature over the previous five generations, *J. Exp. Bot.*, **33**, 1147–60.

Ockendon, D.J., and Walters, S.M. (1968). Linaceae, *Flora Europaea,* **2**, 206–11.

Price, H.J., Chambers, K.L., Bachmann, K., and Riggs, J. (1983). Geographic and ecological distribution of genomic DNA content variation in *Microseris douglasii* (Asteraceae), *Amer. J. Bot.*, **70**, 1133–8.

Schimke, R.T. (1980). Gene amplification and drug resistance. *Sci. American*, **243**, (5), 50–9.

Walbot, V. (1985). On the life strategies of plants and animals, *Trends Genet.*, **1**, 165–9.

Weismann, A. (1883). *On Heredity*, Clarendon, Oxford.

Evolutionary Processes and Metaphors
Edited by M.-W. Ho and S.W. Fox

CHAPTER 5

New genetic mechanisms and their implication for the formation of new species

JEFFREY W. POLLARD

ABSTRACT

Recent advances in the analysis of chromosome organization both in animals and plants have shown that DNA is able to undergo variation by many non-Mendelian mechanisms. These include the promiscuous behaviour of movable genetic elements, gene amplification, the dynamics of genetic interchange within gene families, gene conversion and interspecies gene transfer. In this chapter I suggest that the frequency, and the dramatic effects on phenotype, of some of these non-Mendelian events demands a new theory for the genetic basis of evolution.

Two themes distinguish this new theory from the Neo-Darwinian explanation of evolution. The first is that major changes in evolution (macroevolution) have little to do with the accumulation of point mutations in structural (or regulatory) genes causing gradual phenotypic effects upon which natural selection may act (microevolution). Macroevolution, instead, involves rapid genomic restructuring that produces dramatic alterations of developmental plans independent of natural selection. However, once these changes have occurred, microevolution will hone up the fit between the organism and its environment. The second theme is that the concept of Weismann's barrier (the inviolability of the germline from its environment) is outmoded and needs substantial modification. It now seems that the germline DNA has strategies for coping with environmental changes either by generating genetic variation in response to the modified condition or by gene conversion from somatic variants that have been selected during a period of environmental upheaval.

A theory of speciation based on these themes is proposed. This theory suggests that, during periods of environmental stress or following the invasion of foreign genetic material, rapid genomic restructuring will occur with its concomitant effects on phenotype. If this happens in a small isolated population, new forms will be fixed over a few generations by inbreeding such that the population will be reproductively and phenotypically incompatible with its parent population, i.e. it will be a new species. Such a model is consistent with saltatory speciation.

> History shows that the human mind, fed by constant accessions of knowledge, periodically grows too large for its theoretical coverings, and bursts them asunder to appear in new habiliments,as the feeding and growing grub,at intervals,casts its too narrow skin and assumes another, itself but temporary. . . . A skin of some dimensions was cast in the 16th century, and another towards the end of the 18th, while, within the last fifty years, the extraordinary growth of every department of physical science has spread among us mental good of so nutritious and stimulating a character that a new ecdysis seems imminent. (T.H. Huxley (1863). Reprinted 1959, pp. 72–3.)

INTRODUCTION

Two decades ago, Haldane (1964, p.49) was able to write with some confidence, 'In my opinion, beanbag genetics, so far from being obsolete, has hardly begun its triumphant career'. His meaning was that numerical analysis of changes in gene frequencies (the replacement of one gene by a variant gene) throughout populations would, when properly applied, give a substantial, if not complete, understanding of the process of evolutionary change. At that time such optimism was well founded since the theory of evolution developed by Haldane and others, known as the synthetic theory (Huxley, 1942), provided a coherent hypothesis for the mechanism of evolutionary change. It was founded upon apparently secure facts and its base was built upon the rigour of the 'superior' science, mathematics. The data included observations of selection in action in nature (e.g. melanic moths in smoke blackened woods) and in the laboratory (e.g. antibiotic resistance in bacteria), as well as a voluminous literature on the nature of mutations, their patterns of inheritance and their dynamics in populations. Two decades later, however, the whole edifice of the synthetic theory is being attacked from disciplines ranging from palaeontology to molecular biology (see for example Ho and Saunders, 1984; Pollard, 1984a, b; Temin and Engels, 1984) and a consensus is emerging that the theory has collapsed as a sufficient description of evolutionary change (although opinions differ as to whether it needs to be replaced entirely).

One of the major reasons for this dissatisfaction with the synthetic theory is the recent realization that genomic organization is infinitely more complex and much more liable to dramatic changes than previously conceived. Furthermore, there exists a complete set of enzymes capable of restructuring the genome both within the life-span of an organism and throughout evolutionary time. These findings, in my opinion, demand the development of a new and more comprehensive genetic theory for the mechanism of evolutionary change.

This new theory needs to address at least four major questions: (1) How do novel forms arise? (2) How are major genetic changes integrated into previously existing systems in order to increase the complexity of the organism in a harmonious way? (3) Are the types of genetic changes now documented sufficient to produce a new species *ab initio* or do they simply produce variation within an old species which is followed by selection to produce a new species? (4) Are these genetic events compatible with both change and stasis within species?

Of course, complete answers to all these questions, and most particularly to the first, which is the crux of the matter, are beyond the scope of present knowledge. Nevertheless, advances in the cloning of chromosomal sequences involved in developmental regulation, such as the homeo boxes (McGinnis *et al.*, 1984) and the plethora of recent observations on the dynamics of DNA change are very encouraging. It is my purpose in this chapter to review some of the data demonstrating the flexibility of the genome and comment on their relevance to evolution. I will discuss the major types of processes, other than point mutations, which generate genetic variation, and how the environment may affect the organization of the genome through these processes.

SEQUENCE REITERATION: RECURRING MOTIFS IN EVOLUTION

Gene duplication and the production of new functions

It has always been a problem to understand how organisms gained completely new functions without losing pre-existing ones. Ohno (1970) suggested that this problem could be solved by the process of gene duplication. Such a process would allow the new copy of the gene to accumulate function-altering mutations while the original copy maintained its function. By means of this mechanism, batteries of genes could be built up which are similar in structure and overlapping in function. Many examples of multigene families are now known that appear to have been derived by such a process; these include the chorion proteins of the silk moth (Jones and Kafatos, 1980), the globin gene cluster (Jeffries, 1982), the immunoglobulin gene family (Hunkapiller *et al.*, 1982; Hunkapiller and Hood, 1986) and the serine protease genes (Rogers, 1985). Families range from those containing only a few gene copies to those with thousands of copies. The genes within a family can be virtually identical (e.g. rRNA genes), or they can vary significantly, as, for example, the immunoglobulin genes (Hood, Campbell and Elgin, 1975). In some cases, lineages can be constructed linking different multigene families, suggesting that the whole lineage may have evolved from a single primordial gene, e.g. the gene families for α-microglobulins, immunoglobulins, T-cell receptors and transplantation antigens (Hunkapiller *et al.*, 1982, Ohno, Mori and Matsunaga

1986, Kaufman *et al.*, 1984, Hunkapiller and Hood, 1986). Indeed, it has been suggested that nearly all genes can be considered as members of different gene families (Ohno, Mori and Matsunaga, 1986) and probably, in higher organisms, no gene exists in a single copy.

Recently, a very elegant demonstration of the process of gene duplication evolving to give a functional protein has been documented in the case of human colour vision (Nathans, Thomas and Hogness, 1986). Colour vision is based on three light-sensitive pigments. Initially it appears that there was duplication of the ancestral rhodopsin pigment gene to give a dichromatic visual system of blue and green absorbing pigments. This was followed by further duplications of the green-sensitive gene to produce a battery of genes. One of these at the 3′ end of the cluster underwent sufficient mutation to give a gene more similar to the green gene than the blue (96 per cent versus 43 per cent identity) but with sufficient variation to encode a novel red-sensitive protein. This led to the development of a three-pigment colour system in humans and Old World monkeys (although there must also have been other changes in the brain to process such information). Because New World monkeys do not have this gene, the duplication event that was to result in red and green pigments occurred somewhere in the last 30–40 million years. Thus, the original duplication giving rise to the family of green pigment genes was most probably genetically neutral and was followed by the accumulation of mutations either as the result of genetic drift or by selection of more visually acute individuals.

It is of interest to know whether the rate determining step was the original duplication or the point mutations altering the duplicated gene's function? This is a difficult question to answer because we observe only the frequency of occurrence and not the rate. Molecular clock studies indicate that the nucleotide substitution rate is, for example, in the non-functional region of fibrinopeptides, about 1 per cent amino acid sequence change per 1 to 1.7×10^6 years enabling an estimate for the evolution of a new gene by sequence divergence of about $25–40 \times 10^6$ years (Ohno, Mori and Matsunaga, 1986). If gene duplication occurs more frequently, the rate determining step for the production of the new gene function would be the accumulation of point mutations. Gradual accumulation of mutations is the mechanism of evolution predicted by the synthetic theory. However, in the case of the pigment genes the initial duplication was also absolutely necessary for the production of a new function and such gene duplications are not accounted for by the gradualist model.

Thus, the rate and pattern of occurrence of gene duplications needs to be further explored in order to assess their potential evolutionary impact. If this mechanism is important, the restructuring of duplicate genes by point mutations or by other processes to produce new proteins should correlate with the patterns of speciation found in the fossil record.

Members of gene families engage in genetic exchanges by means of transposition, unequal crossing over and gene conversion with and without RNA intermediates (Smith, 1973; Dover and Flavell, 1984). Thus, mutated sequence can spread rapidly throughout a family in a sexually reproducing population, quite independently of natural selection (Dover, 1982). The result is the concerted evolution of gene families (see Dover, 1986). Dover and Flavell (1984) and Dover (1986) have further argued that a gradual evolution of functionally complementary gene families occurs by stepwise selection within the genome which could be observed as species-specific distributions of differently homogenized gene families. Although it is not readily apparent from this scheme how gene families can evolve away from one another to produce complex multigene families, there is no doubt that homogenization of gene families occurs. But all these processes stem from the biochemical dynamics involving the DNA itself and are not necessarily the consequence of natural selection for gradual alterations in phenotype. The latter undoubtedly plays a role in the process of fixation and maintenance of such genetic changes within species. As survival and spread of a variant through a gene family is a matter of chance rather than selection, it disrupts the predicted patterns of gene segregation and cannot be accommodated within orthodox population genetics theory (Dover and Tautz, 1986).

Novel functions may also arise by duplication of exons due to unequal crossing over. This process, termed 'exon shuffling' (Gilbert, 1978, 1985), alters functions efficiently because RNA splicing ensures that the new mRNA is in the correct reading frame. Thus, either the same sequence motif may be duplicated to give different proteins a common function, e.g. the DNA binding region in steroid receptor proteins and *erbA* (Weinberger *et al.*, 1985), or different exons can be used to build up complex genes, e.g. the low density lipoprotein receptor and the epidermal growth factor receptor (Sudhof *et al.*, 1985 a, b). Apparently, genes may always have been in pieces (Doolittle, 1978), as demonstrated by the conservation of intron positions in the triosephosphate isomerase gene in both plants and animals (Marchionni and Gilbert, 1986). Genes may also acquire new introns (Rogers, 1985) or lose them (Perler *et al.*, 1980); this means that 'shuffled exons' undergo further evolution, generating yet more diversity. Variability may also be increased by alternative splicing events that produce different mRNAs from a single DNA sequence. Many examples of this have now been documented (Sharp, 1985). Thus exon shuffling and alternative splicing appear to be fundamental processes for generating new functions and increasing variability within gene families.

Oligonucleotide repeats and the origin of new functions

A variation on the themes of duplication and exon shuffling has been proposed to be the source of truly new functions within organisms (Ohno, Mori and Matsunaga, 1986). It is suggested that duplications of short oligonucleotide sequences produce repeating periodicities in the DNA and that these repeats provide the primordial building blocks for modern genes (Ycas, 1972; Ohno, 1984). In support of this model, many genes have been shown to have a residual oligonucleotide periodicity such as the collagens, serum albumin, immunoglobulins, involucrin, RNA polymerases, etc. (Ohno, 1984; Allison *et al.*, 1985; Ohno, Mori and Matsunaga, 1986; Eckert and Green, 1986) and also some 'new' genes have been shown to be built up on this type of pattern (Ohno, Mori and Matsunaga, 1986). An example of the latter are the anti-freeze proteins in Arctic and Antarctic species of teleosts. These novel proteins, based on different repeating units in, for example, hakes and flounders, are only found in species inhabiting cold water, and are unknown in temperate species (Ohno, Mori and Matsunaga, 1986). Ohno (1984) has suggested that all genes may ultimately be derived from oligonucleotide repeats and that novel genes are formed by the sequestration of these repeats from the non-coding reiterated parts of the genome. More sophisticated functions may be derived by mutation of these captured sequences. Furthermore, by the insertion of an oligonucleotide repeat into a pre-existing gene, a new protein may be created either by opening up an alternative reading frame or by causing transcription from the opposite strand (Ohno 1984).

Genome size variation and species divergence

Repetitions of genetic sequences within the genome can, therefore, involve non-coding as well as coding regions of the genome. The non-coding sequences may be repeated from a few to millions of times. There has been considerable debate about the function of such repeats. Hypotheses include the views that they merely replicate selfishly, that they are the progenitors of all new proteins, or are important in physiological regulation (for arguments, see Cavalier-Smith, 1985). The size of the genome has some effect on phenotypes such as the length of the cell cycle, the organisms' size and morphology. Furthermore, it can alter the position of chromosomes within the nucleus which may in turn affect their transcriptional activity (Flavell, 1982; Cullis, 1986). All these effects may be subject to selection during evolution. In this chapter, however, I will concern myself only with the origin of molecular morphology and discuss possible effects that variation in this morphology may have on evolution and, in particular, on the formation of new species.

Many of the mammalian repeats that have been analysed appear to have originated relatively recently, post-dating the mammalian radiation. The divergence among repeats has occurred at rates that are compatible with the neutral accumulation of mutations (Daniels *et al.*, 1983; Deininger and

Daniels, 1986). The important evolutionary question is, therefore, whether the spread of repeats is involved in the speciation of various groups (perhaps by causing genomic incompatibility) or is secondary to the speciation process. This question has hardly been considered in mammals but has begun to be addressed in flowering plants where DNA contents may range from 5×10^7 to over 8×10^{10} base pairs (Bennet and Smith, 1976; Flavell, 1986). Some of the variation in genome size in flowering plants is due to tetraploidy (another important method of wholesale gene duplication), but much is due to changes in haploid genome size. In fact, most of the differences in chromosome size and structure between species can be attributed to changes in repetititve DNA content (Flavell, 1982). This variation in size may make chromosomes incompatible during sexual reproduction and so contribute to establishing interspecific barriers. However, species vary in their reproductive tolerance of genome size variation. For example, maize is quite tolerant (Rose and Doolittle, 1983) whereas *Nicotinia* is not (McClintock, 1984). Of course, subtle effects on fertility may be overlooked by simple short-term breeding experiments. Nevertheless, as species are most often distinguished by chromosome structure, this seems to be a better indicator for speciation than primary sequence divergence (Wilson, Sarich and Maxson, 1974). It is quite possible, therefore, that one major genetic mechanism driving speciation is chromosomal incompatibility caused by processes such as amplification, deletion, divergence and translocations which reorganize repetitive gene families. These, in turn, produce major structural differences between chromosomes in different populations, which are sufficient to isolate species. As variation in the number of repeated sequence numbers can occur very rapidly (Flavell, 1982; Cullis. 1984) it may account for saltatory speciation events and provide an explanation for the C-value paradox (the large discrepancies between haploid DNA contents of closely related species). Dramatic variations in haploid genome size might be the outcome of chance events which force subspecies apart by making them reproductively incompatible. Therefore, the large variation in DNA content between closely related species may have no other consequence than to maintain them as separate entities.

Gene amplification during development and in cultured cells

Experimental observation and sequence comparisons have shown that gene amplification is a common biological phenomenon. It has been observed during development, where it apparently fulfils the need for large amounts of a given gene product within a short time, as for example, the chorion proteins during oogenesis in *Drosophila melanogaster*, the extrachromosomal rDNA in the oocyte of *Xenopus laevis* (Stark and Wahl, 1984), and as a mechanism of conferring resistance to particular toxins in organisms ranging from protozoa to mammals (Alt *et al.*, 1978; Coderre *et al.*, 1983; Stark and Wahl, 1984; Chasin, 1985). Interestingly, in some insects, such as the silk moth, the chorion genes are permanently amplified (Jones and Kafatos, 1980) suggesting the fixation of amplified genes during evolution.

In many cases, the gene amplification that confers resistance to a toxin through an increased cellular concentration of an approximate enzyme involves the wild type allele (Schimke, 1984). However, there are cases when a variant allele has been amplified (Lewis, Davide and Merlera, 1982). Many gene amplifications result in visible alterations to chromosome structure, giving homogeneously staining regions or extrachromosomal double minutes. Analysis of these amplified regions has shown that not only the selected gene but also large sections of the genome on either side of it have been amplified (Cullis, 1984; Bostock, 1986). For example, the dihydrofolate reductase gene is approximately 32 kilobases but the amplified regions in different clones range from 500 to 3000 kilobases (Cowell, 1982). Thus, unrelated or related genes may be amplified in concert with the genes whose product confers a selective advantage (Wahl, Vitto and Rubnitz, 1983) resulting in unexpected effects on phenotype. During amplification, the particular sequence amplified at each successive step can differ (Zieg *et al.*, 1983) and novel rearrangements can also occur which may bring together disparate parts of the genome (Bostock, 1986). In the CAD gene, these mechanisms apparently have resulted in a new mRNA species (Meinkoth *et al.*, 1986). Thus, once a cell has begun to amplify a particular gene, rapid amplifications and dramatic rearrangements can occur in a single or a few cell generations (Schimke, 1982; Bostock, 1986). These changes may be genetically stable, though deletions and further rearrangements can also occur (Bostock, 1986) suggesting that the amplified sequences exist as dynamic structures that may change independently of the gradual changes introduced as a consequence of molecular drive. It would be extremely interesting if exon shuffling was also found to occur during the process of amplification since this mechanism could generate dramatic new functions within one cell generation.

In cultured mammalian cells, the rate of gene amplification has been estimated at between 10^{-3} and 10^{-4} events/cell/generation (Johnston, Beverley and Schimke, 1983; Stark and Wahl, 1984) in the absence of the selective agent. Using this figure and the assumptions that gene amplification occurs randomly throughout the genome, and that only those events whose function can be selected are observed, it has been calculated that, at any one time an amplification is occurring in the genome of one in every three cells of a population (Bostock, 1986). However, these cells are transformed and are characteristically aneuploid and exhibit a high degree of chromosomal instability. Unfortunately, from an evolutionary perspective, comparable estimates for the rate of amplification have not been made for mammalian diploid cells *in vivo* for the germline cells.

Mechanisms for sequence duplication

The mechanisms for gene and sequence duplication appear to be diverse and may differ for different types of sequences under different circumstances. They include local alterations in DNA metabolism, the replication of movable genetic elements (transposons) or the consequences of reverse transcription of expressed genes. Apart from transpositions (see below, p. 73) other mechanisms proposed for gene duplication and gene amplification include unequal crossing over, rolling circle mechanisms, slippage replication and aberrant or differential replication (Schimke, 1982; Bostock, 1986); with perhaps the latter mechanism being favoured for large-scale amplification of sections of the genome (Schimke, 1984; Cullis, 1984; Bostock, 1986). However, it is by no means certain, and appears unlikely, that gene duplication, gene amplification and simple sequence repetitions are produced by the same mechanisms. Furthermore, an important evolutionary question is whether the limited number of observations on gene amplification during development and in somatic cells described above are relevant to the alteration in germline amplifications that must have occurred through evolution. It is interesting in this context that in *Xenopus laevis*, prior to the 2500-fold amplification of the rDNA genes in the fertilized oocyte, there is a 40-fold amplification of these genes in the primordial germ cells (Gall, 1969; Gall and Pardue, 1969). Similarly, in *Drosophila melanogaster* the rDNA genes can undergo dosage compensation in the germline (Tartoff, 1973). The presence of large numbers of reverse transcribed sequences in the genome also suggests a substantial degree of this enzyme's activity in cells of the germline. Thus, the necessary enzymes for altering sequence number are present in the germ cells. It remains a challenge to ascertain the frequency of such events in the germline; perhaps estimate of this may be obtained from experiments involving transgenic mice.

THE NATURE OF THE GERMLINE AND WEISMANN'S BARRIER

Much of the above discussion has attempted to relate genetic events that can be observed in somatic cells to the variations in genome organization in the germline of different species. However, it may be argued that due to the existence of Weismann's barrier, the soma and germline are distinct and separate and, therefore, it is not legitimate to draw inferences about the germline from events that can occur in the soma. But how real is this distinction between soma and germline? The concept of the continuity of the germline and its isolation from the soma was developed in the latter part of the nineteenth century by Nussbaum and Weismann. Weismann (1893), in particular, argued for a molecular basis for the distinction between germplasm and soma, and furthermore, he argued that the germplasm was immune from both variation in the soma and environmental influences such as stress, nutritional alterations or somatic usage. This hypothesis came to be known as Weismann's doctrine, and the separation between the soma and the germline as Weismann's barrier. It gained favour because it allowed rejection of so-called Lamarckian inheritance (the inheritance of acquired characters) and because it conformed with embryological observation on germ cell sequestration in vertebrates. It, therefore, soon passed into dogma and became one of the foundation posts of the synthetic theory (Pollard, 1984a, 1987). Recently, however, a fact which has been known since the first two decades of this century came to be appreciated once more: that germline sequestration has little or no relevance for the majority of phyla of animals and does not apply to the whole of the plant kingdom (Buss, 1983; Klekowski, Mohr and Kazarinova-Fukshansky, 1986). In these groups, the germ cells are formed from somatic tissue and, therefore, any somatic mutation can in theory be inherited. For example, in the flowering plants apical meristems reset their developmental pathways from the production of leaf primordia to differentiate into all the components of the flowers including the gametes (Walbot and Cullis, 1985). Thus, given the number of flowers that some plants produce, a very large number of somatic lineages can be represented in the germline of one plant and mutation or genomic reorganization within these lineages can, therefore, become incorporated in the germline. This process, in principle, also allows the mutations to be tested in the environment prior to their incorporation into the germline, in complete contrast to Weismann's doctrine. Indeed, selection has been documented between somatic lineages (Stewart, 1978; Klekowski, Mohr and Kazarinova-Fukshansky, 1986).

Bearing this in mind, the work on environmentally induced genomic reorganization in flax is extremely important to one of the main themes of this chapter. In a series of experiments, rapid quantitative heritable DNA changes were generated by growth in different, characterized environments within one generation (Durrant, 1971; Cullis, 1983, 1984, 1986 and this volume). The progeny of the selfed plants known as genotrophs differed from one another in size, height, nuclear DNA amount, hairiness of the capsules' septa, as well as in isozyme patterns for peroxidase, acid phosphatase and esterase (Cullis, 1984). Interestingly, once the new isozyme patterns were generated, in subsequent breeding they appeared to be controlled at a single locus by a dominant and recessive allele (Cullis, 1979). At the extremes, the genotrophs differed by 15 per cent in DNA content which was largely made up of an altered copy number of rRNA genes, 5S genes and in highly and middle repetitive sequences (Cullis, 1984). Using the rDNA genes as marker genes, it was shown that their number varied during the period under inducing conditions (Cullis and Charlton, 1981). Similar genomic changes could also be induced by growth of cells through tissue culture and by intraspecific crosses (Cullis, 1986).

The inducing condition, therefore, had a direct effect on the genome of somatic lineages within the flax which by virtue of their reprogramming into gametes became inherited. This process contravenes Weismann's doctrine, since under this doctrine the germline is considered to be immune from direct environmental effects such as stress and nutritional status. It was known, of course, that environmental mutagens cause DNA damage and in the strictest sense also contravene Weismann's doctrine. But in the case of flax, the DNA does not appear to be chemically altered by the inducing conditions. Instead, the inducing conditions have elicited the process of sequence *amplification*, for which, in the majority of cases, the function is unknown but, none the less, is accompanied by many phenotypic effects. It must be considered, therefore, that this mechanism is part of the adaptive response of the organism generating genetic diversity under environmentally changing conditions. Although the mechanism whereby the genome senses changes in the environment is unknown, such dramatic effects in plants beg the question of the generality of the phenomenon, particularly to those animal phyla that do not show germline sequestration, and its role in evolution. In this context, it is interesting to note that exposure of Chinese hamster ovary cell culture to hypoxic conditions dramatically enhances the frequency of dihydrofolate reductase gene amplification (Rice, Hoy and Schimke, 1986) suggesting a wider occurrence of 'stress' induced gene amplifications.

These data suggest to me a mechanism for evolutionary change that fulfils many of the criteria outlined in the introduction. If an isolated population of organisms were exposed to an altered environment rapid variation of genomic organization might occur with concomitant effects on morphology, size and patterns of gene expression. There may also be a degree of selection involved since phenotypic effects of the genomic alterations might be expressed in somatic lineages resulting in the selection of one lineage over another and, therefore, its representation in the germline. Once these variations had occurred within individuals in a small isolated population, inbreeding would tend to fix them in the entire population. Furthermore, the newly amplified sequences would be exposed to all the possible changes that occur within gene families and during gene amplification that were discussed above. These processes could result, in a very short space of time, in a sub-population that is reproductively incompatible with and phenotypically different from its parent population. This model suggests that there would be bursts of speciation occuring as new ecological opportunities open up, as appears to have happened during the adaptive radiations which characterize evolutionary history (Eldredge and Gould, 1972; Stanley, 1981). It also suggests that the pattern of speciation discussed in the 'punctuated equilibrium' or 'saltatory' speciation models would be the expected pattern of large-scale evolution (Stanley, 1981; Gould, 1982).

TRANSPOSITION: A 'FREE-LIVING' MODE OF GENE DUPLICATION

Transposition, when replicative, is a gene duplication involving non-contiguous parts of the genome and therefore is part of the continuum of processes described in the sections on gene duplication. In fact, there are also several models that evoke the role of transposable elements in gene amplification (de Cicco and Glover, 1983; Stark and Wahl, 1984). However, transposable elements appear to be evolutionarily related to 'free living' viruses (Temin and Engels, 1984) which suggests that some of these sequences may be able to move between cells thereby violating Weismann's barrier. In this section I will consider the possible effects that a promiscuous life-style may have on evolution.

Transposable elements fall into two classes: those that transpose through a DNA intermediate and those that transpose through an RNA intermediate. The latter group is also divided into two groups. The first are those elements that have homology with retroviruses (retrotransposons) and probably transpose by similar mechanisms and may be able to move between cells, e.g. copia in *Drosophila* and Ty elements in Saccharomycetes (Mount and Rubin, 1985; Mellor *et al.*, 1985; Fink, Boeke and Garfinkel, 1986). These elements may account for 10 to 20 per cent of the genome (Temin and Engels, 1984). The second group are the retrotranscripts consisting of cDNA sequences homologous to the mRNA of protein coding genes (processed genes) and the reverse copies of polymerase III transcripts, e.g. Alu sequences (Vanin, 1984; Jelinek and Schmid, 1982; Sharp, 1983; Temin and Engels, 1984; Temin, 1985; Deininger and Daniels, 1986). These polymerase III elements have an internal promotor giving them replicative autonomy but they do not code for reverse transcriptase. They presumably rely upon a host or virally encoded enzyme in the germline to copy the primary transcript back into DNA. The retrotranscripts appear more common in mammals than in other animals (Wagner, 1986). In mammals these processed genes represent approximately 10 per cent of the genome (Temin, 1985) and taken together, the retroposons and retrotranscripts can account for over 500 000 insertions into the mouse genome (Temin, 1985). Thus, these sequences are a major source of new mutations; they provide sites for recombination or conversion with other genes in a gene family, act as a means of dispersing sequences to other chromosomes and increase genetic variability by their mode of replication (Steele and Pollard, 1987). There is also one case where retrotransposition has provided a new functional gene, the rat preproinsulin 1 gene (Soares *et al.*, 1985).

The role of DNA movable genetic elements also remains a mystery with suggestions for their function ranging from their having none (being simply genomic parasites) to their holding key regulatory roles in the operation of gene batteries (Orgel, Crick and Sapienza, 1980; Davidson and Posakony, 1982; Shapiro and Cordell, 1982; Doolittle, 1985). At least some functional roles during development have been documented in lower eukaryotes, e.g. mating strain switching in yeast (Shapiro and Cordell, 1982) but in the majority of cases no functional role has been assigned. What is established, however, is that they do provide a substantial source of genetic variation that can occur in any generation. They can cause germline instability in chromosomes and large-scale chromosomal rearrangements of up to 10 per cent per generation (Never and Saedler, 1977). They cause mutations by inserting into coding or regulatory sequences and as the consequence of imprecise excision when they leave the gene, which may account for the majority of mutations at a particular locus (Syvanen, 1984). They can also produce re-setting of developmental pathways (Shapiro and Cordell, 1982; McClintock, 1978, 1984; Temin and Engels, 1984; Deininger and Daniels, 1986). These effects and the ubiquity of these elements suggest that they must play a major role during evolution by increasing the range of genetic variability upon which selection can act and also in causing genomic incompatibility between sub-populations. If, on top of this, they (or some sub-classes) play a role in the developmental regulation of gene batteries (Davidson and Posakony, 1982; Sutcliffe *et al.*, 1984) then germline transposition could result in the re-setting of developmental pathways and their timing of expression, perhaps giving new forms *ab initio*.

Returning to the theme of the previous section, it has been shown that stress (i.e. environmental alterations) can increase the rate of transposition of DNA elements in maize (McClintock, 1984). This led to the proposal that environmental stress, causing previously silent transposable elements to transpose, may be a major feature of evolutionary change (McClintock, 1984). This process may not be confined to plants or to DNA elements since environmental stress (e.g. heat shock) has also been shown to induce copia transposition in *Drosophila* cells in culture (Strand and McDonald, 1985), Ty in yeast (Paquin and Williamson, 1984) and the rate of insertional mutagenesis in *Drosophila melanogaster* (Bregliano and Kidwell, 1983).

The correspondence of some transposable elements to free-living retroviruses also suggests that they may contravene Weismann's doctrine by being able to move between different organisms' germlines. It is well documented that retroviruses can infect across species and enter the host's germline after which they are vertically transmitted (Temin, 1985). These insertional events can cause alterations in functions and developmental mutations (Jenkins *et al.*, 1981; Varmus, Quintrell and Ortiz, 1981) and may, therefore, be a major part of the mutational load of an organism. Retroviruses are also able to recombine with host genes and presumably transduce them between members of a species and also between species. This may result in pathology as seen for the oncogenic retroviruses (Weinberg, 1980) or possibly as a means of bringing a new functional protein *de novo* into an organism (Pollard, 1984a, 1986; Syvanen, 1986). Retroviruses are also able to capture host RNA sequences in their virions (Ikawa, Ross and Leder, 1974) and these can be reverse transcribed (Taylor and Cywinski, 1984) giving a further possible mechanism for transferring information between species. These mechanisms may provide a means for supplying organisms with new functions without the need for laboriously building genes by point mutations and thus this process may be a major factor in evolution (Syvanen, 1985, 1986; Pollard, 1987). However, despite the homology between retroposons and retroviruses (Temin and Engels, 1984) it has not yet been demonstrated that retroposons can move between cells although extracellular copies of these elements have been found (Temin, 1985).

The transfer of genetic information within endogenous retroviral vectors is the means of information transfer between the soma and the germline proposed by Steele in his somatic selection hypothesis (Steele, 1979, 1981; Steele, Gorczynski and Pollard, 1984). In this theory any mutation in somatic cells, selected because it conferred an advantageous phenotype on the organism and, expressed as an RNA, could become inherited by a gene conversion event after transfer of this RNA from somatic to germline cells by a retroviral vector. Such a process was seen by Steele to occur frequently, thus acting as a means of feedback between the environment and the germline. The somatic selection hypothesis' separate proposals for the capture of representative cytoplasmic RNAs in retroviral virions and their reverse transcription (Faras *et al.*, 1973; Ikawa, Ross and Leder, 1974; Taylor and Cywinski, 1984), germline and generalized reverse transcription (Temin, 1976, 1985; Baltimore, 1986) and homologous recombination (Smithies *et al.*, 1985; Thomas, Folger and Capecchi, 1986) have all been confirmed (see Steele, Gorczynski and Pollard, 1984; Pollard, 1984a, 1987; Steele and Pollard, 1987). However, direct proof that a cytoplasmic RNA can be transferred from the soma to the germline causing a gene conversion has yet to be definitively obtained. It should be noted, however, that reverse transcription of an RNA conferring a mutant phenotype on a somatic lineage that in turn develops into a gamete or a ramete would constitute a prototype of this process and may of itself be extremely important in evolution.

Although definitive molecular evidence for the somatic selection hypothesis is still to be obtained, experiments by Gorczynski and Steele (1980, 1981), which showed that experimentally manipulated neonatal tolerance (a somatically established trait) could become inherited strongly, suggest that such a process can occur. However, there is considerable controversy over these results (Brent *et al.*, 1981; Steele, Gorczynski and Pollard 1984; Pollard, 1984a). Recently, independent data have been obtained confirming the general phenomenon of paternal influence on the immune status of progeny (Gorczynski *et al.*, 1983; Mullbacher, Ashman and Blanden, 1983; Cooper-Willis *et al.*, 1985). If the proposed genetic mechanism does exist, it will provide a means whereby organisms can track the environment and, furthermore, if it involves not only structural genes but regulatory sequences, it may be a means whereby altered functional patterns could become inherited in one generation (Steele, 1979, 1981; Steele, Gorczynski and Pollard, 1984; Pollard, 1984a, 1987). If this were the case it would be a major mechanism for directed variation during evolution and the establishment of new species during periods of environmental change.

For transposable elements to have evolutionary consequences, it is not necessary for them to carry 'foreign genetic' information, because they can cause substantial genetic effects on their own over and above their mutational capacity. Indeed, one of the most intriguing observations for evolutionary biology to come from the recent analysis of the genome, is the case of the P-transposon in *Drosophila melanogaster*. This transposable element is a classic DNA-transposon that, because of a tissue-specific splicing event to give an element encoded transposase, is only active in the germline (Laski, Rio and Rubin, 1986). When a female lacking this element is crossed with a male containing it, hybrid dysgenesis occurs because the P-elements are reactivated in the progeny. This results in female sterility, large numbers of genomic rearrangements and a high rate of mutation (Kidwell, Kidwell and Sved, 1977). The P-transposon is not found in laboratory stocks collected some 30 years ago but is now present in wild populations. Furthermore, its pattern of distribution in other *Drosophila* species is not correlated with their evolutionary relationship nor with the pattern of distribution of another dysgenic system, the I-R system (Bucheton *et al.*, 1986). This suggests that the P-transposons have spread horizontally among *Drosophila* species (Kidwell, 1983) by an as yet unknown mechanism (Lansman *et al.*, 1985).

The consequence of the activation of the P-transposon is reduced fitness of hybrids, amounting to something like an interspecific barrier. It is possible, therefore, to envisage a mechanism of speciation (Rose and Doolittle, 1983; Pollard, 1984b, 1987) whereby an isolated subspecies is invaded by a transposable element which initially destabilizes the genome causing substantial rearrangements, mutations and restructuring of developmental pathways which generate new forms (Pollard, 1987). After a period of time and substantial inbreeding, however, the element would become harmonized within its host genome, as was seen for the P-element. A new species is created because other members of the original species do not have these elements and, therefore, an effective reproductive barrier is set up due to both phenotypic and genotypic incompatibility. A particularly attractive feature of this mechanism for speciation is that it fulfils the criteria outlined in the introduction. Firstly, genomic alterations can occur which rapidly result in dramatic phenotypic effects. Secondly, species can be created in small populations giving a punctuated mode of speciation. Thirdly, after the period of genetic upheaval the element can become hormonized with the genome resulting in evolutionary stasis until a new element invades the species (Pollard, 1984b, 1987).

CONCLUSION

In the *Logic of Life* Jacob compared DNA with a computer program but made the important distinction that 'one [the computer program] can change at will according to the results obtained, the nucleic acid structure, on the contrary, is inaccessible to acquired experience and remains unchanged throughout life' (Jacob, 1973: 9). However, one of the themes of this chapter is that DNA is a 'metabolic molecule' capable of expanding, contracting, moving around and responding to the environment with an element of feedback regulation. This is not to say that instruction from the environment occurs in a true Lamarckian sense, but that in contravention to Weismann's doctrine, the germline has strategies for increasing genetic variability under periods of environmental upheaval. Furthermore, gene conversion events of the kind envisaged by Steele, or the selection of particular somatic mutant lineages during the life-span of organisms followed by conversion of these lineages into gametes or rametes, can lead to a type of feedback between the environment and the germline.

The synthetic theory of evolution is based on the predictable behaviour of mutations spreading through populations causing gradual phenotypic shifts. One of the other themes of this chapter, however, is that the genetic changes that produce major shifts in form during evolution are qualitatively different from, and occur more rapidly than, the 'beanbag genetics' envisaged by the synthetic theory. Moreover, the mechanism of beanbag genetics plays a minor role in evolutionary change, perhaps honing up the fit between the organism and its environment. This distinction between macro- and micro-evolutionary events at a genetic level is reminiscent of the writings of earlier geneticists such as Goldschmidt (1940) and prior to the development of the synthetic theory, of de Vries and Bateson (see Stanley, 1981). It is concerned, therefore, less with change in structural genes (such as those involved with colour vision discussed above) than with the reorganization of the genome to produce altered developmental pathways.

The mechanisms of genetic change discussed in this chapter are very diverse and complex, and although the behaviour of some of these events can be analysed using mathematical models (see Lewontin, 1985), the element of historical accident and the organisms' responsiveness to environmental alteration may put the realistic representation of evolution beyond such models. However, it is necessary to determine the dominant means of genetic variation that produces new structures during evolution. Unfortunately, to date, we have little understanding of how the genetic events we observe in chromosomal DNA have influenced alterations in form. It does seem possible to me, however, that a major feature of evolution is the reprogramming of gene batteries (maybe by moving regulatory sequences) in response to dramatically altered environments with the result that organisms are pushed down different developmental pathways, perhaps alternative pathways in the epigenetic landscape in much the manner that Waddington envisaged (Waddington, 1957). How the central plans for structures such as the amnion and pentadactyl limb were originally laid down, however, will be the challenge for biology over the next few decades (see Goodwin, 1984; Ho, 1984; Pollard, 1987 for further discussion).

ACKNOWLEDGEMENTS

I would like to acknowledge the continuous and stimulating discussion with Dr Ted Steele over the mechanisms of evolution and to thank Dr Pam Stanley for constructive criticism of the manuscript.

REFERENCES

Allison, L.A., Moyle, M., Shales, M., and Ingles, C.J. (1985). Extensive homology among the largest subunits of eukaryotic and prokaryotic RNA polymerase, *Cell*, **42**, 599–610.

Alt, F.W., Kellems, R.F., Bertino, J.R., and Schimke, R.T. (1978). Selective multiplication of dihydrofolate reductase genes in methotrexate-resistant variants of cultured murine cells, *J. Biol. Chem.*, **253**, 1357–70.

Baltimore, D. (1986). Retroviruses and retrotransposons: The role of reverse transcription is shaping the eukaryotic genome, *Cell*, **40**, 481–2.

Bennett, M.D., and Smith, J.B. (1976). Nuclear DNA amounts in angiosperms, *Phil. Trans. R. Soc. Lond.*, **274**, 227–74.

Bostock, C.J. (1986). Mechanisms of DNA sequence amplification and their evolutionary consequences, *Phil. Trans. R. Soc. Lond.* **312**, 261–73.

Bregliano, J.C., and Kidwell, G.M. (1983). Hybrid dysgenesis determinants. In J. Shapiro (ed.), *Mobile Genetic Elements*, Academic Press, New York. pp. 363–410.

Brent, L., Rayfield, L.S., Chandler, P., Fiertz, W., Medawar, P.B., and Simpson, E. (1981). Supposed Lamarckian inheritance of immunological tolerance, *Nature*, **290**, 508–12.

Bucheton, A., Simonelig, M., Vaury, C., and Crozatier, M. (1986). Sequences similar to the I transposable element involved in I-R hybrid dysgenesis in *D. melanogaster* occur in other *Drosophila* species, *Nature*, **322**, 650–2.

Buss, L.W. (1983). Evolution, development and the units of selection, *Proc. Nat. Acad. Sci. USA*, **80**, 1387–91.

Cavalier-Smith, T. (1985). *The Evolution of Genome Size*, John Wiley, Chichester.

Chasin, L. (1985). The dihydrofolate reductase locus. In M.M. Gottesman (ed.), *Molecular Cell Genetics*, John Wiley, New York, pp. 449–88.

Coderre, J.A., Beverley, S.M., Schimke, R.T., and Santi, D.V. (1983). Overproduction of bifunctional thymidylate synthetase — dihydrofolate reductase in methotrexate-resistant *Leishmania tropica*, *Proc. Nat. Acad. Sci. USA*, **80**, 2132–6.

Cooper-Willis, C.A., Olson, J.C., Brewer, M.E., and Leslie, G.A. (1985). Influence of paternal immunity on idiotype expression in offspring, *Immunogenetics*, **21**, 1–10.

Cowell, J.K. (1982). Double minutes and homogeneously staining regions: gene amplification in mammalian cells, *Ann. Rev. Genet.* **16**, 21–59.

Cullis, C.A. (1979). Segregation of the isozymes of flax genotrophs, *Biochem. Genet.*, **17**, 391–401.

Cullis, C.A. (1983). Environmentally induced DNA changes in plants, *CRC Crit. Rev. Pl. Sci.*, **1**, 117–31.

Cullis, C.A. (1984). Environmentally induced DNA changes. In J.W. Pollard (ed.), *Evolutionary Theory: Paths into the Future*, John Wiley, Chichester, pp. 203–16.

Cullis, C.A. (1986). Plant DNA variation and stress, *Stadler Genet. Symp.* **17**, 143–55.
Cullis, C.A., and Charlton, L.M. (1981). The induction of ribosomal DNA charges in flax. *Plat. Sci. Lett.*, **20**, 213–17.
Daniels, G.R., Fox, G.M., Loewensteiner, D., Schmid, C.W., and Deininger, P.H. (1983). Species specific homogeneity of the primate Alu family of repeated DNA sequences, *Nucleic Acid Res.*, **11**, 7579–93.
Davidson, E.H., and Posakony, J.W. (1982). Repetitive sequence transcripts in development, *Nature*, **297**, 633–5.
de Cicco, D.V., and Glover, D.M. (1983). Amplification of rDNA and type 1 sequences in *Drosophila* males deficient in rDNA, *Cell*, **32**, 1217–25.
Deininger, P.L., and Daniels, G.R. (1986). The recent evolution of mammalian repetitive DNA elements, *Trends Genet.*, **2**, 76–80.
Doolittle, W.F. (1978). Genes in pieces: were they together? *Nature*, **272**, 581–2.
Doolittle, W.F. (1985). Middle repetitive DNAs. In T. Cavalier-Smith (ed.), *The Evolution of Genome Size*, John Wiley, Chichester, pp. 443–87.
Dover, G.A. (1982). Molecular drive: A cohesive mode of species evolution, *Nature*, **299**, 111–17.
Dover, G.A. (1986). Molecular drive in multigene families: How biological novelties arise, spread and are assimilated. *Trends Genet.*, **2**, 159–65.
Dover, G.A., and Flavell, R.B. (1984). Molecular co-evolution: DNA divergence and the maintenance of function, *Cell*, **38**, 622–3.
Dover, G.A., and Tautz, D. (1986). Conservation and divergence in multigene families; alternatives to selection and drift, *Phil. Trans. R. Soc. Lond. B*, **312**, 275–89.
Durrant, A. (1971). Induction and growth of flax genotophs, *Heredity*, **27**, 277–98.
Eckert, E.L., and Green, H. (1986). Structure and evolution of the human involucrin gene, *Cell*, **46**, 583–9.
Eldredge, N., and Gould, S.J. (1972). Punctuated equilibrium: An alternative to phyletic gradualism. In T.J.M. Schopf (ed.), *Models in Paleobiology*, Freeman Cooper, California, pp. 82–115.
Faras, A.J., Garapin, A.C., Levinson, W.E., Bishop, J.M., and Goodman, H.M. (1973). Characterisation of the low molecular weight RNAs associated with the 70S RNA of Rous sarcoma virus, *J. Virol.*, **12**, 334–42.
Fink, G.R., Boeke, J.D., and Garfinkel, D.J. (1986). The mechanism and consequences of retrotransposition, *Trends Genet.*, **2**, 118–24.
Flavell, R.B. (1982). Amplification, deletion and re-arrangement: Major sources of variation during species divergence. In G.A. Dover and R.B. Flavell (eds), *Genome Evolution*, Academic Press, London, UK, pp. 301–34.
Flavell, R.B. (1986). Repetitive DNA and chromosome evolution in plants, *Phil. Trans. R. Soc. Lond. B*, **312**, 227–42.
Gall, J.G. (1969). The genes for ribosomal RNA during oogenesis, *Genetics*, **61**, suppl., 122–31.
Gall, J.G., and Pardue, M.L. (1969). Formation and detection of RNA-DNA hybrid molecules in cytological preparations, *Proc. Nat. Acad. Sci., USA.*, **63**, 378–83.
Gilbert, W. (1978). Why genes in pieces? *Nature*, **271**, 501.
Gilbert, W. (1985). Genes in pieces revisited, *Science*, **228**, 823–4.
Goldschmidt, R. (1940). *The Material Basis of Evolution.*, Yale University Press, New Haven.
Goodwin, B.C. (1984). Changing from an evolutionary to a generative paradigm in biology. In J.W. Pollard (ed.), *Evolutionary Theory: Paths into the Future*, John Wiley, Chichester, pp. 99–120.
Gorczynski, R.M., and Steele, E.J. (1980). Inheritance of acquired immunological

tolerance to foreign histocompatibility antigens in mice, *Proc. Nat. Acad. Sci. USA*, **77**, 2871–5.

Gorczynski, R.M., and Steele, E.J. (1981). Simultaneous yet independent inheritance of somatically acquired tolerance to two distinct H-2 antigenic haplotype determinants in mice, *Nature*, **289**, 678–81.

Gorczynski, R.M., Kennedy, M., MacRae, S., and Ciampi, A. (1983). A possible maternal effect in abnormal hyporesponsiveness to specific alloantigens in offspring born to neonatally tolerant fathers, *J. Immunology*, **131**, 1115–20.

Gould, S.J. (1982). The meaning of punctuated equilibrium and its role in validating a hierarchial approach to macroevolution. In R. Milkman (ed.), *Perspectives on Evolution*, Sinauer Associates, Mass., pp. 83–104.

Haldane, J.B.S. (1964). A defense of beanbag genetics. Reprinted in D. Garber (ed.), *Genetic Perspectives in Biology and Medicine*, University of Chicago Press, Chicago, pp. 35–51.

Ho, M-W. (1984). Where does biological form come from? *Rev. Biol.*, **77**, 147–79.

Ho, M-W., and Saunders, P.T. (1984). *Beyond Neo-Darwinism: An Introduction to the New Evolutionary Paradigm*, Academic Press, London.

Hood, L., Campbell, J.H., and Elgin, S.C.R. (1975). The organization, expression and evolution of antibody and other multigene families. *Ann. Rev. Genet.*, **9**, 305–53.

Hunkapiller, T., and Hood, L. (1986). The growing immunoglobulin gene superfamily. *Nature*, **323**, 15–16.

Hunkapiller, T., Huang, H., Hood, L., and Campbell, J.H. (1982). The impact of modern genetics on evolutionary theory. In R. Milkman (ed.), *Perspectives on Evolution*, Sinauer Associates, Mass., pp. 164–89.

Huxley, J.S. (1942). *Evolution: The Modern Synthesis*, Allen and Unwin, London.

Ikawa, Y., Ross, J., and Leder, P. (1974). Association between globin messenger RNA and 60S RNA derived from Friend leukaemia virus. *Proc. Nat. Acad. Sci., USA*, **71**, 1154–8.

Jacob. F. (1973). *The Logic of Life: a History of Heredity*, Pantheon books, New York.

Jeffries, A.J. (1982). Evolution of globin genes. In G.A. Dover and R.B. Flavell (eds), *Genome Evolution*, Academic Press, London, pp. 157–75.

Jelinek, W.R., and Schmid, C.W. (1982). Repetitive sequences in eukaryotic DNA and their expression, *Ann. Rev. Biochem.*, **51**, 813–44.

Jenkins, N.A., Copeland, N.G., Taylor, B.A., and Lee, B.K. (1981). Dilute (d) coat colour mutation of DBA/2J mice is associated with the site of integration of an ectopic MuLV genome, *Nature*, **293**, 370–4.

Johnston, R.N., Beverley, S.M., and Schmike, R.T. (1983). Rapid spontaneous dihydrofolate reductase gene amplification shown by fluorescence-activated cell sorting, *Proc. Nat. Acad. Sci. USA.*, **80**, 3711–15.

Jones, C.W., and Kafatos, F.C. (1980). Structure, organization and evolution of developmentally regulated chorion genes in a silk moth, *Cell*, **22**, 855–67.

Kaufman, J.F., Auffray, C., Korman, A.J., Shackelford, D.A., and Strominger, J. (1984). The class II molecules of the human and murine major histocompatibility complex, *Cell*, **36**, 1–13.

Kidwell, M.G. (1983). Evolution of hybrid dysgenesis determinants in *Drosophila melanogaster*, *Proc. Nat. Acad. Sci. USA.*, **80**, 1655–9.

Kidwell, M.G., Kidwell, J.F., and Sved, J.A. (1977). Hybrid dysgenesis in *Drosophila melanogaster*: a syndrome of aberrant traits including mutation, sterility and male recombination, *Genetics*, **86**, 813–33.

Klekowski, E.J., Mohr, H., and Kazarinova-Fukshansky, N. (1986). Mutation, apical meristems and developmental selection in plants. *Stadler Genet. Symp.*, **17**, 79–113.

Lansman, R.A., Stacey, S.N., Grigliatti, T.A., and Brock, H.W. (1985). Sequences homologous to the P mobile element of *Drosophila melanogaster* are widely distributed in the subgenus Sophophora, *Nature*, **318**, 561–3.

Laski, F.A., Rio, D.C., and Rubin, G.M. (1986). Tissue specificity of *Drosophila* P-element transposition is regulated at the level of mRNA splicing, *Cell*, **44**, 7–19.

Lewis, J.W., Davide, J.P., and Merlera, P.W. (1982). Selective amplification of polymorphic dihydrofolate reductase gene loci in Chinese hamster lung cells, *Proc. Nat. Acad. Sci. USA.*, **79**, 6961–5.

Lewontin, R.C. (1985). Population genetics, *Ann. Rev. Genet.*, **19**, 81–102.

Marchionni, M., and Gilbert, W. (1986). The triosephosphate isomerase gene from maize: Introns antedate the plant–animal divergence, *Cell*, **46**, 133–41.

McClintock, B. (1978). Mechanisms that rapidly reorganize the genome, *Stadler Genet. Symp.*, **10**, 25–48.

McClintock, B. (1984). The significance of responses of the genome to challenge, *Science*, **226**, 792–801.

McGinnis, W., Garber, R.L., Wirz, J., Kuroiwa, A., and Gehring, W.J. (1984). A homologous protein-coding sequence in *Drosophila* homeotic genes and its conservation in other metazoans, *Cell*, **37**, 403–8.

Meinkoth, J.L., Legouy, E., Brison, O., and Wahl, G.M. (1986). New RNA species is produced by alternate polyadenylation following rearrangement associated with CAD gene amplification, *Somatic Cell Mol. Genet.*, **12**, 339–50.

Mellor, J., Fulton, S.M., Dobson, M.J., Wilson, W., Kingsman, S.M., and Kingsman, A.J. (1985). A retrovirus-like strategy for expression of a fusion protein encoded by yeast transposon Tyl. *Nature*, **313**, 243–6.

Mount, S.M., and Rubin, G.M. (1985). Complete nucleotide sequence of the *Drosophila* transposable element Copia: Homology between Copia and retroviral proteins, *Mol. Cell. Biol.*, **5**, 1630–8.

Mullbacher, A., Ashman, R.B., and Blanden, R.V. (1983). Induction of T-cell hyporesponsiveness to Bebaru in mice and abnormalities in the immune responses of progeny of hyporesponsive males, *Aust. J. Exp. Biol. Med. Sci.*, **61**, 187–91.

Nathans, J., Thomas, D., and Hogness, D.S. (1986). Molecular genetics of human color vision: The genes encoding blue, green and red pigments, *Science*, **232**, 193–202.

Never, P., and Saedler, H. (1977). Transposable genetic elements as agents of gene instability and chromosomal rearrangements, *Nature*, **268**, 109–15.

Ohno, S. (1970). *Evolution by Gene Duplication*, Springer-Verlag. Heidelberg–Berlin, New York.

Ohno, S. (1984). Repeats of base oligomers as the primordial coding sequences of the primeval earth and their vestiges in modern genes, *J. Mol. Evol.*, **20**, 313–21.

Ohno, S., Mori, N., and Matsunaga, T. (1986). *Conditio sine qua non* for *de novo* emergence of new genes and the concept of primordial building blocks, *Stadler Genet. Symp.*, **17**, 157–74.

Orgel, L.E., Crick, F.H.C., and Sapienza, C.A. (1980). Selfish DNA. *Nature*, **288**, 645–6.

Paquin, C.E., and Williamson, V.M. (1984). Temperature effects on the rate of Ty transposition, *Science*, **226**, 53–5.

Perler, F., Efstratiadis, A., Lomedic, P., Gilbert, W., Kolodner, R., and Dodgson, J. (1980). The evolution of genes: the chicken preproinsulin gene, *Cell*, **20**, 555–6.

Pollard, J.W. (1984a). Is Weismann's barrier absolute? In M.-W. Ho and P.T. Saunders (eds), *Beyond Neo-Darwinism: An Introduction to the New Evolutionary Paradigm*, Academic Press, London, pp. 291–314.

Pollard, J.W. (1984b). Paths into the future. In J.W. Pollard (ed.), *Evolutionary Theory: Paths into the Future*, John Wiley, London, pp. xv-xxii.

Pollard, J.W. (1987). The moveable genome, Weismann's doctrine and new models for speciation, *Biol Forum*, **80**, 11–54.
Rice, G.C., Hoy, C., and Schimke, R.T. (1986). Transient hypoxia enhances the frequency of dihydrofolate reductase gene amplification in Chinese hamster ovary cells, *Proc. Nat. Acad. Sci. USA*, **83**, 5978–82.
Rogers, J. (1985). Exon shuffling and intron insertion in serine protease genes, *Nature*, **315**, 458–9.
Rose, M.R., and Doolittle, W.F. (1983). Molecular biological mechanisms of speciation, *Science*, **220**, 157–62.
Schimke, R.T. (ed.) (1982). *Gene Amplification*, Cold Spring Harbor Lab., New York.
Schimke, R.T. (1984). Gene amplification in cultured animal cells, *Cell*, **37**, 705–13.
Shapiro, J.A., and Cordell, B. (1982). Eukaryotic mobile and repeated elements, *Biol. Cell*, **43**, 31–54.
Sharp, P.A. (1983). Conversion of RNA to DNA in mammals: Alu-like elements and pseudogenes, *Nature*, **301**, 471–2.
Sharp, P.A. (1985). On the origin of RNA splicing and introns, *Cell*, **42**, 397–400.
Smith, G.P. (1973). Unequal crossover and the evolution of multigene families, *Cold Spring Harb. Symp. Quant. Biol.* **38**, 507–13.
Smithies, O., Gregg, R.G., Boggs, S.S., Koralewski, M.A., and Kucherlapati, R.S. (1985). Insertion of DNA sequences into the human chromosomal α-globin locus by homologous recombination, *Nature*, **317**, 230–4.
Soares, M.B., Schon, E., Henderson, A., Karathanasis, S.K., Cate, R., Zeitlin, S., Chirgwin, J., and Efstratiadis, A. (1985). RNA-mediated gene duplication: the rat preproinsulin I gene is a functional retroposon, *Mol. Cell. Biol.*, **5**, 2090–103.
Stanley, S.M. (1981) *The New Evolutionary Time Table*. Basic Books., New York.
Stark, G.R., and Wahl, G.M. (1984). Gene amplification, *Ann. Rev. Biochem.*, **53**, 447–91.
Steele, E.J. (1979). *Somatic Selection and Adaptive Evolution: On the Inheritance of Acquired Characters*, 1st edn, Williams-Wallace Productions International, Toronto.
Steele, E.J. (1981). *Somatic Selection in Adaptive Evolution: On the Inheritance of Acquired Characters*, 2nd edn, University of Chicago Press, Chicago.
Steele, E.J., Gorczynski, R.M., and Pollard, J.W. (1984). The somatic selection of acquired characters. In J.W. Pollard (ed.), *Evolutionary Theory: Paths into the Future*, John Wiley, Chichester, pp. 217–37.
Steele, E.J., and Pollard, J.W. (1987). Somatic hypermutation by gene conversion via the error prone DNA→RNA→DNA information loop. *Mol. Immunol.*, **24**, 667–73.
Stewart, R.N. (1978). Ontogeny of the primary body in chimeral forms of higher plants. In S. Subteiny and I.M. Sussex (eds), *The Clonal Basis of Development*, Academic Press, New York, pp. 131–60.
Strand, D.J., and McDonald, J.F. (1985). Copia is transcriptionally responsive to environmental stress, *Nucleic Acids Res.*, **13**, 4401–9.
Sudhof, T.C. Goldstein, J.L., Brown, M.S., and Russell, D.W. (1985a). The LDL receptor gene: a mosaic of exons shared with different proteins, *Science*, **228**, 815–22.
Sudhof, T.C., Russell, D.W., Goldstein, J.L., Brown, M.S., Sanchez-Pescador, R., and Bell, G.I. (1985b). Casette of eight exons shared by genes for LDL receptor and EGF precursor, *Science*, **228**, 893–5.
Sutcliffe, J.G., Milner, R.J., Gottesfeld, J.M., and Lerner, R.A. (1984). Identifier sequences are transcribed specifically in brain, *Nature*, **308**, 237–41.
Syvanen, M. (1984). The evolutionary implications of mobile genetic elements, *Ann. Rev. Genet.*, **18**, 271–93.
Syvanen, M. (1985). Cross-species gene transfer: Implications for a new theory of evolution. *J. Theor. Biol.*, **112**, 333–43.

Syvanen, M. (1986). Cross-species gene transfer: a major factor in evolution? *Trends Genet.*, **2**, 63–6.

Tartof, K.D. (1973). Regulation of ribosomal RNA gene multiplicity in *Drosophila melanogaster*, *Genetics*, **73**, 57–71.

Taylor, J.M, and Cywinski, A. (1984). A defective retrovirus particle (SE21Q1G) packages and reverse transcriber cellular RNA, utilizing tRNA-like primers, *J. Virology.*, **51**, 267–71.

Temin, H.M. (1976). The DNA pro-virus hypothesis, *Science*, **192**, 1075–80.

Temin, H.M. (1985). Reverse transcription in the eukaryotic genome: retroviruses, pararetroviruses, retroposons and retrotranscripts, *Mol. Biol. Evol.*, **2**, 455–68.

Temin, H.M., and Engels, W. (1984). Movable genetic elements and evolution. In J.W. Pollard (ed.), *Evolutionary Theory: Paths into the Future*, John Wiley, Chichester, pp. 173–201.

Thomas, K.R., Folger, K.R., and Capecchi, M.R. (1986). High frequency targeting of genes to specific sites in the mammalian genome, *Cell*, **44**, 419–28.

Vanin, E.F. (1984). Processed pseudogenes, *Biochem. Biophys. Acta*, **782**, 231–41.

Varmus, H.E., Quintrell, N., and Ortiz, S. (1981). Retroviruses are mutagens: insertion and excision of a non-transforming provirus alters expression of a resident transforming provirus, *Cell*, **25**, 23–6.

Waddington, C.H. (1957). *The Strategy of the Genes*, George Allen and Unwin, London.

Wagner, M. (1986). A consideration of the origin of processed pseudogenes, *Trends Genet.*, **2**, 134–7.

Wahl, G.M., Vitto, L., and Rubnitz, J. (1983). Co-amplification of rRNA genes with CAD genes in N-(phosphonoacetyl)-L-aspartate-resistant Syrian hamster cells, *Mol. Cell. Biol.*, **3**, 2066–75.

Walbot, V., and Cullis, C.A. (1985). Rapid genomic change in higher plants. *Ann. Rev. Pl. Physiol.*, **36**, 367–96.

Weinberg, R.A. (1980). Origins and roles of endogenous retroviruses, *Cell*, **22**, 643–4.

Weinberger, C., Hollenberg, S.M., Rosenfeld, M.G., and Evans, R.M. (1985). Domain structure of human glucocorticoid receptor and its relationship to the V-erbA oncogene product, *Nature*, **318**, 670–2.

Weismann, A. (1893). *The Germ Plasm: A Theory of Heredity*, Scott Publishing, London.

Wilson, A.C., Sarich, V.M., and Maxson, R.L. (1974). The importance of gene rearrangement in evolution: Evidence from studies on rates of chromosomal, protein and anatomical evolution. *Proc. Nat. Acad. Sci. USA*, **71**, 3028–30.

Ycas, M (1972). *De novo* origin of periodic proteins, *J. Mol. Evol.*, **2**, 17–27.

Zieg, J., Clayton, C.E., Ardeshir, F., Giulotto, E., Swyryd, E.A., and Stark, G.R. (1983). Properties of single-step mutants of Syrian hamster cell lines resistant to N-(phosphonoacetyl)-L-aspartate, *Mol. Cell. Biol.*, **3**, 2089–98.

Evolutionary Processes and Metaphors
Edited by M.-W. Ho and S.W. Fox

CHAPTER 6

Operationalism, optimality and optimism: suitabilities versus adaptations of organisms

JACK P. HAILMAN

ABSTRACT

The impression that organisms are suitably designed for the environmental problems they face is a proposition in need of a clear methodology for evaluation of specific cases. The apparent reality of such suitability has been cited as evidence for the design of organisms both by the hand of God and by natural selection, the two remarkably similar arguments sharing the logical flaw of inferring processes from the end results the processes produce. The task is to specify a process-independent methodology for identifying suitabilities, a term employed here to avoid the selectionist implications of words like 'adaptations' and 'fitness'. Communication among evolutionary biologists will be improved if the sought for methodology is constructed in terms of measurable variables (operationalism).

Two methods have commonly been employed for identifying adaptations, and these provide the principal candidates for operational modification to identify suitabilities. One method seeks to establish, commonly by quantitative techniques derived from mathematical 'operations research', that a particular species is optimally constructed (or behaves optimally) relative to its environmental problems; this methodology is optimality theory, broadly speaking. The other method seeks to establish the existence of character–environment correlations among unrelated species; this is the comparative method.

Scrutiny of logical and practical problems attendant to adapting these methodologies for the identification of suitabilities reveals that optimality theory cannot be made operational whereas the comparative approach encounters no insurmountable hurdles. The two methodologies are compared point by point, and then additional criticisms of the comparative method are considered. These latter criticisms are all traceable to incomplete consideration of the minimal requirements of a comparative study; namely, the comparison of species in at least two independent taxa such that some members of each taxon inhabit the same environment as some members of the other taxon and each taxon also contains other species inhabiting different environments. Only when unrelated species living in the same environment share characteristics (convergent evolution) and differ in these characteristics from their relatives living in other environments, is the reality of evolutionary suitabilities established. The comparative method, appropriately applied, appears to constitute the only viable approach towards the operational identification of evolutionary suitabilities.

One of the most pervasive marvels of nature is that each kind of living organism seems so eminently suited to its own particular way of life. These apparent suitabilities constitute a focal point of evolutionary biology, usually treated under the heading of adaptations. In this chapter I argue that the notion of adaptations undesirably binds the end results (the suitabilities) to a hypothesis of the evolutionary mechanism that brings them about (natural selection). I further argue that a divorce of results from posited mechanism is strengthened if the results can be defined operationally. I then consider two of the most promising approaches to operational definitions of suitabilities — the optimality method and the comparative method — finding the former to fall short of needed attributes but the latter to engender optimism that suitabilities can be operationally defined in a satisfactory way.

THE PROBLEM

Myths of many cultures attempt to account in various ways for the apparent suitability of organisms to their environments, the subject evoking comment from Biblical times through the ancients of Western civilization right down to the present day. The issue was introduced strongly into biology by the great pre-Darwinian English naturalist John Ray (1627–1705), whose studies of anatomy suggested that parts of the body work harmoniously like parts of a man-made instrument. Ray saw in this apparent 'design' the hand of a Designer, and published his viewpoint in a book entitled *The Wisdom of God in the Works of Creation.* The great French biologist, Jean Baptiste Pierre Antoine de Monnet, Chevalier de Lamarck (1744–1829), put forward the bold notion that characteristics of organisms were not specially created by God but evolved over time within lineages. As everyone knows from the term that adopts his name, Lamarck conceived of the evolutionary process as one of individual change towards environmental appropriateness that was passed on by parents to their offspring. The English theologian William Paley (1743–1805) was appalled by evolutionary thinking, and just a few years before his death published his most important work, *Natural Theology* (1802), in which he roundly attacked Lamarck's views, and resurrected those of Ray with great clarity and ingenuity. It was Paley who argued that animals are constructed as intricately as watches, and since so complex an instrument was designed with its purpose in mind, so must animals have been designed purposely by the ultimate Designer.

One might think that with these lively intellectual events already history the entrance of Charles Robert Darwin (1802–82) would be anticlimactic, but of course we know otherwise. Although the notion of organic evolution was old hat to the intelligentia, Darwin provided evolution with a simple and potentially pervasive mechanism: natural selection. This hypothesis convinced those on the fence and provoked renewed attack from theologians on the other side. The important point here, however, is the shift to thinking about suitabilities not merely as evolved, but as evolved *through natural selection*. This shift is today so pervasively complete that many contemporary biologists have unwittingly created a tautology by parading the existence of suitabilities as *evidence* for the process of natural selection. Even Paley might have been amused by the irony: uncritical biologists have taken this view that 'suitabilities are evidence for creation by God' and simply substituted 'natural selection' for 'God'.

In so far as I can determine, no one has even begun to attempt to define adaptations operationally. In fact, the situation seems to be worse than that, for in a cursory survey of some books with 'adaptation' in the title I could not find any kind of definition of adaptations as evolutionary results (e.g. Burnett and Eisner, 1964; V. Grant, 1963; Mossakowski and Roth, 1982; Wallace and Srb, 1961). One book does define the process of adaptation as 'the hereditary adjustment to the environment' (V. Grant, 1963: 93), which leads by implication to the idea that adaptations as results are the manifestations of this process. However, as we cannot observe the past processes we obviously cannot define their products in terms of them. A sentence from another book adequately summarizes the viewpoints of many: 'The authors believe that a neat, well-pruned definition of adaptation fails to convey to the reader a feeling for what adaptation is, just as a definition of an easy chair fails to describe the satisfaction one feels when one sinks his tired body into it' (Burnett and Eisner, 1964:3). Are we to believe in, and teach our disciples to believe in, something that we cannot or will not attempt to define, much less define operationally? Science and pedagogy should be made of sterner stuff.

One needs merely to thumb the contents of the present volume in order to realize that evolutionary theory is undergoing pervasive scrutiny. Many of the present authors have been struggling for years to distinguish what is and what is not adequately established in evolutionary biology (e.g. Bateson, 1982, 1984; Fox, 1984; Gray, 1986; Hailman, 1982b; Ho, 1984; Ho and Saunders, 1979, 1982, 1984; Kitcher, 1984; Oyama, 1985; Saunders and Ho, 1982) but most of the issues they deal with are by no means new ones. Instead, new doubts are being raised about problems that many workers had considered solved and new scrutiny is being given to problems that once seemed intractable. We are, collectively in evolutionary biology, struggling to distinguish what is knowable from that which is not — and to know the knowable.

Most of the scrutiny of evolutionary biology is focused on processes and mechanisms, such as speciation and natural selection, rather than on the results they are posited to bring about. If mechanisms are not distinguished logically from observable phenomena, an intellectual muddle ensues. A good example of this muddle is an argument marshalled by 'creationists' in the United States that reduces to this: if it can be shown that animals did not evolve by natural selection then one has proved that they did not evolve. Yet we know from the fossil record that animals did evolve, and this fact will remain even if it be shown that natural selection was not the mechanism of evolution. Similarly, the suitability of organisms to their environments needs to stand in its own right on its own evidence; otherwise this suitability constitutes just as persuasive an argument for the hand of a Designer as envisioned by Ray and Paley as it does for adaptation by natural selection.

Therefore, the problem I undertake is to find a way of identifying suitabilities of organisms that is independent of the processes or mechanisms that may bring about such suitabilities. Specifically, I argue that posited suitabilities (like any other scientific hypothesis) must be empirically falsifiable in order to be admitted into the corpus of science, and that they therefore require operational definition.

OPERATIONALISM

There are probably a number of reasons that organismic and populational biology have been so slow to adopt the operationalism that has characterized physics for most of the twentieth century. For one thing, whole plants and animals are entities on our own scale of size: things we can observe directly, unlike the subatomic particles of physics and the objects in the far reaches of the cosmos that so captivate contemporary astronomers. Scientists in many other disciplines cannot 'see' their objects without the aid of instruments. Therefore, physical scientists are more or less forced to define entities and causal relationships among them on the basis of things they can measure, whereas it often seems 'obvious' to an ethologist (for example) when an animal is showing 'aggression'.

Another reason is that biology is exceedingly rich in variables. The natural phenomena of interest to biologists are always fraught with influences, so that even if one could measure all of them it would be a defeating task to attempt such measurements simultaneously and to characterize all the interactions. Instead, biologists have always relied more heavily on intuition than they commonly like to admit, groping to find key variables that account 'pretty well' for natural phenomena.

American Nobel laureate Percy W. Bridgman (1882–1961) consolidated the operational practices of physics into a philosophical outlook. Paraphrased and summarized almost to a caricature, Bridgman's (e.g. 1927, 1938, 1945) viewpoint was this: entities and causes have no meaning apart from the operations by which we measure them in practice. Therefore, if we wish to communicate in science we must specify the measuring operations that give meaning to the objects of our discussion.

Consider the problem of 'travel time' over immense distances such as between star systems. Suppose you time your rocket flight from our solar system to some other star system in the Milky Way; a friend with the same kind of timing device watches your flight from some third star system; you both start timing at liftoff and at setdown as you each perceive them; finally you compare a measured 'flight times'. Except under certain circumstances you will have measured different times because under these conditions 'flight time' is not well defined. The difference lies in the two different sets of measuring operations employed, specifically the fact that you were on board the rocket and your friend was observing from the outside.

What physicists quickly understood, and Bridgman encapsulated philosophically, was that measurements have no absolute meaning, but rather depend upon the measuring operations. If a definition is to be operational, therefore, the quantities on which it depends must be determined by explicit (or adequately implied) measuring operations. When this condition is fulfilled, then criterion values for the measured variables may be used to define logical sets, and anyone technically qualified to make the requisite measurements will arrive at the same answer as to whether something does or does not belong in the criterion set.

It is a tall order to make biology, especially evolutionary biology, operational — although attempts have been made in some areas of evolutionary biology (e.g. Hailman, 1976b; see also 1982a). Consider the notion of a species as 'a group of interbreeding individuals that are reproductively isolated from individuals of other such groups'. Operations need not be specified if they are clearly implied, but where is the implication of how one should measure 'interbreeding'? Clearly even long-lived animals never mate with all other members of the species of opposite sex, so 'interbreeding' is a slippery notion. Although definitions such as this one commonly provide no intellectual problems for the practising biologist, they should. It is simply not clear how one should go about assessing 'interbreeding' and 'reproductively isolated'. So that even if all the other aspects of the definition are clarified, these two key elements of the definition cannot be made operational until one provides the instructions for measuring these two variables.

In view of the high demands made by operational definitions and the subtle ways in which apparently reasonable definitions fail to qualify as operational, the task of finding or devising an operational definition for the suitabilities of organisms to their environments is bound to be a difficult and probably controversial endeavour. However, we can look forward to two commonly used approaches to the notion of adaptations — namely, the optimality and comparative approaches — and ask whether either of these can be modified appropriately for use as a definition for suitabilities.

OPTIMALITY CONSIDERED AND REJECTED

It is certainly a straightforward notion that has occurred to many persons independently that the apparent suitability of organisms to their environments is manifested in some kind of 'match' between the two. It might, therefore, be possible to devise a criterion for the degree of match, and if this degree is high ('optimal') to take the match as identifying a suitability.

The potential use of optimality notions in the quest for an operational definition of suitabilities is fraught with problems. Most (although not all) empirical optimality studies either (a) aim to show that an organism is optimal in order to demonstrate that it possesses an adaptation, or (b) assume that it is optimally adapted to find out how this adaptation works in practice. Therefore, 'optimality' itself carries adaptational baggage that we must be careful to discard if we are to divorce suitabilities from the mechanisms that bring them about. Furthermore, the notion of an optimal match between organism and environment may be called a fit, and this term too implies selectionist mechanisms. The implication is historical rather than etymological, largely due to Herbert Spencer's catch-phrase 'survival of the fittest'. If those organisms that fit their environments best also survive best, then fitness becomes synonymous with survival (or survival and reproduction) — a tautological use that is today widespread in evolutionary biology. These problems with optimality are truly semantic, however, and so long as one uses 'suitability' to avoid the mechanistic baggage of 'adaptation' and 'match' to avoid the mechanistic baggage of 'fit', it is possible to reason clearly.

Optimality theory in the broad sense has enjoyed immense popularity in recent years. This popularity is due in no small part to the trickling into biology of quantitative techniques that grew up during the Second World War under the rubric 'operations research' (there being no direct connection with Bridgmanian operationalism). These techniques were devised to manage the material and personnel of war more efficiently, and operations research included such topics as the expansion of statistics, network analysis, queue theory, Monte Carlo simulation, feedback control, information theory, and two logically related techniques of optimization: linear programming and the theory of zero-sum games of strategy (e.g., Singh, 1972). Many of these techniques, linear programming in particular, have come to biology principally through the social sciences, where they enjoy great popularity (e.g., Rachlin *et al.*, 1981).

It is only the mathematical sophistication that is new, however, not the notion of optimality in evolution. Darwin used this notion in Chapter 8 of the *Origin* with reference to the hexagonal packing of cells in the honey bee's (*Apis mellifera*) hive: 'Beyond this stage of perfection in architecture, natural selection could not lead; for the comb of the hive-bee, as far as we can see, is absolutely perfect in economizing labour and wax' (p. 353 from the 6th edn., Darwin, 1897). I shall have more to say about hexagonal packing below, but here need only note that use of such words as 'perfection' and 'absolutely perfect' to emphasize the point that optimality is an old notion in evolutionary theory.

Three genres of optimality studies

In considering below the logic of optimality criteria as a potential operational definition for suitabilities of organisms to their environments I will cite some empirical examples. I want to make clear at the outset that the cited studies were not necessarily undertaken to show optimality, and the degree to which individual authors argue that their (real or imagined) optimal results demonstrate adaptations varies greatly. Furthermore, finding (as I do) that optimality criteria are worthless for an operational definition of suitability does not imply that optimality studies *per se* have no worth. To make this latter point clearer, consider the spectrum of optimality studies as represented by three genre of investigations.

First, there are studies of optimality that find usefulness wholly apart from any evolutionary considerations whatever. For example, Hailman and Jaeger (e.g. 1976) use extensively an optimality notion in phototactic behaviour, defining the 'optimum ambient illumination' (OAI) of a species' eye as the infection point of its intensity–response curve. It is this point where a given difference in physical intensity produces the largest difference in perceived brightness and hence optimizes a variety of visual tasks such as detection of form and movement. This optimality concept led to the discovery that animals are not 'photopositive' or 'photonegative' as the literature implies, but rather that a given animal seeks light (behaves photopositively) when ambient is below its OAI and retreats from light (behaves photonegatively) when ambient is above this value. Furthermore, the spectral response curve of the phototactic animal differs markedly depending upon whether the ambient light is above or below the OAI. These are purely phenomena of the stimulus–control of behaviour and, albeit that other aspects of this research can be used to argue traditional adaptational questions, the basic optimality concept is free of adaptational considerations.

A second kind of optimality study assumes that the animal is doing something optimally, such as foraging, and uses this assumption to explore how the behaviour works in practice. In principle, the assumption optimality need not imply adaptationalist thinking although in practice such viewpoints are almost always present. This kind of optimality study is often referred to as the 'functional' approach, which Jamieson (1986) has recently criticized primarily for the myopia it causes in practitioners who fail to ask certain kinds of relevant questions about the animal.

The second kind of optimality study grades into a third in which the objective is to show that an animal is behaving optimally as a demonstration of its adaptation. That the two kinds of study do grade into one another, both embedding the assumption of natural selection having brought about the optimality in question, is clear from works on behavioural ecology (e.g. Krebs and Davies, 1984; see especially the chapter by Krebs and McCleery, 1984). Ironically, it is this third, explicit search for optimality that comes closest to providing a candidate for operationally defining suitabilities of organisms to their environments. By neglecting the assumptions and implications of natural selection, and concentrating instead on the apparently straightforward question of whether organisms are or are not doing something optimally, it is possible to see if the optimality approach is sufficiently tight to yield operational criteria for suitabilities.

1) Subtle circularity: lack of an optimal goal

We may begin by reiterating the commonly encountered reasoning that argues, *a posteriori*, that a particular value found is optimal essentially because it was the value found. Often this tautological reasoning is quite subtle, embedded in an interplay of results and discussion. For example, Kaufmann (1975, p. 221) does not use the term 'optimal', but does say that whereas 'no macropod has achieved a very high level of social development', the eastern grey kangaroo, *Macropus giganteus*, 'has a relatively impressive combination of related adaptations', one of which apparently is group size (to which most of the discussion was devoted). Kaufmann disagreed with previous authors 'that grey kangaroos have no persistent group structure. Caughley [a previous author — J.P.H.] concluded that group sizes depended on an essentially random process . . .and that there was no tendency for groups of a given size . . . to occur more often than expected from random association' (p. 220).

Kaufmann studied two 'mobs' of grey kangaroos, charting the frequency of various sizes of subgroups found in each. He believed the data for mob B to be the more realistic, and found a mean of 3.7 individuals per subgroup (Caughley had found 3.3). Is this the optimal size of a kangaroo subgroup? Is about 3.7 individuals some kind of suitable match to environmental problems? Perhaps, but the variation from one to 15 individuals in a subgroup suggests that 'average

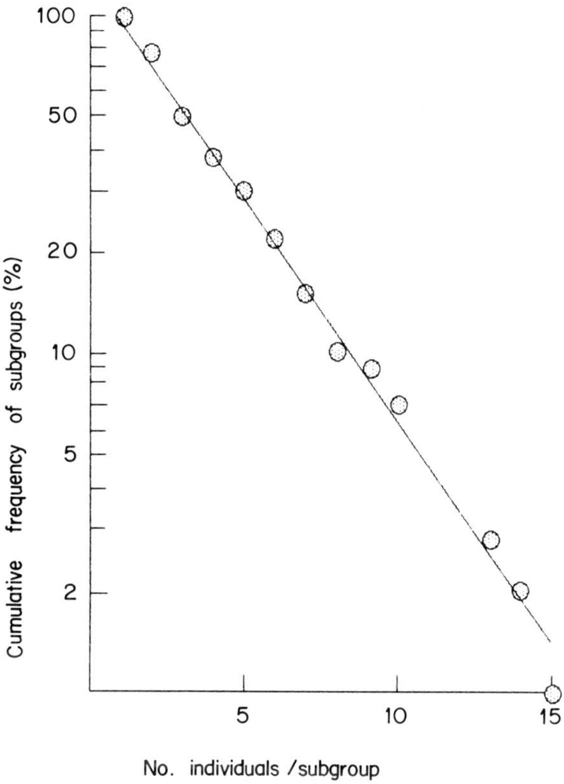

Figure 1 Survivorship curve of the size of subgroups of the grey kangaroo (see text). The fit to a straight line (exponential decay) suggests that group size is randomly determined

size' of a group is not even a very important measure of social behaviour. Furthermore, if one plots the log cumulative distribution of subgroup sizes for mob B the resulting graph (Figure 1) is remarkably similar to an exponential decay function. That is, the survivorship plot that results approximates a straight line (which I have fitted by eye rather than calculating a best fit by least-squares regression) of random expectation that occurs, for example, when kangaroos join groups with a constant probability regardless of the size of the existing group.

I re-emphasize that the purpose of this example is neither to show that the study cited is faulty nor to imply that studies of optimality in group size generally have problems. There are, for example, some quite sophisticated studies of this kind (e.g., Cohen, 1971: Pulliam and Caraco, 1984). The point made here is simply that kangaroo group size appears to be simply random rather than approaching some meaningfully average value, much less an optimal one. I shall have more to say below about the lack of a null hypothesis in optimality studies; the point here is that the *a posteriori* designation of a suitability based on quantitative data can be nothing more than a tautology showing no optimality at all.

2) Choice of the reputedly optimized variable

If we wish to see whether some measurable characteristic of an animal (or plant) attains or even approaches optimality it is evident that we also need to know the variable that is being optimized. For example, much of the literature on optimal foraging theory stimulated by the pioneering study of MacArthur and Pianka (1966) is cast in terms of optimizing time or energy (with more thorough examples considering other variables as specific nutrients or risk of predation). The idea is to optimize the net gain by specifying the cost-to-benefit ratio. Gray (1986) has provided a critical review of optimal foraging theory, but the only point that need concern us here is that the investigator must choose the variable supposedly being optimized, and that choice is of course subjective.

Be that as it may, one might argue that the subjective choice of the possibly optimized variable is no problem because all one is doing is testing whether or not the animal really is optimizing that particular variable. In other words, if one finds that the animal is optimizing that variable, then we have identified a suitability; if it is not optimizing the variable, then we may have chosen the wrong variable and it may be some other variable that is being optimized. A problem with this argument is that it removes the decision concerning suitabilities from manifestly two-valued logic. Ideally we should like to decide whether or not a given characteristic is a suitability, but here the decision leads to the alternatives of (a) yes it is because optimization has been proven, or (b) it might or might not be a suitability that is optimizing some variable not yet considered. On the basis of an optimization test we are unable to separate characteristics into the set of suitabilities and its complement (the set of traits that are not suitabilities).

Choice of a variable is in principle not a problem, however, for the choice is simply a hypothesis to be tested empirically. If the chosen variable is not the right one, then the investigator should choose a new one and try again. In practice the test for optimality is usually so involved that investigators do not in fact try again. Perhaps this problem is really not critical: if there are no other problems with the optimalization criterion for suitabilities, and no alternative criteria available, then we could live with the ambiguity of a negative result and at least identify some suitabilities. However, there are other problems.

3) Failure to consider alternative hypotheses

A large body of cost/benefit maximization argument has arisen with regard to 'parental investment' (e.g. Andersson, Wiklund and Rundgren, 1980; Barash,

1975; Boucher, 1977; Dawkins and Carlisle, 1976; Maynard Smith, 1977; Trivers, 1972). Studies of nest defence in birds have consistently interpreted results in terms of parental investment optimization, in particular the expectation that defence will increase through the breeding season. The argument is straightforward and compelling: as there is high mortality in young offspring, the older the offspring the greater being its worth in terms of producing breeders for the next generation. Therefore, parental defence should increase (in intensity, frequency and so on) in proportion to the age of the offspring. This expectation has been overwhelmingly confirmed in at least 12 empirical studies cited by Knight and Temple (1986a: 324).

The 'catch' is simply this: investigators are so convinced, *a priori*, that the increase in nest defence is an adaptation that they do not design their studies to *test* that hypothesis. All of the empirical studies entailed repeatedly visiting nests through the course of the breeding season. Knight and Temple (1986a) divided nests of two species into two groups: in one the nests were visited repeatedly as in previous studies and in the other each nest was visited only once, but visits were made at various times during the nesting season. In the former group they found the oft-replicated result of increased parental defence, but in the latter they found no difference through the nesting cycle in variables such as call rates, closest approach to the predator, number of dives and strikes and so on. It appears that increased nest defence is not due to optimizing parental investment at all, but rather is explicable in terms of experience of the parents. It may be that repeated visits are sensitizing the parents, or they may be positively conditioned by the fact that nest defence 'works' in making the human predators leave the nest without taking the young. Knight and Temple (1986a) further suggest that the parents increasingly lose their fear of the human intruders with repeated visits so that aggressive defence becomes more dominant in their behaviour; in another study they found that familiar nest visitors provoked stronger responses than novel ones (Knight and Temple, 1986b: 564, table III).

The lesson from this extended example is clear. The search for optimization can so blind investigators to alternative hypotheses that one after another adds to the body of empirical evidence reputedly demonstrating the expected result. Without the study by Knight and Temple (1986a) we would today be strongly tempted to term increased nest defence through the breeding season as a well-demonstrated behavioural adaptation predicted by the optimization of parental investment. This lesson should at least raise the flag of caution about using optimization criteria for identifying suitabilities.

4) Lack of a null hypothesis

A persistent problem in claims for optimality lies in the lack of comparison between the empirical findings and those expected on some random basis. The null hypothesis may be unknown to anyone or simply unknown to the author, who tries to show that his or her results approach a specified optimal condition. But approach how closely? If optimization is not absolutely perfect, how do we know that the empirical result is not simply random unless the random expectation is specified? This situation easily leads to claims for optimality when no such thing has been demonstrated, hence specifying a reputed adaptation that may in fact be imaginary.

To begin, consider what is likely the best-known case of optimality: hexagonal packing. It was probably D'Arcy Thompson's (1917) classic compendium that popularized the notion that biological materials (such as batteries of cnidarian nematocysts, mammalian epithelial cells and the like) are hexagonally packed. The well-known example of a honey bee's colony cells, argued by Darwin in the *Origin*, was used by Batchelet (1975) for an elegant mathematical demonstration that an hexagonal array is indeed the optimal use of space when packing circular objects of the same size. For example, if one places a monolayer of chequers on the bottom of a shallow but very large box, the maximum number of chequers can be contained in the box if they are arranged in an hexagonal array (instead of placing them in rows and columns, or any other arrangement).

Hexagonal packing may be visually obvious in structures such as bee hives, but if one considers entities such as animal territories the situation is not so simple. P.R. Grant (1968) reasoned that in an homogeneous environment with crowded conditions 'if territory sizes are similar and neighbors equally spaced, then the shape of each will be hexahedral' (p. 75). ('Hexagonal' would seem the more appropriate term; Grant was expressing the fact that avian territories are three-dimensional, but the only point of interest is their boundaries on the surface.)

How does one judge that territorial packing forms an hexagonal array? P.R. Grant (1968), using data gathered by R.T. Holmes on the pectoral sandpiper, *Calidris melanotos*, measured angles of incompletely mapped boundaries, as these should be 120° for regular hexagons. Actual angles ranged from a low of 70–79° to a high of 160–169°, with means for two years at about 117°, from which Grant concluded 'that territories are, on the average, hexahedral (or hexagonal), a result which is consistent with the reasoning given in the introduction' (p.78). The variation in angles indicates territories are not strictly optimally packed, but do they approach optimality? How can we decide when we do not know what average angle would occur under the null hypothesis of random packing?

A related example comes from Barlow (1974), who presented a photograph of the pit territories of a captive group of cichlid fish, *Tilapia mossambica* (see also Wilson, 1975:271–2, who relates the apparent joint discovery of these polygonal territories and reproduces Barlow's photograph as figure 12–9). Barlow counted sides rather than measuring angles, and found under two different densities of fish the modal number was five (not six). However, he noted that under the higher density 'the proportion of hexagonal territories nearly doubled' (p.877), yet it is clear from his figure 2 that the proportion of territories with *seven* sides nearly doubled as well. He concluded that 'the shift toward hexagons, therefore, resulted in more efficient packing for territories', but how can we tell without knowing what kind of distribution of side-numbers would occur under the null hypothesis of random packing?

Buckley and Buckley (1977) appear to be the first authors to recognize that without a null hypothesis, any claim for optimal packing is logical nonsense. Suppose one scatters points randomly on a surface and then encloses them with polygons formed by drawing perpendiculars through the midpoints of the lines

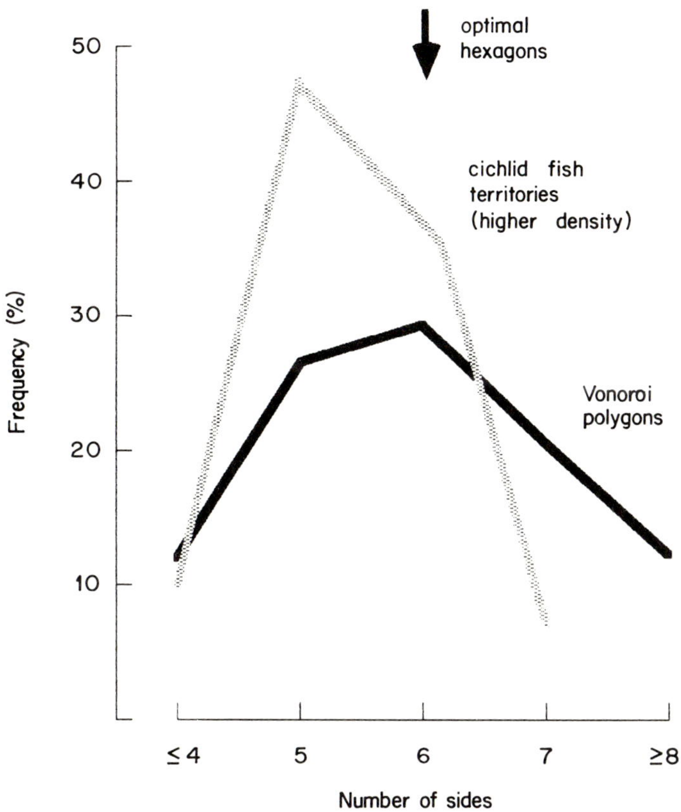

Figure 2 Frequency distribution of the number of sides of cichlid fish territories compared with the distribution of Voronoi polygons (darker curve) and optimally packed hexagons (arrow) all of which have six sides (see text). There is no indication that the fish territories are tending towards optimally packed hexagons, and in fact they are less efficiently packed than random Voronoi polygons

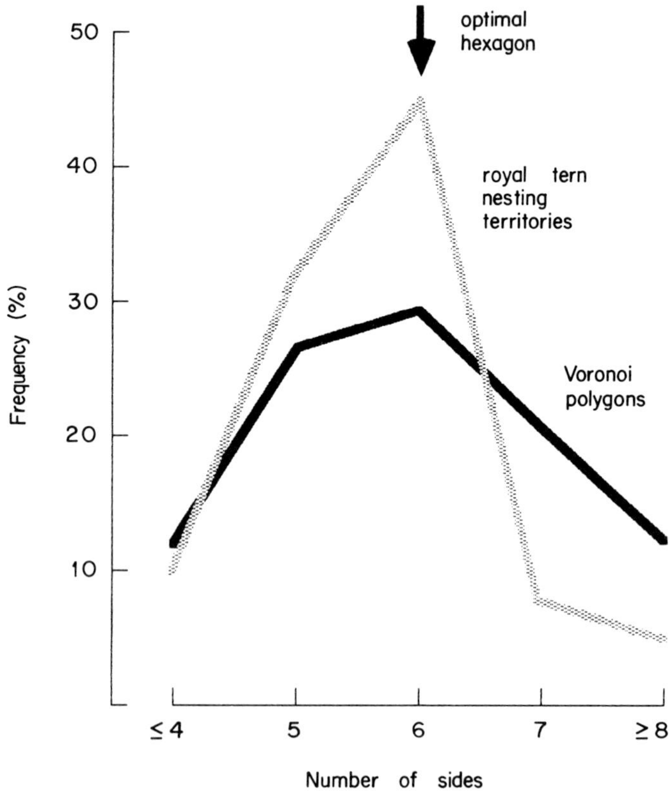

Figure 3 Frequency distribution of the number of sides of royal tern nesting territories compared with the distribution of Voronoi polygons (darker curve) and optimally packed hexagons (arrow) all of which have six sides (see text). The tern territories are more efficiently packed than those of the cichlid fish shown in Figure 2 (above), and the tern territories tend towards optimal packing. However, it is not obvious whether the tern territories are closer to optimal hexagons or to random Voronoi polygons

connecting the points. Under this condition of random packing, the resulting 'Voronoi polygons' have an average of six sides and a modal value of six (with five being nearly as frequent). A formula for generating the frequency distribution of side number in Voronoi polygons has not been derived, but I.K. Crain did a Monte Carlo simulation of 11 000 polygons with which Buckley and Buckley compared their data on nest territories of birds (considered in the next section).

Figure 2 compares the frequency distributions of the numbers of Voronoi polygons with those of high-density cichlid territories (the first from table 3 in Buckley and Buckley, 1977:40, with the other from figure 2 in Barlow 1974:877, which may entail some lack of precision in my reading of the published graph). It is evident that the high-density cichlid distribution comes nowhere near approaching hexagonal packing, and whereas it does depart from the Voronoi distribution, the departure is *not* in the direction of being more hexagonal like. Whether or not this result suggests that the fish are not packing themselves even as efficiently as random is irrelevant to our point here, which is simply that once the null hypothesis is known any illusion of approaching optimal packing quickly vanishes.

5) The unknown spectrum between null and optimal

Suppose the null expectation is well characterized, so that departures from it can be analysed statistically to satisfy the usual comments of journal reviewers. Do we not then have a situation that allows operational assessment of reputed optimality? No. Specification of the null expectation is not necessarily sufficient for judging the degree to which an empirical set of data tends toward optimality. I demonstrate this point by returning to the case of optimal packing of animal territories, this time considering the nesting territories of royal terns, *Sterna maxima*, studied by Buckley and Buckley (1977).

Figure 3 plots from Buckley and Buckley (1977:40, table 3) the distribution of sides of tern nesting territories compared with those of Vonoroi polygons. (The result is similar to their figure 2 on p. 43, but is here constructed to match my Figure 2 above and to show the data with the boundaries Buckley and Buckley used in their statistical test.) It can be seen from Figure 3 that the tern territories have a modal distribution at six sides, which is markedly stronger than the mode of the Voronoi polygons. Furthermore, Buckley and Buckley (1977) showed that the tern distribution was statistically different from the distribution of Voronoi polygons by a goodness-of-fit test. From this and other comparisons such as nearest neighbour analyses Buckley and Buckley (1977:36) conclude that the royal tern 'has achieved an exceedingly close approach to maximal or hexagonal packing of its nests, probably the first firm demonstration of this effect in any aspect of a vertebrate social system'.

But *how* close is this 'exceedingly close approach' to hexagonal packing? Actual hexagonal packing would yield 100 per cent six-sided territories (arrow in Figure 3), and it is evident that less than half of the tern territories have six sides. What we really need to know is where the tern territories lie with respect to the difference between random and optimal packing, for only with such an assessment can labels such as 'exceedingly close approach' become meaningful. In this case, thorough statistical analysis is frustrated by the lack of a generating function for Voronoi polygon distribution, and even a simple goodness-of-fit test is not possible because the hexagonal distribution has zero expected frequencies for all but the six-side class. In sum, there just seems to be no clear way to tell how far the tern distribution departs from random towards optimal hexagonal packing. The best we can say is that the departure from random is in the direction of optimality, which at least is an improvement over the situation represented in Figure 2 (above). The tern example shows that even when the null hypothesis is specified, it may be exceedingly difficult or impossible to judge whether the empirical data lie closer to random or optimality.

6) Unacceptability of a criterion for suitability

Suppose all the foregoing problems could be overcome: we avoid *a posterori* reasoning, somehow identify the variable that is reputedly optimized, adopt a thorough-going empiricist's programme to formulate and rule out alternative hypotheses, specify the null hypothesis of random expectation, and even measure objectively the difference between the random and the optimal. As variation is to be expected in most biological systems, very few truly optimal systems such as bee combs will be found, but by being able to measure where the system lies on the continuum between random and optimal we should be able to identify quantitatively the approach to optimization.

In this case we are confronted with a final judgement: how close does the trait in question have to approach optimality before it is termed a suitability? Is 'reliably different from random' (and in the right direction) sufficient? Or is simply 'closer to optimal than to random' a criterion that all would agree to? I doubt it. I suspect instead that whatever arbitrary criterion one author selected, other authors would disagree, and in the end the literature would become filled with wrangling. I suspect that the reason no criterion would be agreed upon is not that the choice is subjective or arbitrary, but rather because of the nature of the entire optimality approach. That is, we accept arbitrary criteria commonly in science, such as the alpha level of 0.05 indicating an acceptable statistical difference. The reluctance to accept an arbitrary criterion for approach to optimality lies more subtly in the uncertainty that the chosen variable is that which is really being optimized.

Consider this issue further with reference to the example in Figure 3 (above). Suppose we could measure where the tern nesting territories fell on the spectrum between the random expectation of Voronoi polygons and the optimal hexagons. The following criterion would not be a very good one because it does not take into account the whole distribution of tern data, but note that about 28 per cent of Voronoi polygons are six-sided, about 45 per cent of tern territories are six-sided, and 100 per cent of hexagonally packed territories would be. We could then say that tern territories lie at the point $(45 - 28)/(100 - 28) = 24$ per cent along the distance from random to optimal. Buckley and Buckley (1977) showed that the tern distribution differed significantly from the distribution of Voronoi polygons, and the difference does *seem* to be in the right direction towards hexagonal packing (cf. Figure 2, above), so why should we be reluctant to accept this as sufficent proof that tern territories are approaching optimum? I believe that the reason for such reluctance is that we have no assurance that we are on the correct continuum.

For example, the cichlid fish territories in Figure 2 (above) are probably also statistically different from Voronoi polygons, but clearly not approaching optimal hexagons, so we would not accept this result as showing an approach to optimality. The difficulty with the tern situation is that the packing of territories may not be tending towards optimal hexagons at all, but rather towards some other packing distribution that no one has yet even thought of.

In other words, unless a variable is truly optimized we are unlikely to accept any arbitrary criterion for an approach to optimum because we have no assurance that the conceived optimum is that which is being approached. It is not the arbitrariness of the criterion that engenders the reluctance, but rather the uncertainty of the optimal goal itself which was initially selected subjectively rather than being generated by empirical evidence.

Conclusions: optimality rejected

Optimality will not do as a criterion for the operational definition of suitabilities. Even in those extremely rare cases where an optimal condition really does seem to have been achieved by a species, as in hexagonal packing of the honey bee's comb-cells, there is no operational way to specify the 'perfect fit' to the environment. As Darwin himself noted, the hexagonal array optimizes use of space, use of materials (for common walls of cells), use of energy (in constructing cells) and so on. One can view this as a happy coincidence that many variables are simultaneously optimized, but if we are to engage in a rigorous search for the processes that brought about this result we do need to specify the result. If hexagonal packing of bee cells optimizes these things, perhaps it also optimizes a lot of other things that no one has yet thought of, so the search for the processes bringing about hexagonal packing will be frustrated by lack of identification of what is being optimized by those processes.

Most characteristics will not show optimization at all, but merely 'approach' it. Gray (1986) has reviewed published papers claiming to show optimal foraging and found that few actually show achievement of optimality. As noted in the subsection immediately above, even if the 'approach' to optimality can be measured quantitatively, some arbitrary line must be drawn as to how close the approach need be in order to term the character in question a suitability. In the end, even if all the other problems with optimality studies can be avoided, the finding of an optimal condition or an approach to optimality both entail terminal problems that cannot be overcome without leaving the realm of operationalism. In sum, optimality cannot be made an operational criterion for a suitability, so we must turn elsewhere.

OPTIMISM: THE COMPARATIVE METHOD

If attempting to show an optimal match between organism and environment is not sufficient to define suitabilities operationally, then perhaps showing a correlation between the two might do instead. The notion of comparison among animals is an old one, used in some way or another by virtually all authors writing about adaptation, but the notion carries less selectionist baggage than many other terms in contemporary evolutionary biology and so is less likely to engender subtle confusion. Part of the reason that the notion is fairly 'neutral' lies in the term 'comparative method', which connotes a plan of investigation rather than a conclusion reached. Furthermore, biologists compare organisms for a variety of reasons, such as tracing phylogeny, assessing the principles of ritualization that make behavioural patterns socially communicative, and simply finding just the right species with which to attack some particular problem (e.g. Hailman, 1976a, 1981). Here we are concerned with another use, namely to demonstrate convergent evolution, which began in the pre-Darwinian writings of Ray and others as the notion of analogy (Hailman, 1976b). Clutton-Brock and Harvey (1984) term this use of the comparative method an approach 'to investigating adaptations'. Comparative arguments, however, frequently suffer from an incompleteness that is best laid bare by being explicit about the minimum requirements for a useful comparison.

Minimum requirements for the comparative method

Table 1 sets out what I consider to be the minimum requirements of the comparative method, namely at least one species of each of two taxa that inhabits environment **e** and at least one other species of each taxon that inhabits some different environment **ê** (read 'not-**e**'), totalling four species. Of course, one species each in the four tabular cells is hardly a convincing sample size, but that issue is a question of statistical reliability rather than the basic methodological logic being addressed here first.

TABLE 1 Minimum requirements for the comparative method (see text)

	Environments	
	e	ê
Taxon 1	e1	ê1
Taxon 2	e2	ê2

What we seek in a convincing convergence is a consistent similarity between the measurable characteristics of the unrelated species in cells **e1** and **e2** *plus* a difference between species in cells **e1** and **ê1**, and those in **e2** and **ê2**. The environment **e** must be the same for all species in cells **e1** and **e2**, but the environments of species in cells **ê1** and **ê2** may differ, even within a taxon, so long as they are not the same as environment **e**.

Consider a simple application of this comparative method by asking whether it establishes convincingly the suitability of oarlike appendages to the marine environment. We shall say that an animal lives in environment **e** if it obtains all its food beneath the surface of the ocean by moving in the water column (as opposed to moving along the bottom), never flies or feeds in the air, and never feeds on land (although it may haul out on land for reproduction). A species that does not meet these environmental requirements is in environment **ê**, which varies widely among species. Many taxa of animals have species in both **e** and **ê**, but we here consider only the vertebrate classes Aves and Mammalia. The notion of an oarlike appendage can be made operational in principle by stipulating that it has one long dimension and is flattened in another dimension more at its distal end than its proximal end. Penguins are the only birds meeting the requirements of environment **e**, but mammals have two such species-groups, namely the whales (cetaceans) and seals (pinnipeds); all other birds and mammals inhabit some other environment **ê**, which varies enormously. As indicated in Table 2, penguins, cetaceans and seals all have oarlike appendages whereas no other birds or mammals do.

TABLE 2 Oarlike appendages as suitabilities

	Environments	
	Ocean	Elsewhere
Birds	*Penguins	Other birds
Mammals	*Cetaceans	Other mammals
	*Pinnipeds	

* Only forms marked with an asterisk possess oarlike appendages.

One important value of the comparative method illustrated by the facile example in Table 2 is that it constitutes an ordinary scientific hypothesis that is falsifiable by new evidence that does not show the correlation between oarlike appendages and the ocean environment. Conversely, our confidence in this hypothesis can be strengthened by finding new groups in which the correlation continues to hold. For example, sea turtles also have oarlike appendages whereas terrestrial chelonians do not. New evidence, however, may prompt some further distinctions in the operational definitions of environments or measurable characters, or both. For example, freshwater turtles have appendages somewhat intermediate between those of sea turtles and terrestrial tortoises, so one might modify the hypothesis in order to distinguish among three environments (say, $\mathbf{e}_1$, $\mathbf{e}_2$ and the wastebasket category **ê** meaning neither of the first two). Or, a different approach might be to modify the definition of original environment **e** to include freshwater existence. Whatever the modification, the objective is to achieve as high a correlation as possible between measurable characters of animals and measurable aspects of the environments in which they live.

I now turn to scrutinizing this comparative method for identifying suitabilities. The goal is to see if the comparative approach avoids the problems of optimality criteria, or possibly generates some new ones.

1) Tautology impossible

There is a seductive tendency in the engineering approaches to adaptation, optimality included, to say that something 'must' be an adaptation simply because it occurs: for example, if animals occur in certain sized social groups that size 'must' be the optimal one. There seems to be no sense in which the comparative method as outlined above can fall prey to subtle circularities. If one were looking for the suitability of group size to the environment, it would be necessary to identify measurable environmental characteristics that might correlate with the size of social groupings. The very method prevents false reasoning from characters alone by demanding both differences in characters between related species of two environments *and* similarities in characters between unrelated species in the same environment.

2) Choosing variables for correlation

The problem with selecting a reputedly optimized variable in optimality studies is that if the animal turns out not to optimize that variable we have 'no decision' and do not know where to turn. As pointed out above when this problem was considered in optimality approaches, it is not a serious one but rather an inconvenience. To an extent the analogous problem arises when using the comparative method, for it is necessary to have some environmental variable (or cluster of variables) that may correlate with measurable characteristics of organisms. There is, however, a useful difference between the two approaches in this regard: straightforward morphological and behavioural descriptions of organisms do not often suggest a hypothesis about what variable might be optimized whereas such descriptions often do immediately suggest character–environment correlations.

Although this difference is one of degree rather than kind, let me try to secure it with a brief example. Many animals have deposits of melanin somewhere in their bodies, very frequently in the integument. If we were to ask whether such pigmentation is optimizing a solution to some environmental problem it would require a good deal of detailed knowledge about a given species even to generate a viable candidate variable for optimization. However, straightforward descriptions of animals suggest immediately an environmental variable to be investigated comparatively as a starting place because all sorts of unrelated animals living in desert regions are extensively melanistic. At the very least such a convergence suggests that we should try correlating melanism with the amount of destructive ultraviolet radiation in the environment, the environmental temperatures produced by infrared radiation, the degree to which aridity could produce dehydration and so on (Hailman, 1977).

In sum, the comparative method (like the optimality approach) requires the investigator to try environmental variables to see which are relevant. However, the comparative method offers two advantages: basic descriptions of animals often suggest a cluster of relevant variables, and comparative testing by correlation is generally easier than repeated attempts to test for optimality.

3) Alternative hypotheses

A problem encountered in optimality studies (and in 'functional' approaches in general, as stressed by Jamieson, 1986) is that the investigator is often so convinced that some variable is critical that alternative hypotheses are not considered. Entire literatures, like that on nest defence, are generated that seem to support the functional hypothesis, whereas in fact the results are due to a consistent confounding of variables. An analogous problem arises in the comparative method, where correlations between characters of organisms and measurable variables of the environment may be spurious. As Clutton-Brock and Harvey (1984) point out, such correlations may be secondary in the sense that the variables measured may be correlated with others where the more direct causality really lies. For example, brain size in small mammals correlates with food habits but other variables are also important: habitat, vertical stratification, activity times and so on. One accounts for these multiple correlations in numerous ways, including standardization (e.g. comparing only nocturnal species with one another) and multivariate analyses.

The possibility of failing to consider alternative hypotheses seems less serious in the comparative method than it is in optimality and other functional approaches for several related reasons. One is that perfect character–environment correlations that are spurious are unlikely to arise by chance. Unlike functional studies where results may correspond closely with some false prediction, correlations are never perfect; one is always encouraged to improve them by further empiricism where extraneous variables are better accounted for or new independent variables are tried. Furthermore, the comparative method is more 'neutral' in generating hypotheses than is the optimality approach because the former generates hypotheses about correlations from empirical descriptions; the latter usually generates hypotheses largely from selectionist models about what organisms 'should' be doing in order to survive and reproduce optimally. The more empirically oriented comparative method is therefore more likely to get an investigation on the 'right track' than on a spurious functional siding.

In sum, the comparative method is less likely to fall prey to ignoring alternative hypotheses than are functional approaches because early phases of investigation are more empirically grounded, initial results are unlikely to produce seductively high but false correlations, and several practical options are available for improving correlations in ways that suggest alternative hypotheses.

4) The null hypothesis is specified

A problem more serious than the foregoing issues was identified in optimality studies as the lack of a null hypothesis. For example, until recently workers dealing with packing arrays intuitively felt that their empirical data approached optimal hexagonal packing without benefit of knowing what a random alternative might be. Null hypotheses might be unknown to anyone (e.g. the situation before mathematical work on Voronoi polygons as in Figure 2, above) or just unknown to a particular investigator (e.g. exponential decay function for group size as in Figure 1, above). This problem of lack of a null hypothesis does not intrude into the comparative method because the null state is simply no correlation, and an array of statistical techniques exists for determining the reality of associations among variables.

5) Data location in the spectrum of possibilities

A related problem that arises in optimality studies is that even when the null hypothesis is known it may be difficult or even logically impossible to specify where empirical results lie on the continuum between the random and optimal conditions. This problem does not arise in comparative studies because the spectrum of possibilities is the degree of association between character variables and environment variables. The null hypothesis, as noted above, is always 'no association' or correlation whereas the other end of the spectrum is 'complete association' or perfect correlation. Statistical measures such as correlation coefficients express precisely the degree of association and hence inform the investigator directly where his or her data lie on the spectrum of possibilities.

6) Criterion for suitability

It might seem at first that although the comparative method proves more satisfactory than the optimality approach at every juncture to this point, it must ultimately founder for the same reason that optimality foundered: unacceptability of a criterion for suitability. However, there is an important difference between the two approaches. In order to bring out this difference let us assume in both cases all the foregoing kinds of problems were satisfactorily solved. We then end up with (a) a statement as to where the data fall between null and optimal in optimality studies, or (b) a statement about the strength of association between characters and environment in comparative studies. In both cases we are confronted with devising an arbitrary criterion for defining suitability.

It is not the arbitrariness of the criterion that matters, for we use arbitrary criteria frequently in science (e.g. alpha levels in statistical testing) and they are perfectly operational. What matters is the willingness of the scientific community to accept a given arbitrary criterion. We accept an arbitrary alpha level of 0.05 because it is intuitively reasonable, and also because it can be shown in some statistical tests to balance the chances of committing type I and type II decision errors.

The optimality and comparative approaches for establishing suitabilities both require arbitrary criteria, but the two approaches differ in the acceptability of establishing a criterion. In the optimality approach the criterion must tell us where on the spectrum between null expectation and optimal condition the data must lie to certify reality of approach towards optimality. Scientists are unlikely to agree on this locus because there is no way to ensure that a departure from null, regardless of how empirically reliable, indicates a tendency towards the subjectively selected optimal goal. I believe that the scientific community is more likely to accept a criterion of the degree of association between characters and environments because any correlation above chance level signifies an approach to underlying causal relationships. Of course, the character variable and environment variable being correlated may have only indirect causal links, as mentioned in a previous section, but totally spurious correlations with no indirect causal links at all are unlikely to arise.

This distinction can also be cast in somewhat different terms. In optimality studies the goal has been picked more or less subjectively by the investigator for testing, and only if the organism shows nearly full optimality are we assured that the right goal was picked. Therefore, partial approaches to this goal could be due to entirely fortuitous circumstances about which we have no clues. In comparative studies the ideal goal of complete character–environment correlation is generated as an empirical observation. Relatively weak but statistically reliable correlations suggest that we do not have precisely the correct variables, but there is a likelihood that the correlation is not completely fortuitous. Furthermore, because these weak correlations are likely due to causal chains, the outcome suggests how to modify the approach to yield stronger correlations.

For these reasons it becomes far more satisfying intuitively to accept an arbitrary criterion for suitabilities in the comparative method than in functional approaches including optimality. The obvious candidate for an arbitrary criterion is a statistically significant association or correlation between character and environment. Such a statistically significant result in comparative studies is highly likely to 'be on the right track' whereas a similarly significant difference between data and null expectation in optimality studies offers no similar assurances.

Interlude: optimality versus comparative approaches

In review, the optimality and comparative approaches for establishing an operational definition of suitabilities of organisms to their environments have been compared in six respects. (1) Optimality can fall prey to assuming that the modal case is optimal whereas such circularity is impossible with the comparative method. (2) Both approaches require selection of variables for testing, but these are likely to be arbitrary choices in optimality studies rather than deriving from empirical generalizations as in comparative studies. (3) The failure to consider alternative hypotheses is likely in optimality studies, much less so in comparative approaches. (4) The null hypothesis of random expectation can be undefined or even undefinable in optimality studies whereas no such problem exists with the comparative method. (5) The location of the data on the spectrum between null expectation and full suitability may be undefinable in optimality approaches but is always explicit in comparative studies. Finally, (6) arbitrary criteria for approach to optimality are unlikely to enjoy agreement among scientists whereas statistically significant association is an intuitively acceptable operational criterion in the comparative method. These differences constitute a clear-cut rejection of optimality approaches for operationally defining suitabilities while endorsing the comparative method as a promising candidate. However, it is necessary to ask whether the comparative approach does not generate new problems heretofore not considered.

7) Convergence without environmental diversity

A common short-cut in comparative studies of adaptation is simply to show that unrelated animals living in the same environment have similar characteristics, a situation usually referred to as convergent evolution. For example, nightjars (Family Caprimulgidae), also called goatsuckers, are nocturnal (or crepuscular) birds that have no bright coloration (reds, oranges, yellows, greens, blues) but are basically black and white with some brown and grey. Barn owls (Family Tytonidae) have similar coloration and similarly are all nocturnal. It is tempting to conclude from such evidence that black and white coloration is suited to nocturnal activity, and we could even speculate why this might be so (e.g. insufficient light for colour vision, white more visible as signal coloration and so on). Table 3 warns us, however, that the comparative method has not been followed completely because neither family has diurnal representatives, so there is no control for phylogenetic correlation (cf. Table 1, above).

In this case the convergence of black and white coloration in nocturnal birds

TABLE 3 Coloration of some birds

	Environments	
	Nocturnal	Diurnal
Nightjars	*67 species	[No species]
Barn owls	*10 species	[No species]

* These species have essentially black and white plumage with some grey or brown (no red, orange, yellow, green, or blue).

does not qualify as a suitability from the evidence. To rectify this situation the evidence must be expanded to include other groups that have both nocturnal and diurnal members; for example, typical owls (Family Strigidae). If sufficient variation in habits cannot be found within at least two groups of birds, the comparison must be further enlarged. (In the extreme, for instance, one could note that the flowers of higher plants are white in night-blooming species, and so look to see if they have day-blooming relatives with more colourful flowers.)

8) Differences without taxonomic diversity

The 'reverse' of the incompleteness represented in Table 3 also occurs. Clutton-Brock and Harvey (1984: 13–15) decompose comparative studies into three types, the first being 'comparisons between a small number, usually a pair, of closely related species'. They cite the example of E. Cullen's (1957) classic study of the black-legged kittiwake (*Rissa tridactyla*), a type of cliff-nesting gull that shows many morphological and behavioural differences from typical ground-nesting species such as the black-headed gull (*Larus ridibundus*) studied extensively by Tinbergen's research group at Oxford. For example, the kittiwake does not remove the eggshell from the nest after the chick has hatched as does the black-headed gull (Tinbergen *et al.*, 1962), so that Table 4 represents the situation that could be drawn up for a host of other differences between these two species.

TABLE 4 Behaviour towards eggshells

	Environments	
	Cliff nesting	Ground nesting
Gulls	*Black-legged kittiwake	Black-headed gull
[No taxon]	—	—

* This species does not remove eggshells.

There are two problems evident in Table 4. The first is simply the small sample size of species, namely two. As pointed out by Clutton-Brock and Harvey (1984) a large sample is desirable, which leads to their second category of comparative studies 'considerable numbers of species, which are allocated to different ecological categories'. In the case of gulls, Tinbergen and E. Cullen already knew that in every ground-nesting gull that had been studied the species always showed the same traits as the black-headed gull as opposed to the kittiwake, so the implied comparison actually had a far greater number of species in cell **ê1** (see Table 1, above) than implied by Table 4. Furthermore, a subsequent study of the Galapagos swallow-tailed gull (*Larus furcatus*) showed that this cliff-nesting species had traits similar to those of the kittiwake (Hailman, 1965). Furthermore, the cliff habitat of this Galapagos species differs from that of the kittiwake in a number of ways that make it more like that of ground-nesting gulls. This situation predicts that the characters of the swallow-tailed gull should in some respects be intermediate between the two species shown in Table 4, and that proved to be the case. Further cliff-nesting gulls would be desirable to increase the sample size cell **e1** of Table 4, but unfortunately the only other cliff nester in this group is the very closely related red-legged kittiwake (*Rissa brevirostris*).

The more important problem with Table 4, however, is the lack of a second taxon minimally required for the operational definition of a suitability (see Table 1). Fortunately, subsequent to the study by E. Cullen (1957) on the kittiwake, J.M. Cullen and Ashmole (1963) extended the comparison of cliff- and ground-nesters to terns (see also J.M. Cullen, 1960); terns are confamilial but in a different subfamily from gulls. Later ornithological work has explored characteristics of cliff-nesting species in wholly different families (e.g. auks, Family Alcidae) and even different orders (e.g. boobies, Family Sulidae of Order Pelecaniformes). Each new taxon studied provides a further test of the hypotheses that various characteristics of cliff-nesting birds are suitabilities for the cliff-nesting environment.

9) Parsing the phylogenetic factor

Clutton-Brock and Harvey (1984:15) recommend that a 'third category of interspecies comparison, which involves the use of formal statistics to test associations between interspecific differences in behaviour and ecology, is usually preferable'. This recommendation emphasizes the statistical nuts and bolts of quantitative comparison while ignoring the importance of cross-taxon comparisons. Indeed, they even found a primate study to be 'less satisfactory' because 'it included all species within an order (instead of within a subfamily), thus increasing the effects of phylogenetic influences' (p.14). The real criticism here should be that each taxon was not treated as an independent replicate so that the phylogenetic factor could not easily be parsed.

Later (pp.19–21) these authors consider the breadth of taxonomic array to be considered in a comparative study, but instead of treating each taxon as a useful independent replicate of some character–environment correlation they are more concerned with whether lumping the data provides better regression fits than do separate calculations for each taxon. What all this amounts to, I believe, is oversight of the fact that the comparative method as I have outlined it in Table 1 allows for factoring out of phylogenetic influence.

Consider the method of drinking water by birds. We know from a large literature that most species drink by scooping water into the lower mandible and tilting back the head so that water flows down through the throat (the 'scoop-and-tilt' method). Some finches in the Subfamily Estrildinae drink by inserting the bill into water and somehow sucking the water with head lowered (the 'pump' method). All these finches that have been reported drinking in this way are desert species, whereas other estrildines drink by the scoop-and-tilt method (e.g. Immelmann, 1962). The only other kinds of birds known to me that also use the pumping method are pigeons and doves (Order Columbiformes). Species living in the arid southwestern United States drink by this method, but so do all other species that have been reported, regardless of their habitat. We, therefore, have the situation shown in Table 5. Obviously the data do not show a tight character–environment correlation because all pigeons and doves drink by the pumping method; and, worse yet, apparently all birds other than the two groups shown in Table 5 drink solely by the scoop-and-tilt method. The comparison does, however, reveal the phylogenetic constraint: of all the groups in Class Aves only the estrildine finches have differences in drinking methods among the component species.

There is no viewpoint of evolutionary theory of which I am aware that does not recognize that lineages have evolutionary constraints. Indeed, the conservatism of lineages is that property most important in establishing phylogenies. As the vast majority of the nearly 9000 species of birds probably drink by the scoop-and-tilt method, one may assume that (for whatever reason) the pumping method is difficult to evolve even when it might be suitable to some environmental condition. Only the estrildine finches and the pigeons seem to have evolved the pumping method. In the estrildine finches this method

TABLE 5 Drinking methods of some birds

	Environments	
	Arid regions	Other
Estrildine finches	*Several species	Many species
Pigeons and doves	*Several species	*Many species

* Only these species drink by the pumping method.

appears to be suitable to the arid environment, but we cannot establish that point by Table 5 because no other taxon in Class Aves has a diversity of drinking methods among its species. The situation is similar to that of the previous subsection (see Table 4, above) when we eliminate the pigeons and doves from Table 5 because of lack of diversity in drinking methods. That is, not only do we need at least two different taxa to establish a suitability, but moreover each taxon must have species that live in both of the minimal two environments and each taxon must show diversity in the character being scrutinized as a reputed suitability.

10) 'Biological relevance' is irrelevant

Clutton-Brock and Harvey (1984:15) state that 'The first question that any attempt to make quantitative comparisons needs to ask is whether the variables are biologically relevant'. For example, they cite comparative studies of the brightness of male plumage in birds and ask for assurances that the birds themselves (or the species with which they interact) 'judge the brightness on a similar scale'. Such 'biological relevance' is irrelevant to our operational definition of a suitability, as Clutton-Brock and Harvey (1984) were considering the use of the comparative method to establish adaptations, which is to say suitabilities with all the traditional selectionist implications attached. In identifying suitabilities we are concerned solely with establishing a link between animal characteristics and types of environmental variables so that we know precisely what it is that we want to account for by selectionist or other explanations. To introduce 'biological relevance' into this process would be to rob the notion of a suitability of its operational status.

One apparent way out of this predicament would be to establish a greatly expanded approach to defining suitabilities in which 'biological relevance' is also made operational. I doubt that this is desirable or practical, and it may be logically impossible. It is not at all clear that it would be possible to make a vague notion like 'biological relevance' operational. Numerous problems suggest themselves immediately upon considering any concrete example, such as bright plumage of birds. Is it merely the brightness judgement that must be similar to ours? How about the colour vision of the birds, which we know from human vision literature interacts with brightness perception? How about flicker–fusion thresholds, which also interact with brightness perception? Indeed, how can we be certain that we are choosing the 'biologically relevant' issues? Most importantly, how can we know when we have assured ourselves of *all* the 'biologically relevant' variables?

What the establishment of unencumbered suitabilities does for us is to divorce 'functional' considerations from character–environment correspondences, so that we can begin on a sturdy foundation with operationally established phenomena. It is a persistent temptation to adulterate this operational purity by allowing subtle functional or selectionist issues to creep into the epistemology, and to allow this to happen would destroy the operational foundation being sought in this chapter.

Conclusions: optimism about the comparative approach

There is never any assurance when attempting to establish a consistent way of looking at natural phenomena that one has guarded against all serious problems. In the foregoing subsections I have shown that the comparative method of establishing an operational definition for character–environment suitabilities does much better than optimality approaches (points 1–6) and survives a number of other potential criticisms (points 7–10). Clutton-Brock and Harvey (1984) consider several other specific issues about interspecific comparisons, in their use for establishing adaptations, and these all seem to fall into one of five categories: (a) specifics of quantitative methods, which are either detailed questions that do not raise logical issues or else do raise issues that fall into one of the following categories; (b) problems that arise because convergence without environment diversity has been considered (see point 7, above); (c) problems that arise because differences have been considered without the control of taxonomic diversity (see point 8, above); (d) problems that arise because the taxon considered does not have the phenotypic diversity required (see point 9, above); and (e) problems that arise due to the introduction of 'functional' (and hence for our purpose, irrelevant) issues (see point 10, above). In sum, we have encountered no insurmountable problems in using the comparative method, as outlined, to establish character–environment suitabilities in an operational fashion. This fact does not mean that no problems exist, but it does promote optimism about this step in establishing an operational evolutionary biology.

ACKNOWLEDGEMENTS

I am grateful to Russell Gray, who warned at the outset that I could not effectively employ the term 'adaptation' as a process-free word without confusing readers; I refused initially to follow his good advice — a mistake that required unusually extensive revision of the first draft of the manuscript after it had twisted some very good minds. For excellent comments on that draft I thank Robert Bleiweiss, Millicent S. Ficken, Gray, Mae-Wan Ho, Ian G. Jamieson, Peter H. Klopfer, Susan Oyama, Jonathan R. Reed and Hugh M. Robertson.

REFERENCES

Andersson, M., Wiklund, G., and Rundgren, H. (1980). Parental defence of offspring: a model and an example, *Anim. Behav.*, **28**, 536–42.

Barash, D.P. (1975). Evolutionary aspects of parental behavior: distraction behavior of the Alpine accentor, *Wilson Bull.*, **87**, 367–73.

Barlow, G.W. (1974). Hexagonal territories, *Anim. Behav.*, **22**, 876–8.

Bateson, P. (1982). Behavioural development and evolutionary processes. In *Current Problems in Sociobiology* (eds King's College Sociobiology Group), Cambridge University Press, Cambridge, pp. 133–51.

Bateson, P. (1984). Sudden changes in ontogeny and phylogeny. In G. Greenberg and E. Tobach (eds), *Behavioral Evolution and Integrative Levels*, Lawrence Erlbaum Associates, Hillsdale, NJ, pp. 155–66.

Batchelet, E. (1975). *Introduction to Mathematics for Life Scientists*, 2nd edn, Springer-Verlag, Berlin.

Boucher, D.H. (1977) On wasting parental investment, *Amer. Natur.*, **111**, 786–8.

Bridgman, P.W. (1927). *The Logic of Modern Physics*, Macmillan, New York.

Bridgman, P.W. (1938). *The Intelligent Individual and Society*, Macmillan, New York.

Bridgman, P.W. (1945). Some general principles of operational analysis, *Psychol. Rev.*, **52**, 246.

Buckley, P.A. and Buckley, F.G. (1977). Hexagonal packing of royal tern nests, *Auk*, **94**, 36–43.

Burnett, A.L. and Eisner, T. (1964). *Animal Adaptation*. Holt, Rinehart and Winston [place of publication not specified in book].

Clutton-Brock, T.H., and Harvey, P.H. (1984). Comparative approaches to investigating adaptations. In J.R. Krebs and N.B. Davies (eds), *Behavioural Ecology: An Evolutionary Approach*, 2nd edn, Sinauer Associates, Sunderland, Mass, pp. 7–29.

Cohen, J.E. (1971). *Causal Groups of Monkeys and Men: Stochastic Models of Elemental Social Systems*, Harvard University Press, Cambridge, Mass.

Cullen, E. (1957). Adaptations in the kittiwake to cliff-nesting, *Ibis*, **99**, 275–302.

Cullen, J.M. (1960). Some adaptations in the nesting behaviour of terns, *Proc. XII Int. Ornithol. Congr.*, Helsinki, pp. 153–7.

Cullen, J.M., and Ashmole, N.P. (1963). The black noddy *Arious tenuirostris* in Ascension Island, *Ibis*, **103**, 423–43.

Darwin, C. (1897). *The Origin of Species by Means of Natural Selection, or the Preservation of Favored Races in the Struggle for Life*, 6th edn, D. Appleton, New York. [In two volumes.]

Dawkins, R., and Carlisle, T.R. (1976). Parental investment, mate desertion and a fallacy, *Nature*, **262**, 131–3.

Fox, S.W. (1984). Morphogenesis within evolution: morphomolecular evolution. In J. Mlikovsky and V.J.A. Novak (eds), *Evolution and Morphogenesis*, Academia, Prague, pp. 263–78.

Grant, P.R. (1968). Polyhedral territories of animals, *Amer. Natur.*, **102**, 75–80.

Grant, V. (1963). *The Origin of Adaptations*, Columbia Univ. Press, New York and London.

Gray, R. (1986). Faith and foraging: a critique of the paradigm argument from design. In A.C. Kamil, J.R. Krebs and H.R. Pulliam (eds) *Foraging Behavior*, Plenum Press, New York.

Hailman, J.P. (1965). Cliff-nesting adaptations of the Galapagos swallow-tailed gull, *Wilson Bull.*, **77**, 346–62.

Hailman, J.P. (1976a). Uses of the comparative study of behavior. In R.B. Masterton, W. Hodos and H. Jerison (eds), *Evolution, Brain, and Behavior: Persistent Problems* Lawrence Erlbaum, Hillsdale, NJ, pp. 13–22.

Hailman, J.P. (1976b). Homology: logic, information and efficiency. In R.B. Masterton, W. Hodos and H. Jerison (eds), *Evolution, Brain, and Behavior: Persistent Problems*, Lawrence Erlbaum, Hillsdale, NJ, pp. 181–98.

Hailman, J.P. (1977). *Optical Signals: Animal Communication and Light*, Indiana University Press, Bloomington and London.

Hailman, J.P. (1981) Comparative studies. In D. McFarland (ed), *Oxford Companion to Animal Behaviour*, Oxford University Press, Oxford, pp. 92–7.

Hailman, J.P. (1982a). Ontogeny: toward a general theoretical framework for ethology. In P.P.G. Bateson and P.H. Klopfer (eds), *Perspectives in Ethology*, vol. 5, Plenum, New York, pp. 133–89.

Hailman, J.P. (1982b). Evolution and behavior: an iconoclastic view. In H.C. Plotkin (ed), *Learning, Development, and Culture*, John Wiley, London, pp. 205–54.

Hailman, J.P., and Jaeger, R.G. (1976). A model of phototaxis and its evaluation with anuran amphibians, *Behaviour*, **56**, 215–49.

Ho, M.-W. (1984). Where does biological form come from? *Rivista di Biol.*, **77**, 147–79.

Ho, M.-W., and Saunders, P.T. (1979). Beyond neo-Darwinism — an epigenetic approach to evolution, *J. Theor. Biol.*, **78**, 573–91.

Ho, M.-W., and Saunders, P.T. (1982). Adaptation and natural selection: mechanism and teleology. In S. Rose (ed), *Towards a Liberatory Biology*, Allison and Busby, London, pp. 87–104.

Ho, M.-W., and Saunders, P.T. (eds) (1984). *Beyond Neo-Darwinism: An Introduction to the New Evolutionary Paradigm*, Academic Press, London.

Immelmann, K. (1962). Beträge zu einer vergleichenden Biologie australischer Prachtfinken (Spermestidae), *Zool, Jb. Syst. Bd.*, pp. 1–196.

Jamieson, I.G. (1986). The functional approach to behavior: is it useful?, *Amer. Natur.*, **127**, 195–208.

Kaufmann, J.H. (1975). Field observations of the social behaviour of the eastern grey kangaroo, *Macropus giganteus, Anim. Behav.*, **23**, 214–21.

Kitcher, P. (1984). Species, *Philos. Sci.*, **51**, 308–33.

Knight, R.L., and Temple, S.A. (1986a). Why does intensity of avian nest defense increase during the nesting cycle?, *Auk*, **103**, 318–27.

Knight, R.L., and Temple, S.A. (1986b). Methodological problems in studies of avian nest defence, *Anim. Behav.*, **34**, 561–6.

Krebs, J.R., and Davies, N.B. (eds) (1984). *Behavioural Ecology: An Evolutionary Approach*, 2nd edn, Sinauer Associates, Sunderland, Mass.

Krebs, J.R., and McCleery, R.H. (1984). Optimization in behavioural ecology. In J.R. Krebs and N.B. Davies (eds), *Behavioural Ecology: An Evolutionary Approach*, 2nd edn, Sinauer Associates, Sunderland, Mass., pp. 91–121.

MacArthur, R.H., and Pianka, E.R. (1966). On the optimal use of a patchy environment, *Amer. Natur.*, **100**, 603–9.

Maynard Smith, J. (1977). Parental investment: a prospective analysis, *Amin. Behav.*, **15**, 1–9.

Mossakowski, D., and Roth, G. (eds) (1982). *Environmental Adaptation and Evolution: A Theoretical and Empirical Approach*, Gustav Fischer, Stuttgart and New York.

Oyama, S. (1985). *The Ontogeny of Information: Developmental Systems and Evolution*, Cambridge University Press, Cambridge.

Pulliam, H.R., and Caraco, T. (1984). Living in groups: is there an optimal group size? In J.R. Krebs and N.B. Davies (eds), *Behavioural Ecology: An Evolutionary Approach*, 2nd edn, Sinauer Associates, Sunderland, Mass., pp. 122–47.

Rachlin, H., Battalio, R., Kagel, J., and Green, L. (1981). Maximization theory in behavioral psychology, *Behav. Brain Sci.*, **4**, 371–417.

Saunders, P.T., and Ho, M.-W. (1982). Is neo-Darwinism falsifiable — and does it matter, *Nature and System*, **4**, 179–96.

Singh, J. (1972). *Great Ideas of Operations Research*, rev. edn, Dover, New York.

Thompson, D'A. (1917). *On Growth and Form*, Cambridge University Press, Cambridge.
Tinbergen, N., Brockhuysen, G.J., Feekes, F., Houghton, J.C.W., Kruuk, H., and Sczulc, E. (1962). Egg shell removal by the black headed gull, *Larus ridibundus* L.: a behaviour component of camouflage, *Behaviour*, **19**, 74–117.
Trivers, R.L. (1972). Parental investment and sexual selection. In B. Campbell (ed), *Sexual Selection and the Descent of Man*, Aldine, Chicago, pp. 139–79.
Wallace, B., and Srb, A.M. (1961). *Adaptation*, Prentice-Hall, Englewood Cliffs, NJ.
Wilson, E.O. (1975). *Sociobiology: The New Synthesis*, Harvard University Press, Cambridge, Mass.

Evolutionary Processes and Metaphors
Edited by M.-W. Ho and S.W. Fox

CHAPTER 7

On not holding nature still: evolution by process, not by consequence

MAE-WAN HO

ABSTRACT

The Neo-Darwinian paradigm of 'evolution by consequence' depends on the conceptual separation between organism and environment, and on the supposed randomness of the variations on which natural selection acts. Both these assumptions have been invalidated in the light of contemporary knowledge.

Organism and environment are intimately interconnected, from the sociocultural domain right down to genomic DNA. Moreover, forms and variations are neither random nor arbitrary. Instead, they taken shape from the dynamic properties of the generative processes involved. The emerging framework of 'evolution by process' sees organisms evolving in unison with nature through processes nested in space and time, and not as the consequence of natural selection. Development is, therefore, continuous with evolution, and hence the organism's own activity is effective in shaping evolutionary history.

The metaphor of natural selection derives from the dominant socioeconomic ideology of the Victorian era, now rejected by nearly all of humanity. The mechanistic conception of life which it inspires is equally outmoded and inappropriate. Why should we cling to this metaphor when it can serve no other purpose than to reinforce those prejudices which gave it birth?

THE UNITY OF NATURE

The theme of this chapter is the integration of the organism with nature, and the logical implications for evolution which flow from that. To many great civilizations past and present, the unity of nature is simply a fact of immediate experience that needs no special pleading. In the West, however, much of the history of science is concerned with separating and reducing this unity into ever smaller and smaller fragments out of which nature has somehow to be glued together again. It is a history not only of fragmentation but of our own alienation from nature. Some will say it began the moment the mind set itself apart as observer on a nature observed. From that vantage point, the drama of nature mysteriously unfolds, and we become unwitting sufferers of unknown fate.

If the inanimate world were reducible to mechanisms and atoms, it would be only a matter of time before the organic world, too, would succumb. Darwinism might be seen as the culmination of mechanical materialism (Barzun, 1958), but it is Neo-Darwinism which dealt the final blow in disintegrating the organism to a mere collection of particles (genes) shuffled by blind selective 'forces'. The extension of Neo-Darwinian principles to social behaviour gave birth to the discipline of sociobiology. Its chief exponents claim to account for social behaviour, culture and even morals in terms of the natural selection of random mutations. Many have recoiled from this final capitulation to crude mechanism by severing their connections with biology altogether; thereby leaving humanity to the dilemma of a disembodied and hence impotent mind pitched against the mindless automaton of a body controlled by genes whose sole imperative is to replicate.

Actually, it is not our biology that is at fault, nor indeed biology in general; it is merely our view of it through the Neo-Darwinian looking glass (Ho *et al.*, 1987). Even in the West, there has always been a rival tradition that resolutely resists fragmentation in favour of integration and process. It includes such distinguished figures as poet–scientist Goethe, evolutionist Lamarck, embryologist Driesch, and closer to our time, D'Arcy Thompson, Alfred North Whitehead, Joseph Needham, Richard Goldschmidt and Conrad Waddington. An integrated nature demands a study of living organization at different levels, each with its own distinctive emphasis, yet always in relation to the whole.

Once we begin to see biology again in the light of nature's unity, mind and body will become reunited through processes embracing every level from the sociocultural to the molecular. The organism itself — its functions, volitions and actions — will then be rightly perceived, not as the sole *consequence* of natural selection, but as a focus of being *immanent to process* and emerging *simultaneously* with it. Thus relocated within nature, the organism becomes both actor and producer of the evolutionary drama.

THE DARWINIAN METAPHOR AND NEO-DARWINIAN MYTH

In the *Origin of Species*, Darwin collected together a mass of observations to support his theory that evolution occurs by the natural selection of random variations. As pointed out by Darwin himself and many others since, neither the theory that evolution has taken place, nor the idea of the survival of the fittest, was new. What was new was in suggesting that one was the mechanism for the other, and so offering a naturalistic alternative to explanations based on vitalism and on divine creation.

The theory of natural selection states that given the presence of heritable variations in individual organisms, their propensity for Malthusian (i.e. geometric) increase, and limited resources in the environment, there will be competition for survival and reproduction in which the fittest variants will eventually win. 'Fitness' is, therefore, the net reproductive capacity associated with particular variants. Natural selection leads to an improvement of the species over a number of generations; and if it is continued for a long time, especially if there is a change in the environment, the species will also change into a new one. To this Darwinian account, Neo-Darwinists add the theory of Mendelian genetics, according to which heredity is mediated through discrete particles, the genes, and variations are caused by random mutations in those genes. 'Fitness' thus becomes 'genetic fitness' — the component that can be attributed to the possession of certain genetic variants, and is conventionally represented as a coefficient which modifies genotype frequencies in a binomial expansion dignified with the appellation, 'Hardy–Weinberg law'.

Young (1985) identifies the three immediate sources of the Darwinian metaphor as Paley's argument from design, artificial selection and Malthus's principle of population. The first posed the *problem* of adaptation: how it could be explained naturalistically. The second yielded the metaphor of 'selection'. But it was Malthus's law, that human population has a natural tendency to increase geometrically while food supply increases arithmetically, which finally provided the mechanism for natural selection. Malthus's law is an 'important step in the series of developments which overcame the belief that man and his environment were in harmony' (Young, 1985). By transposing this law to the rest of nature, Darwin also found the perfect solution to Paley's problem. Whereas Paley stressed adaptation, Malthus stressed conflict; and Darwin synthesized them by proposing that struggle both *explains* and *produces* adaptation.

Indeed, much of the reason for the instant success of Darwin's theory is that it was cut from the very fabric of Victorian English society. Barzun (1958) remarked that Darwin's philosophy was:

> imbibed, whether he knew it or not, from the economic, social and metaphysical speculations of his time. It did not arise spontaneously from his facts. Malthus and Spencer, combined with Lamarck and Positivism, provided him with the needed assumptions and attitudes, both seemingly so simple and so common in their kind as to pass for matter of fact. What brought him rapid victory and prolonged sway over his age was thus the ability of the age to recognize itself in him.

In other words, competition and the struggle for survival were seen to be so much the order of the day that everyone believed they understood the theory, and moreover, that it must be true. Ultimately, natural selection — a metaphor borrowed from life — was most easily mistaken to be an apt explanation for life itself.

To the extent that Darwinism was successful in countering vitalism, it did so by attempting to reduce vital process to a mechanical conception (Ho, 1987). Organisms are conceptualized as *objects*, alienated from the environment, and acted upon by selective *forces*. Darwin himself had great difficulty with the separation between organism and environment that the theory of natural selection requires (see below). It was Weismann's doctrine of the independence of the germplasm, coupled with Mendelian genetics, which finally achieved this. The organism became a collage of *phenotypic* characteristics, a mere shadow of the underlying *genotype* which is insulated from the environment and passed on essentially unchanged from generation to generation.

After the discovery of the double helix, the genotype became identified with nuclear DNA. The latter was deemed to contain the genetic programme encoding all the necessary 'information' that is 'decoded' into phenotype during development. The process of development is thereby reduced into an initial and a final state. So, although natural selection must act via the phenotype, it is only the genotype that matters. In other words, what count are the genes that are selectively propagated. As the genes increasingly take on the illusion of (misplaced) concreteness, so organisms become mere ghosts of departed quality.

NATURAL SELECTION AND VARIATION — ONE PROCESS OR TWO?

It is often said that Darwin's theory of evolution was incomplete because it lacked a mechanism of heredity. Actually, what is really lacking in both Darwinism and Neo-Darwinism is a theory of form and its variation.

Darwin (1859) was aware of this. In Chapter 4 of the *Origin of Species*, he reminded his readers that, 'the term Natural Selection implies only the *preservation* of such variations as arise and are beneficial to the being under its condition of life' (italics mine).

Although Darwin was less concerned with the 'laws of variation' in the formal sense, he went to considerable lengths to document the existence of variation in animals and plants under domestication (Darwin, 1868). On the causes of variability, he stated that 'each modification must have its own distinct cause, and is not the result of what we blindly call accident'.

Darwin recognized and insisted upon the conceptual distinction between natural selection and variation, while freely admitting to the difficulty in practice of separating the two:

> in many cases, it is most difficult to distinguish between the definite result of

> changed conditions, and the accumulation through natural selection of indefinite variation. (*Variations* II, p.281)
>
> in all such cases, it is very difficult to distinguish between the effects of long-continued selection and those which follow from the increased action of the part, or directly from some other cause. (*Variations* II, p.286)

But this difficulty arises precisely because developmental physiology, which produces variations in response to the environment, is itself deeply confounded with heredity, on which selection is supposed to act. Moreover, the presumed selective forces are often none other than the environmental conditions involved in producing the variation (Ho, 1984a).

It is therefore well-nigh impossible to disentangle variation due to developmental physiology from the genetic endowment ascribed to the consequence of natural selection having operated in the past. Despite this, generations of Neo-Darwinists — even those who went out of their way to stress the importance of developmental physiology, such as Goldschmidt (1938) and Waddington (1957) — have all assumed that variation is distinct from selection. This implies that there is an initial state, the genotype, and an end state, the phenotype — jointly the product of genes and environment — that is then subject to selection by the environment.

Not only is genotype regarded as distinct from phenotype, but the generations are assumed to be separate. The genotypes in the offspring are a selected subset of the genotypes in the parents, subject only to recombination and rare mutations. Each generation, as it were, starts from a clean slate; apart from the 'messages' encoded in the genes passed on from the selected parents.

Many of us have argued passionately against the idea that the genetic and environmental components of development can be neatly separated; the weight of evidence being that the 'internal' and 'external' factors are inextricably interwoven in the physiology of development. Yet the theory of natural selection depends on just such a separation between the external environment which selects and internal genetic variations in organisms which can be selected. There is a serious contradiction here for those who argue against the separation between genetic and environmental components of development, but nevertheless persist in the use of the natural selection metaphor (e.g. Oyama, this volume).

EVOLUTION BY CONSEQUENCE

The predominant theme of 'evolution by consequence' that natural selection inspires, has pervaded philosophy, sociology (see Plotkin, 1982, and several articles therein) and even behavioural psychology (Skinner, 1981). One does not ask where variations come from, merely what adaptive *consequence* they have. The fit between organism and environment is thus simply conceived to be the result of past fortuitous variations selected by their consequences and preserved by heredity.

The foundation stones for the thesis of evolution by consequence are Weismann's doctrine and the randomness of variation — both outmoded beliefs recently brought under increasingly sharp scrutiny by the new findings in molecular genetics (see below). Weismann's doctrine of the separation between soma and germline is usually interpreted to mean that physiological interactions with the environment in the course of the organism's development cannot have any heritable effects because they do not lead to changes in germline DNA (Jacob, 1970). The randomness of variation further ensures that genetic mutations in the germline are not directed by the environment or the physiological status of the organism. Adaptive evolution, therefore, can *only* proceed by the selection, *a posteriori*, of genetic variants which happen to be favoured in particular environments. The gift of Weismannism to evolution is that development can be deemed to be safely ignored (Maynard-Smith and Holliday, 1979).

Instead, when we look at organisms as they are: living, breathing, acting, responding, learning, feeling, developing, and in tune with every aspect of their internal and external environments, it is clear that the fit between organism and environment must arise through reciprocal feedback and adjustments occurring on time scales that range from split seconds to hours and years and even generations. In other words, organisms adapt to the environment and adapt the environment to themselves through continuous processes nested in space and time. Organisms are interconnected wholes: psychology with biology, soma with germline, phenotype with genotype. These wholes are themselves in continuity with past and future generations, not only through biological reproduction but also through environmental, cultural and social inheritance (Ho, 1987a; Sinha, 1984; also, this volume, Gray, Oyama, and Kitcher). This intimate interrelationship between organism and environment makes it *necessary* to see adaptation as both immanent and simultaneous with process, and not as the sole consequence of differential survival and reproduction.

In the rest of the chapter, I shall demonstrate that the processes generating form and variation are far from 'random' or arbitrary at the morphological as well as molecular levels. I shall then address the integration of the organism with nature especially in the light of relevant recent findings in molecular genetics, which contrast with the separation fundamental to Neo-Darwinism. A protocol is outlined for the study of behaviour within the emerging paradigm of evolution by process, and contrasted with that based on evolution by consequence. Finally, I suggest that the natural selection metaphor should be abandoned on grounds that it is mostly irrelevant, and often misleading. Its continued usage not only impedes our comprehension of evolutionary phenomena, but also serves to reinforce those ideological norms from which it arose.

THE RATIONALITY OF FORM

The processes generating form and variation exhibit organizational principles which can often be expressed in terms of physics and mathematics. Let me give some examples.

Lessons from the honeycomb

The honeycomb is a beautiful structure, each cell showing a perfect hexagonal cross-section. Darwin (1859) wrote, 'He must be a dull man who can examine the exquisite structure of a comb, so beautifully adapted to its end, without enthusiastic admiration.' This structure Darwin attributed to the hive-making instinct of bees, perfected by natural selection. D'Arcy Thompson (1914) showed, however, that the hexagonal cross-section of the cells, as well as their trihedral pyramidal ends, are both the result of compression due to close packing. In other words, the impressive symmetry of the honeycomb arises automatically from the physics of the hive-making process (see Note, p. 141).

The deception of mimicry

Mimicry refers to the close resemblance between different species. According to convention wisdom, the purpose of mimcry is to deceive. The Monarch butterfly is avoided by predators because it is nauseous. The Viceroy, on the other hand, is more palatable, but is avoided by predators all the same because it pretends to be a Monarch by copying the latter's wing pattern. While this may explain why the mimicry persists, it says nothing as to how it arose in the first place. Furthermore, if selective advantage were the only consideration, the Viceroy should simply have evolved its own nauseous taste instead of copying the complex wing colour pattern of the Monarch (see Saunders, 1984). The former strategy would have conferred an even greater increase in fitness because mimicry becomes less effective as the ratio of mimics to models increases.

The real explanation surely, is that the two species have similar patterning processes, which, in similar environments, will most likely result in convergent wing patterns; and developmental studies have borne this out (Nijhout, 1985). Many patterns can be changed from one to another by different environments of rearing; in other words, they are related by transformation from an underlying dynamical structure. This developmental explanation may also account for the phenomenon of 'pseudomimicry', in which two species of butterflies, often belonging to different families and living hundreds of miles apart, resemble each other as closely as any pair of real mimics. An example of pseudomimicry, originally drawn to my attention by Dick Vanewright, is that

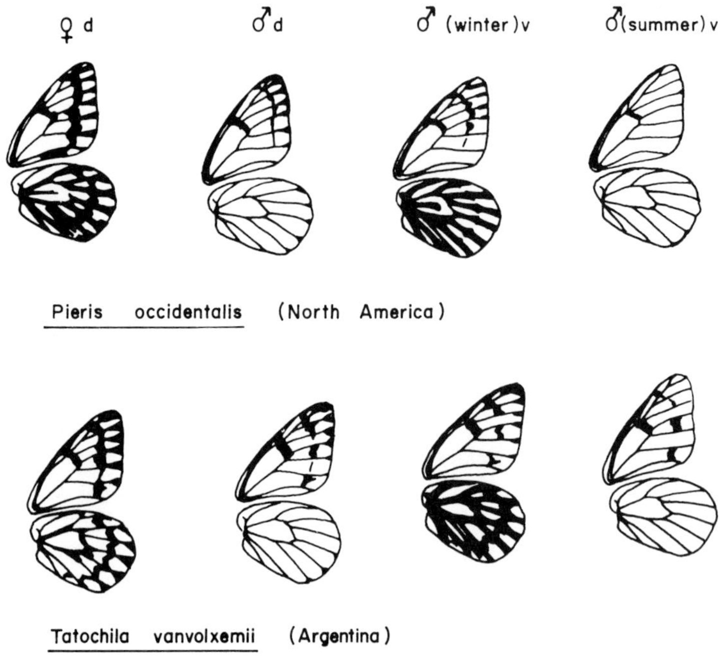

Figure 1 Parallel seasonal polyphenism in two genera of butterflies inhabiting similar but separate ecological environments. (Redrawn from Shapiro, 1984)

between *Anetia cubana*, which lives in Cuba, and *Lexias aeropus*, which lives in the East Indies (see Ho, Saunders and Fox, 1986). It is difficult to see what possible advantage either of them gains by looking alike.

Parallelism and convergence are commonplace in the evolution of butterfly wing patterns (see Figure 1). In fact, they also occur throughout the living world (Ho, 1987b), for the simple reason that the range of forms that can be generated are probably quite limited. The next example will make this clear.

'Mimicry' between physicochemical and biological forms

The cellular slime mould alternates between a multicellular slug which ends in a fruiting body bearing spores and a unicellular amoeboid phase. The amoebae feed and multiply until food runs out, then they aggregate to form a slug. Aggregation is initiated by a pulse of cyclic AMP given off by a single amoeba, which then attracts neighbouring amoebae to move towards it and at the same time to release a similar burst of cAMP. Thus, concentric waves of amoebae move rhythmically towards the centre of attraction, simultaneously relaying the signal outwards to other amoebae. The resulting pattern is very dramatic, but it is the automatic outcome of the relay process and again requires no extra explanation based on natural selection. Furthermore, it happens to be identical to the alternating blue and orange rings created by an oscillating oxidation–reduction system known as the Belousov–Zhabotinskii reaction (see Figure 2).

This remarkable similarity in form occurs in two systems which differ completely with regard to the detailed mechanisms involved. Such convergence between physicochemical and biological forms convinced D'Arcy Thompson (1914) that certain mathematical principles are involved in the generation of all forms. This belief in the underlying rationality of basic forms was also the major motivation behind Rene Thom's catastrophe theory (see Thom, 1975; Saunders, 1980; Ho, 1984b; Ho, 1987b).

THE GENERIC NATURE OF VARIATION

Particular forms and their variations are repeatably generated when conditions are right, because they represent stable states inherent in the dynamics of the processes involved. This is so in the inorganic as in the organic world. Snowflakes, for instance, exhibit a remarkably reproducible range of shapes. For biological forms, however, there is a tendency to regard the genes as the sole agents responsible for their accurate reproduction. And as the genes are supposed to mutate at random, so it is that within Neo-Darwinism, the randomness of variations has the operational meaning that nothing much could be said about the variations because each is effectively unique. Though in defence, Neo-Darwinists will claim it is only to be taken in the weak sense that the variations are not directly correlated with the selective force. In other words, they are supposed to arise within the germline genome without any reference to the physiology of the organism or its external environment. This is the so-called Weismann's barrier which ensures that the organism's experience cannot influence its genes directly.

Actually, the assumption of randomness in both those senses has been violated; especially as the result of recent findings in molecular genetics. The latter also show that Weismann's barrier is far from absolute (see 'The fluid genome' below, p. 132).

In this section, we examine some examples of variations in morphology as well as in proteins and genes, that are repeatably generated in response to specific environmental conditions. They demonstrate most clearly the contradiction of both the randomness postulate and the conceptual separation between organism and environment on which Neo-Darwinism theory is based.

Homeological transformations in *Drosophila*

Although it is undoubtedly true that some nucleotide changes in the DNA may be fortuitous, the resulting variations in the organism are never random because they occur within the context of an epigenetic system which is highly

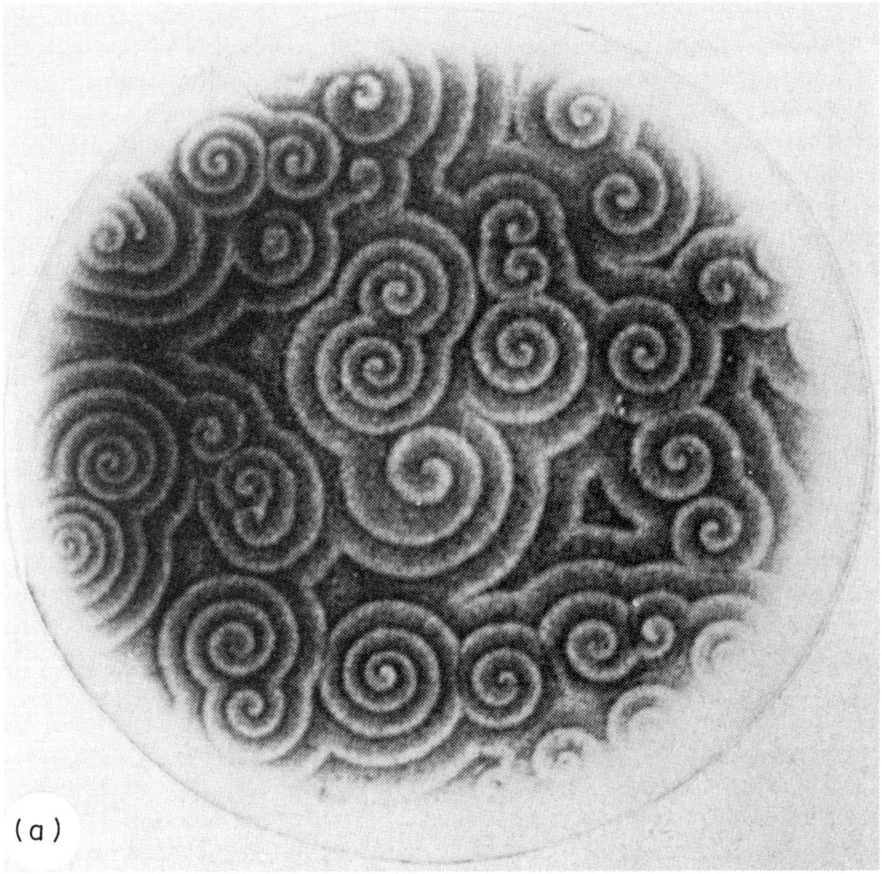

Figure 2a Convergence between biological and physicochemical forms. Pattern of aggregation in the slime mould amoebae. (From Winfree and Strogatz, 1983. Reproduced by permission of the authors)

structured. This dynamical structure in turn *shapes* the variations, such that the same variations can result either from mutations or from the appropriate environmental perturbations. While I do not claim that it is possible to generate so-called phenocopies of every mutation (though it may be true), it is certainly the case that if we knew the timing of particular developmental events and can make a shrewd guess at the mechanism involved, then we can induce the phenocopies we have in mind.

A number of morphogenetic mutants of *Drosophila* have been isolated and molecularly cloned recently (see Coulter and Wieschaus, 1986, and references). These cloned genes have been used most ingenously to locate the temporal and spatial domains of their expression during embryogenesis. The fruitfly's body is made up of about 14 repeated parts or segments, each of which has a different identity. Mutations in the so-called *homeotic* genes scramble the

Figure 2b The Belousov–Zhabotinskii reaction in a Petri dish. (From Winfree and Strogatz, 1983. Reproduced by permission of the authors)

identity of segments, while mutations in the *segmentation* genes lead to incorrect divisions into segments. The phenomenology of embryogenesis has been greatly enriched by molecular genetics, but these studies simply do not address the problem of how the genes become expressed in their particular spatial configurations during development (see Goodwin, this volume). Spatial organization begins with physicochemical processes which set up spatial patterns of cytoplasmic states. The latter, in turn, trigger the differential expression of genes (see Ho, 1984a). This is why a simple physical pertubation — exposure to ether — suffices to induce a range of segmentation defects in *Drosophila* (Ho *et al.*, 1987). Specific homeological transformations (see Goodwin, this volume) were repeatedly induced at precise stages of development. Many of these resemble mutant phenotypes which have been described,

Figure 3 A range of unusual larval phenotypes in *D. melanogaster* resulting from ether treatment. The normal phenotype is at the centre

and so may be regarded as phenocopies. On the other hand, a substantial number of the phenotypes obtained are new, and have never been described as mutants (see Figure 3). It will be of interest to see if *genocopies* of these will eventually be produced by mutation.

Evolution of novel enzyme function in *Escherichia coli*

Wild type *E. coli* metabolizes lactose by first breaking it down with the enzyme β-galactosidase. Special strains in which the β-galactosidase gene is deleted have no enzyme activity and do not metabolize lactose if other carbon sources are present. Experimenting with these strains, Campbell, Lengyel and Langridge (1973) found that as the other carbon sources became exhausted, mutant colonies appeared which possessed lactose-splitting activity. The enzyme responsible was not the β-galactosidase that had been deleted, but another enzyme altogether, called *ebg*, mapping to the other side of the genome. This had undergone mutations that enabled it to metabolize lactose. By itself, this result is unremarkable because it could be interpreted as the artificial selection of a fortuitous mutation. However, the experiment was immediately repeated by other workers (Hall and Hartl, 1974), who isolated 34 different lactose-utilizing strains by the same method. All of these contain enzyme activity identical to *ebg*.

Moreover, in 31 of the strains, the synthesis of the newly evolved enzyme is regulated by lactose, i.e. there has been an additional mutation in another gene which interacts with lactose to regulate *ebg* (Hall, 1982). There is nothing fortuitous in this highly repeatable response to the same environmental challenge, which involves appropriate mutations in two different genes appearing simultaneously (see Opadia-Kadima, 1987).

Remarkable as this example appears, it is really no different from the repeatable production of specific transformations when a given environmental perturbation is applied in *Drosophila* embryos. It is the physiological state of the cell in one case, and the epigenetic system of the organism in the other, that organically 'selects' the appropriate response. Part of that response may involve defined mutations at specific sites in the genome.

Pesticide resistance

Pesticide resistance has been a major and persistent problem in agriculture. As fast as new insecticides are manufactured and brought into use, new resistant populations of the insects evolve. The rapid evolution of insecticide resistance has become a textbook example of the supposed power of natural selection in increasing the frequency of a pre-existing allele which confers resistance.

Actually, insecticide resistance turns out to be a paradigm example of something else: the process of gene amplification (see Hyrien and Buttin,

1986). It is estimated that amplification events occur spontaneously at frequencies of between 10^{-3} and 10^{-4} in cultured cells (see Pollard, this volume). The existing evidence suggests that many, if not all, parts of the genome may be prone to amplification (Schimke, 1984). Gene amplification is thus part of the physiological repertoire of all cells. Toxic chemicals in the environment *stabilize* those cells in which the genes for the appropriate detoxifying enzymes are being amplified. In that respect, the environmental agent is again 'selecting' for a physiological response, but not for a pre-existing genetic variant, since every cell or organism is potentially capable of amplifying its genes. It is possible, of course, that the presence of the toxic chemical itself creates a physiological state that favours the amplification of the appropriate detoxifying genes, but this remains to be investigated.

Directed changes in the genome

There are other cases of directed, non-random changes in the genome (see below). The striking feature in the last two examples is that the changes involved are so obviously adaptive as well as generic. Another comparable situation which comes to mind is the specific and reproducible nature of the immune response to particular antigens. A great deal of antibody diversity is initially generated by DNA rearrangements in which different combinations of genes are spliced together to make the composite antibody gene in each lymphocyte. Subsequently, somatic mutations in the antigen binding site give rise to high affinity antibodies during the maturation of the immune response. Analysis of the antibodies produced against a given antigen reveals that the same somatic mutations recur, not only among different antibodies in the same mouse, but also among antibodies produced in different mice as well (Griffiths *et al.*, 1984).

All these instances of directed variations demonstrate that the organism responds physiologically to the environment in one *continuous* process. The environment is 'selective' in so far as it selects, via the physiological system for an appropriate response, but no real 'selective deaths, have taken place as required by natural selection. The fact that the response may involve genomic changes shows that the strict dichotomy between genotype and phenotype really does not exist as far as the organism is concerned.

As the variations are generated by the physiological interaction between organism and environment, the persistence of the variations over the generations depends not on natural selection but on heredity! The variations would indeed persist in the absence of natural selection, so long as heredity operates in their favour. There is a great deal of conceptual confusion here which some good philosopher of biology should try to sort out in future. For now, it seems most fitting for us to direct our attention to the nature of heredity itself.

HEREDITY BEFORE AND SINCE THE RECOMBINANT DNA REVOLUTION

Until very recently, heredity was supposed to be due to the transmission of something which remains unchanged from generation to generation. The widespread use of the term 'inheritance' is significant, as it brings to mind something akin to the family heirloom. This concept depended on basic assumptions of genetics which practically everyone accepted as true in the 1960s and early 1970s. These are as follows:

1. DNA (and in some viruses, RNA) is the genetic material.
2. Genetic information flows from DNA to RNA to protein, but never in reverse.
3. One polypeptide is specified by one gene locus.
4. Colinearity exists between the base sequence of DNA in the gene and the amino acid sequence of the polypeptide it encodes.
5. The genetic code is universal.
6. The codons are read in one direction, without overlap, and only in one correct reading frame.
7. With very few exceptions, the DNA of all cells remains constant during development, only the genes expressed differ between different cells.
8. Environmentally induced modifications do not affect DNA and cannot be inherited.

Since then, all but the first assumption has become violated. The first assumption remains true only by virtue of definition, as it has long been accepted that certain cytoplasmic components of the oocyte also affect heredity, and therefore would qualify as 'heredity material', if not 'genetic material' (see Ho, 1986, 1987a; Ho and Goodwin, 1987).

By far the most significant picture to emerge out of recombinant DNA research is the dynamism and flexibility of the eukaryotic genome in both its organization and function. This is in striking contrast to the relatively static and mechanical conception which previously held sway (see quotation from Jacob, 1970, in Pollard, this volume).

Some idea of the functional flexibility may be gleaned by considering the status of the one gene–one polypeptide relationship. This relationship has been violated so many times and in so many ways that there appears no longer to be any general rule in the matter. All conceivable mappings between genes and polypeptides have now been found. It really exposes the fallacy of the idea that single genes could be involved in 'determining' any single character (cf. Saunders, this volume).

The functional flexibility of the genome is in part associated with, and indeed fully matched by, its structural fluidity, which in turn locates the genome firmly within the physiological system of the organism as a whole, as we shall see.

I will now concentrate on the two main aspects of the recent findings which are most relevant to my thesis of evolution by process.

THE FLUID GENOME

The first is the fluidity of the genome. This refers to all observations suggesting that genomic DNA is subject to relatively large alterations both during development and in evolution. Genomic fluidity depends on a host of mechanisms which can rearrange DNA (such as that involved in the synthesis of immunoglobulins), move or transpose sequences or contract them, and even sequences, greatly amplify particular sequences or contract them, and even convert sequences from one to another in the same or different part of the genome (see Ho *et al.*, 1987 and refs therein).

It is not known whether entire genomes are potentially fluid, although much actual fluidity is associated with the multigene families whose sequences are dispersed all over the genome. Some parts of the genome are known to change in a predictable way during development. Others have been found to change in both predictable and unpredictable ways under different kinds of environmental conditions.

For example, mammalian somatic cells, both *in vivo* and *in vitro*, undergo non-random multigenic amplifications when challenged with cytotoxic drugs. These changes involve morphologies or specific chromosomes, and the appearance of certain chromatic bodies in the nucleus (Gudkov and Kopnin, 1985). More dramatically, large changes in germline DNA can be induced by environmental manipulations within a single generation. The best studied case is the induction of heritable changes in flax plants treated with different fertilizers (see Cullis, 1983; also this volume). The stable lines produced differ from the parental lines and from one another in morphological characters, isoenzyme patterns, as well as in amounts of nuclear DNA, the number of ribosomal RNA genes and other repeated sequences.

A substantial amount of genomic fluidity with erratic outcomes results from transpositions. Transpositions are found to be particularly frequent during stress in both maize and *Drosophila*. Transposons mutate other genes by inserting into or near them so that they become either activated or inactivated as a result. Transpositions can also cause gross chromosomal rearrangements, thus scrambling the genome in a major way. They occur, not only in somatic cells, but also in germ cells, so that the resultant mutations and scrambled genomes are passed straight on to the next generation.

A comparison of multigene families from related species suggests a relatively high background rate of DNA 'turnover' in all organisms due to cycles of rearrangements, amplifications, deletions and mutations (Figure 4) (Flavell,

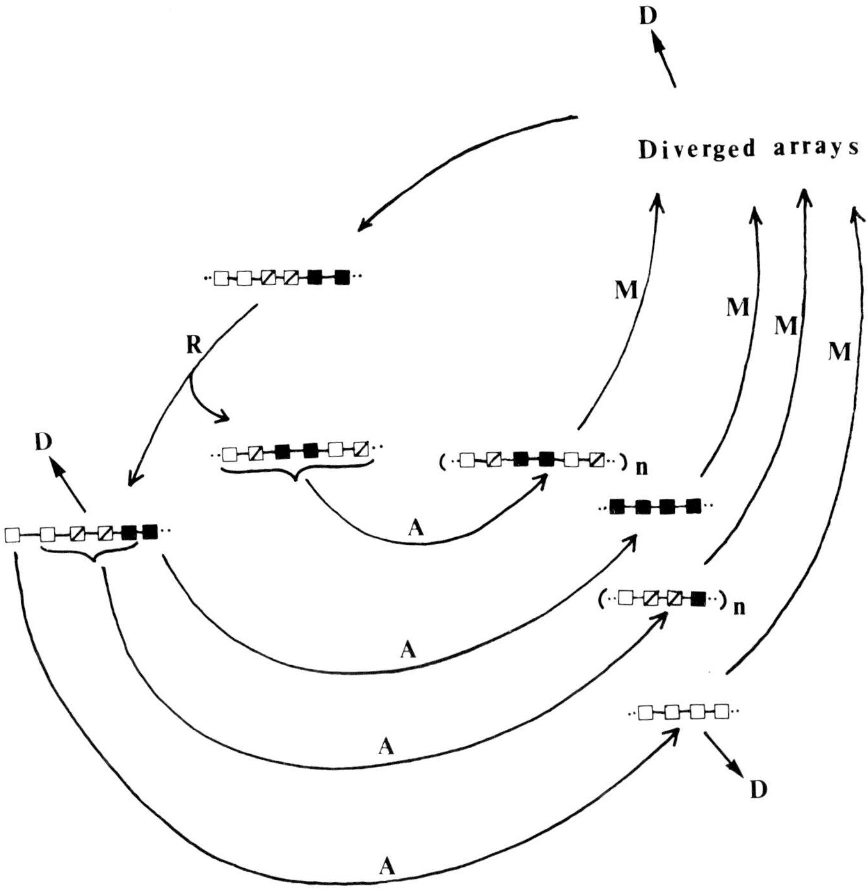

Figure 4 Dynamics of genomic 'DNA turnover'. A, amplification; D, deletion; R, rearrangement; M, mutation. Different square symbols represent different gene sequences. (Redrawn from Flavell, 1982)

1982). This leads to species divergence in the long term, quite independently of natural selection. Furthermore, under certain environmental conditions, the same mechanisms responsible for DNA turnover can cause sufficiently large genetic changes to precipitate the rapid formation of new species (see this volume, Cullis, Chapter 4 and Pollard, Chapter 5). This kind of saltatory speciation is also independent of natural selection.

In summary, DNA may be just as responsive and flexible as the rest of the organism, from behaviour to developmental physiology to protein synthesis (see Ho, 1987a). In fact, genomic fluidity can itself be regarded as a phenotypic character which varies between lines, between species, and according to the environment and the physiological state of the organism (Cullis, this volume).

THE PERMEABILITY OF WEISMANN'S BARRIER

The second aspect of the findings from recombinant DNA research is the permeability of Weismann's barrier (see Pollard, 1984, and this volume; Ho, 1986; Ho *et al.*, 1987). This involves all sorts of processes in which the soma 'talks' back to the germline as part of the functional interactions between different levels within the organism or between the organism and the external environment. I would include (a) the large-scale reverse transcription of processed RNAs into DNA and reinsertion into the germline genome; (b) the non-random changes in genomic DNA in certain environments which can become stably inherited in subsequent generations; (c) biased gene conversion with or without mRNA intermediate, which is based on hypothetical functional feedback to genomic DNA.

I have not included other non-Mendelian processes mentioned above in connection with genomic fluidity, nor the now well-established horizontal transfer of genes between individuals and between species by viruses and plasmid vectors, because they do not seem to be directly related to physiological functions. (It should be noted that in the transfer of plasmids carrying antibiotic resistance genes among micro-organisms, there is an indirect involvement with physiological functions.) But that may only reflect the extent of my ignorance at this point. It may be that as further knowledge becomes available, these non-Mendelian processes will prove to be part and parcel of the organisms's repertoire in the performance of its vital functions.

Reverse transcription plays a large role in shaping the eukaryotic genome (Baltimore, 1985). Since its discovery in retroviruses, the same process has been identified in a wide variety of organisms. For example, certain eukaryotic transposable elements are very similar to retroviruses, and transpose by reverse transcription of processed RNA. Vertebrate cells have cytoplasmic particles which are similar to retroposons (transposons with RNA intermediates). These have active reverse transcriptase and are represented in the genome as proviral DNA, accounting for some 0.3 per cent of the chromosomal DNA of the cell. That reverse transcription occurs frequently for eukaryotic genes is strongly suggested by the widespread presence of so-called *processed pseudogenes* in the genome both for single copy gene sequences and for multigene families. One sequence, the *Alu*, repeated half a million times in the human genome, is reverse transcribed from the cytoplasmic RNA of the 'signal recognition particle' involved in translocating protein into and across cytoplasmic membranes. Most pseudogenes are probably non-functional; but one of the rat insulin genes has been shown to be part of a functional retroposon (Soares *et al.*, 1985).

More intriguing still is the recent report that a certain family of repeated DNA sequences in primates, the L1, may be derived from a sequence encoding a reverse transcriptase-like protein (Hattori *et al.*, 1986). It will be of interest to see whether the protein does have reverse transcriptase activity, and if so, whether reverse transcription is a normal function of the organism.

Reverse transcription violates the central dogma as well as Wiesmann's barrier. Some people argue that the violation of the central dogma may be confined within germ cells, and so for the organisms in which germline and soma are well separated, this does not constitute a violation of Weismann's barrier. As pointed out by Buss (1983), however, there is no early separation between germline and soma in protists, fungi, plants and over half the existing phyla of animals. In addition, the presence of processed pseudogenes in proteins such as the haemoglobins, which are expressed only in specialized somatic cells, suggests that communication from somatic to germ cells can occur, possibly by way of retroviral infection (Temin, 1971; see also Pollard, 1984 and this volume, Chapter 5). Indeed, cellular mRNAs are often found associated with isolated retroviral particles. Ikawa, Ross and Leder (1974) showed that when Friend leukaemia virus was grown on red blood cells that were actively transcribing globin genes, one in a thousand of the virus particles had packaged β-globin mRNA inside.

I have already mentioned the directed changes in genomic DNA on treating flax plants with different fertilizers. These changes are specific to different environments and reproducible for each environment; so they are not just generalized stress responses. The genomic changes occur in the course of development, and involve all cells in the meristem simultaneously. Subsequent to that, the plants improve in growth. Thus, at least some of the changes may constitute an adaptive physiological response which is subsequently stably inherited (see Cullis, this volume).

Biased gene conversion involves the transfer of sequences between homologous genes present in different parts of the genome. This process contributes to so-called 'concerted evolution' in which members of a multigene family become more homogeneous within a given species than between species (see Dover, 1986 and refs therein). Apart from its obvious contribution to genomic fluidity, gene conversion may be related to gene function especially if an RNA intermediate is involved. In that situation, expressed genes will convert non-expressed ones (Doolittle, 1985). Also, if there is some correlation between the state of methylation of a gene and its expression, then methylated genes which are not expressed may become preferential targets for conversion (see Kourilsky, 1986). This would constitute yet another example of an obvious physiological adaptation that involves genomic changes, but in such a way as to stabilize and maintain those gene sequences that are functional.

It is clear that genomic DNA is not immune to change as the result of feedback from the environment and the physiological state of the organism. In fact, DNA changes are involved in the stabilization of gene function as much as in altering it. This only reflects the necessary relationship which DNA has with the rest of the organism. In a sense, there is nothing special in the status of DNA. Gene expression states can be stably inherited without changes in DNA; conversely, alterations in DNA instigated by changes in the environment within one generation can become inherited (see Ho, 1986; Ho, 1987a).

A REFORMULATION OF HEREDITY

The real problem of heredity is to account for the stable and repeatable nature of reproduction. This feature was previously widely attributed to the constancy of DNA. The major consequence of the discovery of the fluid genome is to expose the untenability of this assumption. DNA is functionally and structurally as flexible as the rest of the organism. How then should we see heredity? Where does stability reside if not in the constancy of DNA? In order to answer this question, we have to remind ourselves of some elementary facts of biology.

Living beings engage in the process of living. Process is an *activity* at all levels: from the behaviour of organisms in their ecological and social environment to the expression of genes in their cells. From the beginning of development to maturation and senescence, there is almost nothing that remains static and unchanging. Organisms are life histories and not mechanical objects as conceptualized within Neo-Darwinism (Ho, 1987a). Whereas the stability of mechanical objects depends on static equilibrium, that of organisms is *dynamically maintained*, and is utterly dependent on activity; in other words, on fluidity and change. The cessation of activity spells death.

Heredity — a name given to the observed constancy of reproduction — must ultimately be looked upon as a process, and not as some *material* which is passed on from parent to offspring. Processes, whether biological or physicochemical, have an inherent dynamic which generates patterns and regularities (recall the aggregation of the slime mould and the Belousov–Zhabotinskii reaction mentioned earlier); and here is where at least part of the stability of reproduction resides. Another source of stability is due to the fact that the 'control' of development is *web-like* and *circular* rather than linear and unidirectional (see Figure 5). This means that the 'cause' of development is not just the DNA, but is *distributed throughout the complex interrelationships between the different levels of organism and its environment.* It may be that the fluidity of DNA plays an indispensable role in the maintenance of the organismic system as a whole, for all components of the system must be able to adjust and respond as appropriate to their particular milieu. What is inherited in each successive generation is not only the precise copies of DNA molecules, but an entire experiential repertoire including maternal, cytoplasmic effects, the physicochemical, biotic and social environments (Ho, 1987a), all of which conspire to make development similar to the previous generation. Heredity is therefore inseparable from development.

Similarly, development is directly linked to evolution in two senses. The first

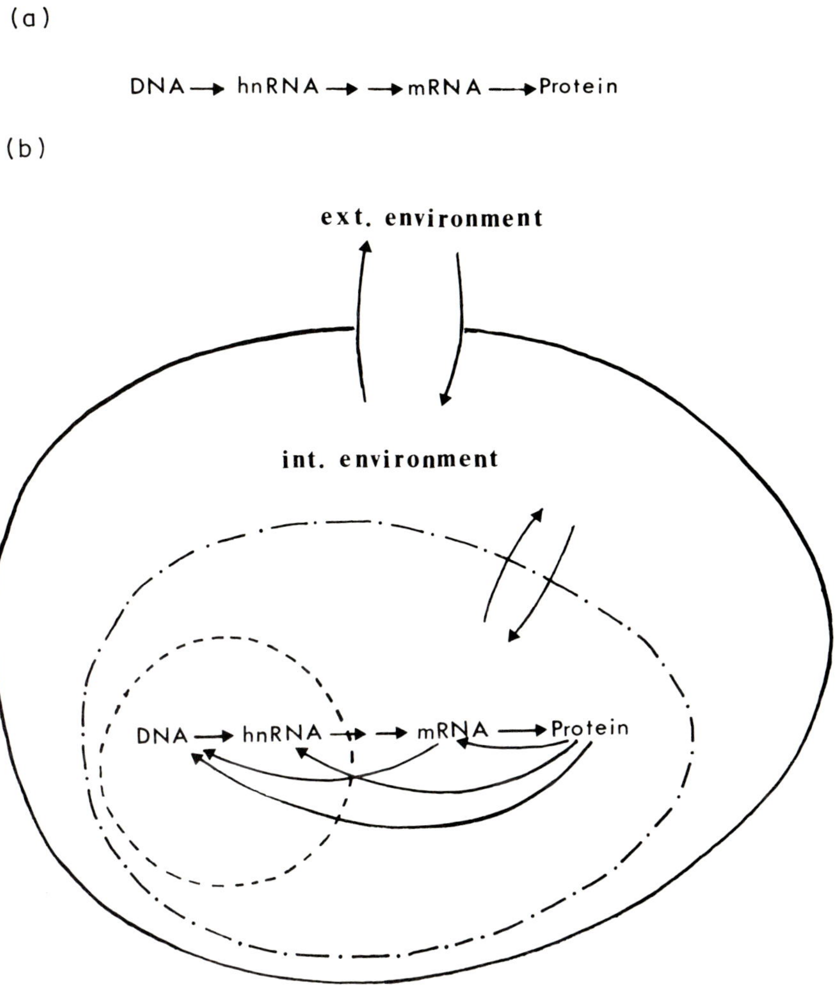

Figure 5 Two models of gene function in development. (a) The central dogma. (b) The process view (this paper). hnRNA, heterogeneous nuclear RNA or primary transcript

is in the formal sense that development, through the dynamical structure of the epigenetic system, defines the sort of changes that can occur under different contingent conditions. The second is by virtue of the fact that generations are not separate and the germline not so inviolable as previously held. This means that the experience of each generation will be transmitted culturally and ecologically as well as via physiological influences which may, or may not, include changes in germline DNA.

EVOLUTION BY PROCESS

How does one see evolution by process as opposed to evolution by consequence? This is really the subject of at least another paper, but I will outline an approach in the study of animal behaviour which a number of workers are already developing, although they themselves may not necessarily think so.

One of the first concerns (Packard, 1987) is to restore the notion of 'fitness' to its former meaning (cf. Henderson, 1913) as an appropriate or harmonious relationship between organism and environment; rather than as a measure of reproductive success. Such fitness arises simultaneously with the organism acting and responding to the environment in continuous processes occurring at multiple levels over a range of time scales. It does not result from random variations which are selected *a posteriori*. The experiencing organism registers change as it is experiencing, and this will influence its future actions and experiences as well as those of future generations. This is where I believe Skinner's analogy between the consequence of reinforcement and the consequence of selective reproduction must break down. The consequence of reinforcement involves the registering of real experience by the organism; as Skinner (1984) has said, an experienced rat is a 'changed rat'. On the contrary, no such registering of real experience is permitted in the Neo-Darwinian theory of natural selection. In fact, this is specifically forbidden on account of Weismann's barrier. If one accepts the arguments presented in this chapter then there is a *continuity*, and not just an analogy, between operant reinforcement within one generation and reinforcement by the process of heredity in successive generations (see below). This is a worthwhile project for future exploration.

One issue that must be raised is the notion of active choice (see Ho, 1984b) which a number of Neo-Darwinist ethologists and biologists have emphasized. I believe the role of choice is at best ambiguous within Neo-Darwinism. While the animal is invested with the capacity to choose, it is the *consequence* of the choice that is supposed to be selected and inherited, thus making the act of choosing meaningless. This contradiction arises most obviously in the juxtaposition of mate *choice* and sexual *selection* as the consequence of the choice made (see also Bateson, this volume).

If choice is to be really exercised, then it is the ability to choose that is inherited, and not the choice itself. Of course, the extent to which choice is 'free' is another matter. Animals are social beings that seek approval, if not love; and experience both pleasure and pain. It has been suggested that such psychological states may play a large role in the evolution of behaviour in influencing the animal's choice (Packard, 1987). This means that the animal itself is evaluating at all times, both its own actions (by a 'sense of satisfaction' (Skinner, 1985) perhaps in relation to its social *milieu*), and the effect of its actions on the environment. It is not 'judged', *a posteriori*, by natural selection.

In order to make this discussion more concrete, let us consider predation. The Darwinian picture is that the prey evolves because the weakest, slowest running prey, and their bad genes, get elimated by the predator. By the same token, the fastest, most cunning predators capture the prey and leave the most offspring, thus preferentially propagating their good genes. The only thing that prevents both predator and prey from evolving towards the speed of light is some vague and timely appeal to 'developmental constraints'.

A more rational view of the whole process may be as follows. The prey, having survived the predator, and reciprocally, the predator, having successfully (or for that matter, unsuccessfully) preyed, both register the experience. The imprints of the experience go from the visual/olfactory systems to ionic currents and physical, biochemical membrane changes in the central nervous system in the short term (Farley *et al.*, 1983; Mason and Rose, 1987). These may then translate into changes in synapses and synthesis of glycoproteins and perhaps neuroreceptor molecules in the longer term (see Rose, 1987). Similarly, the exercise of muscles involved in running either to catch the prey or to escape from the predator, will effect changes in the contractile properties of the muscles (Lømo, Westgarrd and Kahl, 1974), which include the type of nerve endings present, and the expression of different ATPases and myosin genes (see Laing and Lamb, 1985) and refs therein).

These imprints, which are really internalizations of the environment at successive levels in space and time, may then alter heredity in the sense reformulated here (see also Ho, 1987a). At the 'highest' level, cultural inheritance will ensure that the next generation of both prey and predator will be taught to deal effectively with each other. Social cohesion will encourage a conformity of behaviour among individuals all of whom will reinforce it for generations to come. At the 'lowest' level, the change in gene expression states may be inherited cytoplasmically, or maternally. It is not impossible, though not necessary, that changes in genomic DNA may also be involved. This is open to investigation by recombinant DNA techniques now available.

Skinner's hypothesis of operant conditioning and reinforcement in shaping and maintaining behaviour does have the virtue that the entire process can be made quite transparent and explicit. One starts from the organism's initial experience of the external environmental stimuli, and ends in changes at the physiological, and perhaps molecular, genetic level. There is no need for 'genetic programmes' or unknown and unknowable 'internal representation states' to 'control' behaviour. By the same token, there is no need to appeal to 'genes for reinforcement' as Skinner (1981) has done, especially if all organisms are supposed to have them, for genes are primitively defined on *differences*.

The process view of evolution that I have presented is at one with the dynamic holism that the comparative psychologist Schneirla (1966) and others have advocated: a description that loses nothing of the texture and colour of reality. It is necessarily pluralistic becuase it involves all levels and their interconnections (cf. Tobach and Greenberg, 1984).

IN SEARCH OF NEW METAPHORS

Darwinism epitomizes the development of a *Zeitgeist* in nineteenth-century Britain, which in turn lent credence to some of the most pernicious political ideologies in the present century. As Barzun (1958) remarked:

> 'Matter' and 'force', when applied to human beings, found some dangerous simple applications. And when the idea of force is embodied in the notions of Struggle and Survival of the Fittest, it should be expected that men will use these revelations of science as justifications for their own acts.

This completes the positive feedback loop between the dominant sociocultural ideology of the day and a scientific theory to which it gave birth. Social Darwinism and theories of racial inequality formed the backdrop to the rise of the Nazi regime.

Is history repeating itself now with sociobiology? Sociobiology is not just a parlour game for dons. Like Social Darwinism, it has considerable sociopolitical implications. Bateson (1985) points out:

> The emphasis on selfishness and the struggle for existence in evolutionary biology has had an insidious confirmatory effect on the public mind.
>
> The political dangers of representing all human social relationships in terms of competition is that the expectation is self-fulfilling.

Nevertheless, Bateson does not see the need to give up sociobiology. Instead he proposes to de-emphasize competition and to concentrate on cooperation (or co-adaptation) in nature. Of course, this is merely a rehearsal of the many arguments which have been pitched against Social Darwinism in its day, and history has demonstrated that they are woefully inadequate. Within Darwinism as within Neo-Darwinism, competition is primary. Where cooperative behaviour does occur, it has to be explained as a phenomenon arising almost paradoxically out of competition — as by kin selection or reciprocal altruism. Consequently, any attempt to construct a Neo-Darwinist sociobiology in which cooperation is anything more than a relatively minor aspect of behaviour is doomed to failure. Besides, nothing is to be gained by replacing one metaphor with another unless it takes us further ahead.

The major weakness in the critiques of sociobiology, as in those of Social Darwinism, is the almost universal and unquestioned acceptance of the foundations on which they are based: Neo-Darwinism and Darwinism respectively. Our predecessors were combating vitalism on the one hand and Fundamentalism on the other, and therefore could see no real alternative to mechanical determinism. It is not so with us. I have tried to show how natural selection is already discredited by the empirical evidence now available. It presents a reduced and distorted view of the organism and its relationship with nature. It imposes a conceptual framework that does considerable violence to reality, not only in theory but in practice as well. Because science and our perception of the world are interdependent, a distorted science can in turn distort sociocultural reality through the self-fulfilling prophecy.

Evolution by process is premised on the unity and rationality of nature. Organism and environment are interconnected from the genes to the sociocultural domain. Through these interconnections, the organism registers formative imprints of its experience, and at the same time actively creates and structures its environment. In other words, organism and environment engage in continual mutual transformation. The processes of transformation are neither random nor arbitrary. Instead, they proceed according to organization principles which apply equally to the living and non-living realms (Ho and Saunders, 1986).

Evolution by process accords with the empirical findings of science and with our deepest experience of nature. In emphasizing integration and intelligibility, it re-establishes a fertile and harmonious relationship between biology and the exact sciences on the one hand and between biology and the social and human sciences on the other.

ACKNOWLEDGEMENT

A version of this chapter was presented at the Third International Conference in Comparative Psychology in Costa Rica, July 1986, at the invitation and instigation of Ethel Tobach. I am grateful to Professor Skinner for stimulating comments, reprints and preprints; and to Andrew Packard for sharing his considerable insights with me in a preprint. Thanks are also due to Peter Saunders, Brian Goodwin, Jack Hailman and Susan Oyama for helpful comments on earlier drafts.

NOTE

1. Actually, the symmetry is not so perfect as perceived by naturalists eager to impute perfection either to divine inspiration or to the geometric instinct of bees. But the example does illustrate how simple physics often suffices to explain the marvellous and seemingly mysterious handiwork of 'lowly' creatures.

REFERENCES

Baltimore, D. (1985). Retrovirus and retrotransposons: the role of reverse transcription in shaping the eukaryotic genome, *Cell*, **15**, 481–2.

Barzun, J. (1958). *Darwin, Marx, Wagner*, Doubleday Anchor, New York.

Bateson, P. (1985). Sociobiology and human politics. In S. Rose and L. Appignanesi (eds), *Science and Beyond*, Blackwells, Oxford, pp. 79–99.
Buss, L.W. (1983). Evolution, development and the units of selection, *Proc. Nat. Acad. Sci. USA*, **80**, 1387–91.
Campbell, J.H., Lengyel, J.A., and Langridge, J. (1973). Evolution of a second gene for β-galactosidase in *E. coli, Proc. Nat. Acad. Sci. USA*, **70**, 1841–5.
Coulter, D., and Wieschaus, E. (1986). Segmentation genes and the distributions of transcripts, *Nature*, **321**, 472–4.
Cullis, C.A. (1983) Environmentally induced DNA changes in plants. *CRC Critical Reviews in Plant Sciences*, **1**, 117–31.
Darwin, C. (1859). *Origin of Species*, John Murray, London.
Darwin, C. (1868). *Variations of Animals and Plants under Domestication*, John Murray, London.
Doolittle, W.F. (1985). RNA-mediated gene conversion? *TIGS*, **1**, 64–5.
Dover, G.A. (1986). Molecular drive in multigene families: how biological novelties arise, spread and are assimilated. TIGS, **2**, 159–64.
Farley, J., Richards, W.G., Ling, L.J. Liman, J.D., and Alkon, D.L. (1983). Membrane changes in a single photoreceptor cause associative learning, *Hermissender Science*, **221**, 1201–3.
Flavell, R.B. (1982). Major sources of variation during species divergence. In G.A., Dover and R.B. Flavell (eds), *Genome Evolution*, Academic Press, London.
Goldschmidt R. (1983). *Physiological Genetics*, McGraw Hill, New York.
Griffiths, G.M., Berek, C., Kaartinen, M., and Milstein, C. (1984). Somatic mutation and the maturation of immune response to 2-phenyl oxazolone, *Nature*, **312**, 271–5.
Gudkov, A., and Kopnin, B. (1985). Gene amplifications in multidrug resistant cells: molecular and karyotypic events. *BioEssays*, **3**, 68–71.
Hall, B.G. (1982). Evolution on a petri dish. The evolved β-galactosidase system as a model for studying acquisitive evolution in the laboratory, *Evol. Biol.*, **15**, 85–150.
Hall, B.G., and Hartl, D.L. (1984). Regulation of newly evolved enzymes. 1. Selection of a novel lactase regulated by lactose in *Escherichia coli, Genetics,* **76**, 391–400.
Hattori, M., Kuhara, S., Takenaka, O., and Sakaki, Y. (1986). L1 family of repetitive DNA sequences in primate may be derived from a sequence encoding a reverse transcriptase-related protein, *Nature*, **321**, 625–8.
Henderson, L.J. (1913). *The Fitness of the Environment*, Macmillan, New York.
Ho, M.-W. (1984a). Environment and heredity in development and evolution. In M.-W. Ho and P.T. Saunders (eds), *Beyond Neo-Darwinism: Introduction to the New Evolutionary Paradigm*, Academic Press, London, pp. 267–89.
Ho, M.-W. (1984b). Where does biological form come from? *Rivista di Biologia,* **77**, 147–79.
Ho, M.-W. (1986). Heredity as process. Towards a radical reformulation of heredity, *Rivista di Biologia*, **79**, 407–47.
Ho, M.-W. (1987a). Genetic fitness and natural selection: myth or metaphor? In *Proc. 3rd T.C. Schneirla Conference*, Lawrence Erlbaum Assoc., NJ (in press).
Ho, M.-W. (1987b). Evolution by process, not by consequence: implications of the new genetics in development and evolution. *Int. J. Comp. Psychol.* (in press).
Ho, M.-W., and Goodwin, G.C. (1987). *S298 Genetics Unit 1, An Open University Course*, Open University Press, Milton Keynes.
Ho. M.-W., Matheson, A., Saunders, P.T., Goodwin, B.C. and Smallcombe, A. (1987). Ether-induced disturbances in segmentation in *Drosophila melanogaster. Roux Arch. Dev. Biol.*, **196**, 511–521.
Ho, M.-W., and Saunders, P.T. (1986). Evolution: natural selection or self-

organization? In C.W. Kilmister (ed.), *Disequilibrium and Self-Organization*, D. Reidel, Dordrecht, pp. 231–42.

Ho, M.-W., Saunders, P.T., and Fox, S.W. (1986). A new paradigm for evolution, *New Scientist* (27 Feb.), 41–3.

Ho, M.-W., Saunders, P.T., and Fox, S.W. (1987). Through a neo-Darwinian glass darkly, *BioEssays*, **6**, 3–4.

Hyrien, O., and Buttin, G. (1986). Gene amplification in pesticide-resistant insects. *TIGS*, **2**, 275–6.

Ikawa, Y., Ross, J., and Leder, P. (1974). An association between globin messenger RNA and 60s RNA derived from Friend Leukemia virus, *Proc. Nat. Acad. Sci. USA*, **71**, 1154–8.

Jacob, F. (1970). *The Logic of Life: A History of Heredity*, Pantheon Books, New York.

Kourilsky, P. (1986). Molecular mechanism for gene conversion in higher cells, *TIGS*, **2**, 60–3.

Laing, N.G., and Lamb, A.H. (1985). Muscle fibre types and innervation of the chick embryo limb following cervical spinal cord removal, *J. Embryol. Exp. Morph.*, **89**, 209–22.

Lømo, T., Westgarrd, R.H., and Kahl, H.A. (1974). Contractile properties of muscle: control by pattern of muscle activity in the rat, *Proc. Roy. Soc. Lond. (B)*, **187**, 99–103.

Mason, R.J., and Rose, S.P.R. (1987). Lasting changes in spontaneous multi-unit activity in the chick brain following passive avoidance training, *Neuroscience* (in press).

Maynard-Smith, J., and Holliday, R. (1979). Preface. In J. Maynard-Smith and R. Holliday (eds), *The Evolution of Adaptation by Natural Selection*, The Royal Society, London, pp. v-vii.

Nijhout, H.F. (1985). The developmental physiology of colour patterns in Lepidoptera, *Adv. Insect Physiol.* **18**, 181–247.

Opadia-Kadima, G.Z. (1987). How the slot machine led biologists astray. *J. Theor. Biol.*, **124**, 127–35.

Packard, A. (1987). The role of behaviour in evolution: a metaselection theory. Unpublished preprint.

Plotkin, H.C. (ed.) (1982). *Learning, Development and Culture,* Wiley, London.

Pollard, J.W. (1984). Is Weismann's barrier absolute? In M.W. Ho, and P.T. Saunders (eds), *Beyond Neo-Darwinism*, Academic Press, London, pp. 291–315.

Rose, S.P.R. (1987). Neural mechanism of memory. In D. Alkon and G. Woody (eds), *Neural Mechanisms of Conditioning,* Plenum Press, New York (in press).

Saunders, P.T. (1980). *An Introduction to Catastrophe Theory*, Cambridge University Press, Cambridge.

Saunders, P.T. (1984). Development and evolution. In M.-W. Ho and P.T. Saunders (eds), *Beyond Neo-Darwinism: Introduction to the New Evolutionary Paradigm,* Academic Press, London, pp. 243–63, Academic Press, London.

Schimke, R.T. (1984). Gene amplification in cultured animal cells, *Cell*, **37**, 703–13.

Schneirla, R.C. (1966). Behavioural development and comparative psychology, *Quart. Rev. Biol.,* **41**, 283–302.

Shapiro, A.M. (1984). The genetics of seasonal polyphenism and the evolution of 'general purpose genotypes' in butterflies. In K. Wohrmann and V. Loeschcke (eds), *Population Biology and Evolution,* Springer-Verlag, Berlin, pp. 16–30.

Sinha, C. (1984). A socio-naturalistic approach to human development. In M.-W. Ho and P.T. Saunders (eds), *Beyond Neo-Darwinism*: *Introduction to the New Evolutionary Paradigm*, Academic Press, London, pp. 331–62.

Skinner, B.F. (1981). Selection by consequence, *Science*, **213**, 501–4.
Skinner, B.F. (1984). Author's response: representations and misrepresentations (Open peer commentary of 'behaviourism at fifty'), *Behav. Brain. Sci.,* **7**, 655–67.
Skinner, B.F. (1985). News from nowhere, 1984, *The Behaviour Analyst*, **8**, 5–14.
Soares, M.B., Schon, E., Henderson, A., Karathan – asis, S.K., Cate, R., Zeitlin, S., Chirgwin, J., and Estratiadis, A. (1985), *Mol. Cell Biol.*, **5**, 2090–103.
Temin, H.M. (1971). The protovirus hypothesis: speculations on the significance of RNA-directed DNA synthesis for normal development and for carcinogenesis, *J. Nat. Cancer Inst.*, **46**, III–VII.
Thom, R. (1975). *Structural Stability and Morphogenesis* (trans. D.H. Fowler), W.A. Benjamin, Reading, Mass.
Thompson, D'A.W. (1914). *On Growth and Form*, Cambridge University Press, Cambridge.
Tobach, E., and Greenberg, G. (eds) (1984). *Behavioural Evolution and Integrative Levels*, Lawrence Erlbaum Associates, N.J.
Waddington L.H. (1957). *The Strategy of the Genes*, George Allen and Unwin, London.
Winfree, A.T., and Strogatz, S.H. (1983). Singular filaments organize chemical waves in three dimensions. I. Geometrically simple waves, *Physica*, **8D**, 35–49.
Young, R.M. (1985). *Darwin's Metaphor*, Cambridge University Press, Cambridge.

Evolutionary Processes and Metaphors
Edited by M.-W. Ho and S.W. Fox

CHAPTER 8

Morphogenesis and heredity

B.C. Goodwin

ABSTRACT

Morphogenesis and heredity are different aspects of organismic life histories. The former is concerned with the process that generates the distinct morphologies of species, while the latter studies the mechanisms that stabilize these processes from generation to generation. Focus on hereditary relations between organisms results in a genealogical science in which species are seen as members of an historical unity, starting from a common origin of life. A systematic study of morphogenetic relationships between organisms results in a science of space–time transformations in which species are seen as members of a rational unity linked by a common set of generative processes. This chapter explores the evidence for, the structure and the consequences of, the latter type of biological science.

Two quite different traditions of analysis and explanation have developed in biology, identified primarily with the study of development on the one hand and of inheritance on the other. The former addresses the problem of biological form: what processes underlie the generation of the remarkable diversity of organismic forms? Are there generative principles whereby this diversity can be understood as an intelligible unity? The latter is concerned with the particular influences that act within organisms to limit the range of phenotypes expressed in different environments, and how these influences are transmitted from one generation to the next. In seeking a resolution of these traditions, it is useful to examine some observations from which different lines of enquiry may be seen to diverge, and then to consider how these may be reunited.

FORM AND INHERITANCE

In Figure 1 are shown some results of a classical study of form and inheritance in milfoil (*Achillea millefolium*). Multiple cuttings from individual plants were

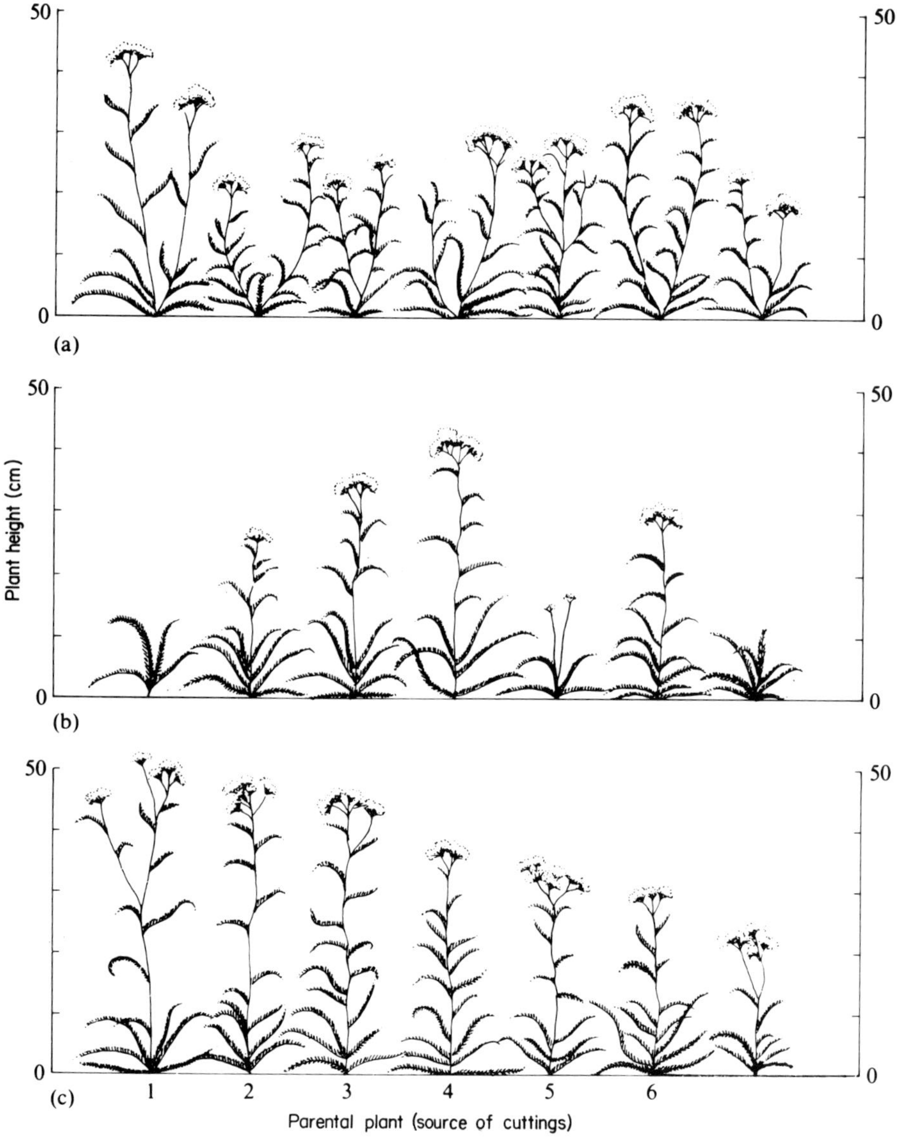

Figure 1 Responses to elevation of seven different plants (*Achillea millefolium*). Cuttings from each plant were grown at: (a) high (3050 m); (b) medium (1400 m); and (c) low (30 m) elevation. Numbers identify plants from the same parent

grown at different altitudes but under the same conditions of soil and watering. It is evident that plants with the same genotype can have quite different forms in different environments, the greatest difference in the whole set being between plants 1(b) and 1(c) (see Figure 1) which grew from cuttings taken from the same parent. What is characteristic of a particular heredity line (a clone in this case) is not the characters generated but the response of individuals with the same genotype to different environments. This can be made quantitative by determining the mean and variance of a particular character, such as height, and plotting it as a function of an environmental variable such as altitude, for plants derived from the same parent. Such a relationship defines the *norm of reaction* of the character in individuals with the same genotype to changes in an environmental variable.

There are other ways in which the relationship between genotype, phenotype and environment can be read. For example, plants 3(b) and 4(c) have very similar forms, but have experienced the influence of different environments and different genotypes. One could say that they are phenocopies of one another, i.e. taking some genotype and environment as standard (parental plant 3 and 30 m, say), there is an environmental influence (growth at 1400 m) that results in the same phenotype with this standard genotype as a particular change of genotype (4) at the standard altitude. Equivalently, one could refer to this relationship as a 'genocopy': the environmental influence that produces the form of plant 3(b) can be mimicked by a change of genotype (4) at the standard altitude. Evidently genotype and environmental influence have equivalent effects in the sense that one can substitute for the other in the generation of a particular phenotype and there is no reason to regard either as primary. This equivalence is in fact embodied in the classical genetic definition of organisms:

> We may start from the fact that all organisms, considered as individuals or as populations distributed in space or in time, represent the outcome of interactions between factors intrinsic to the organism and factors external to it. The usual way of stating this is to say that the individual is a joint production of his heredity and his environment. The intrinsic factors consist of an assemblage of genes derived from the parents. (Dunn, 1963).

It is generally assumed that there are limits to the range of phenotypic variation within a species, but they are limits that cannot be defined because there is a virtually inexhaustible set of possible genetic and environmental variations that could be studied. Hence the statistical nature of the norm of reaction: a particular genotype confers on organisms of a given species a certain phenotypic stability (a norm plus variance) in relation to environmental change. Genetics is in this respect concerned with stability, or equivalently with the frequency of occurrence of characters in populations of defined genetic composition. In population genetics one gets a further notion of stability by introducing the concept of 'fitness' for various characters (characters now seen as aspects of life cycles). The concern then becomes with the capacity of the species to persist, a type of dynamic stability analysis. As gene frequencies and environments vary, the distributions of characters change in populations and hence so does the fitness of the constituent organisms (equivalently, life cycles). Natural selection turns out not to be about selection after all, for nothing is doing any selecting. Population genetics is about the dynamic stability of population–environment interactions. Since both components of the interaction can change in response to the other, we have a dynamic system whose variables are gene frequencies and environments and whose stable states include point attractors, limit cycles, and strange attractors, between which there can be long and complex transients. This view generalizes to the Gaia hypothesis (Lovelock, 1979). Natural selection is a confusingly anthropomorphic concept to use in this context since there is no constant selecting agent.

The emphasis in genetics is thus on the frequencies of occurrence of characters and the stabilities of life cycles, dissected into characters, as functions of genes and environments. However, there is no generative theory that links genes and environments to characters. It is often assumed that, because different genotypes alter the frequencies of characters in organisms, the genes generate the characters. This is to mistake necessary for sufficient conditions, a point that has been extensively discussed elsewhere (Webster and Goodwin, 1982; Goodwin 1984a, b, 1985). In order to define sufficiency, and hence to achieve an exact analysis, it is necessary to proceed along a different route, asking different questions. In this case we do not want to know what factors alter the frequencies of occurrence of different characters and the stabilities of life cycles treated as aggregates of characters. We want to know what type of process it is that generates the space–time forms that are characteristic of organisms, from which characters are abstracted. Such a generative theory would tell us what organismic forms are possible. Actual organisms would then be understood as instances of the possible under certain conditions of inheritance and environment, for example. The diversity of organisms would then be intelligible within a generative unity. Otherwise, organisms remain historical accidents, as in a genetic analysis. The emphasis of such an enquiry is therefore not stability and probability (frequency), but on the generative principles that make organisms possible, i.e. on their *natures* (Webster, 1984). If such a programme can be realized, then biology would become an exact science, with theories of both how its basic entities (organisms and their life cycles) are generated and the stabilities or organism–environment dynamics. These are clearly complementary to one another. It may be that biology cannot become exact, that it is an intrinsically historical science lacking any unifying generative principles. However, the existence of a long and fruitful tradition of rationalism in biology suggests otherwise. To explore this, we return to the primary data and follow an alternative path.

FORM AND TRANSFORMATION

Despite the great diversity of forms in even the small sample of plants shown in Figure 1, they all have common features that allow us to recognize them as members of a set. That is to say, all the individuals are, in some sense, transformations of one another. They are not literal or actual transformations, since there is no way in which plant 2(a), for example, can be actually transformed into the shape of plant 6(c). However, it is perfectly reasonable to assume that a cutting from plant 2(a), grown in an appropriate environment, could resemble as closely as one wished the form of plant 6(c). The cutting from parent 2 grown at 1400 m 2(b) does in fact closely resemble 6(c). The plants could then be regarded as generative transformations of one another.

Consider now the leaf forms shown in Figure 2. These are complete sets of leaves of individual plants, taken at flowering, the first or bottom leaf being at the lower left, the top or last leaf being at the lower right. Tracing out the sequence for each species, we again can understand these forms as members of a transformation set. To be more concrete about these transformations, Figure 3 shows the sequence of forms assumed by each leaf during its development on the plant from a bud, the mature leaves being arranged in the same manner

Figure 2 Leaf sequences taken from mature plants (at flowering) of different species: A, *Circerbita alpina;* B, *Scabiosa lucida;* C, *Souchus oleraceus*. The bottom or first leaf in each sequence is at the lower left, the top or final leaf at the lower right

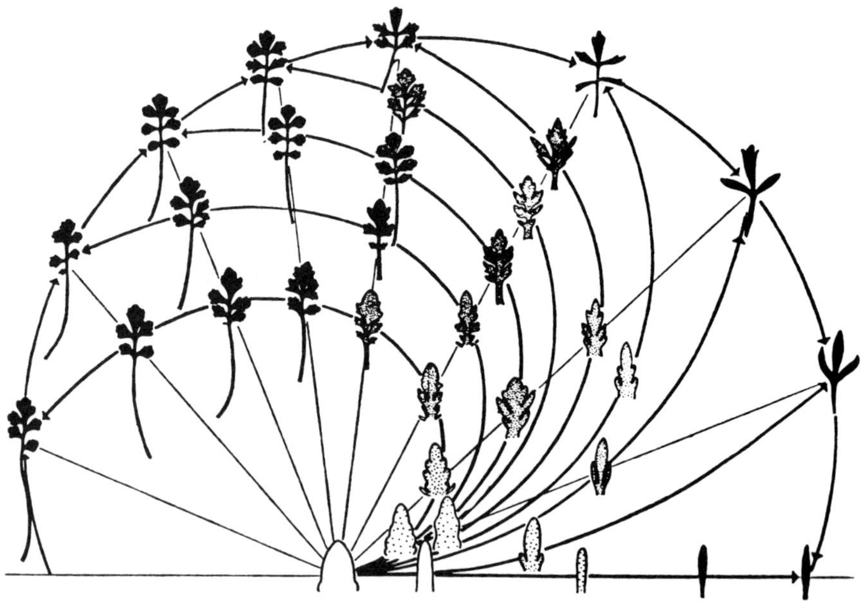

Figure 3 Morphogenetic sequences of leaves from a garden cress plant (*Lepidium sativum*), arranged in the same order as in Figure 2. All leaf buds start with the same basic shape, but progress through a series of changing forms shown by the curving trajectories from initial bud to final leaf shape

as in Figure 2, first leaf at bottom left and last leaf at bottom right of the semicircle. Initially all buds look very similar so they are all shown deriving from the same bud, though on the plant each one arises in a distinct position and at a particular age in the growth of the plant. Each leaf has the same genotype and grows in the same external environment, but now the systematic variations that give rise to the sequence of different forms are developmental, internal to the plant, varying with position and developmental age.

Other parts of the plant can be similarly understood as generative transformations of one another. Petals are similar in form to leaves, the transition occurring via sepals. Anthers, on the other hand, seem to be quite different from petals. However, intermediate forms can be seen in certain species, such as the white waterlily. Actual transformations of structures can be induced experimentally, such as leaves into tendrils in the pea, resulting from surgical manipulation of the bud (Sachs, 1978). It thus emerges that each bud has the potential to realize a larger number of different forms, resulting in the concept of multipotentiality.

The same potential for the transformation of different parts of organisms into structures characteristic of other parts is observed in animals. For example, legs can regenerate in place of lost palps in crabs, wings in place of halteres or legs in place of antennae in *Drosophila*, four-digit limbs (characteristic of fore-limbs) in place of five-digit hind-limbs in newts and salamanders. These are classical instances of homeotic transformations, which can occur spontaneously (congenitally) or as a result of either gene mutation or deliberate environmental perturbation, such as transient heat or chemical disturbance of the embryo at particular stages of development. Such observations all suggest that developing organisms embody certain principles of order which make possible the generation of a range of forms. The question we now face is whether it is reasonable to suggest that all organisms are to be understood, in some abstract sense, as transformations of one another. In Darwinism such a radical unifying generalization is generally accepted as an historical reality, since the major hypothesis of evolution is that all organisms are descended from a common primordial ancestor. Biology thus becomes a unified *historical* science. But can it also become a *rational* science, in the sense of a subject unified by a common set of generative processes whereby organismic forms are produced, those realized historically being intelligible as a particular subgroup of this potential set actualized by specific conditions of inheritance and environment? And, if so what is the nature of these common generative processes? If they exist, then the focus of attention in biology shifts from history to logical order and intelligibility. Instead of organisms being brute historical facts, inexplicable in terms other than their own existence as a result of a series of accidents, they become particular instances of a space–time order that is as logically necessary as liquids or crystals, given the conditions on this planet. That is to say, organisms would be understood as instances of a particular state of organization of matter that is inevitable because it is possible, and the conditions for its manifestations exist on earth. This is simply to say that there is no basic difference between biology and physics, although the ordering principles at work in organisms are different in degree, and possibly in kind, from those that predominate in currently known physical systems. But these are all still speculative issues. The next step in the argument is to examine some recent evidence that suggests what the general nature of the spatio-temporal ordering principles may be in organisms.

GENERATIVE PROCESSES IN *DROSOPHILA*

During the past few years, the role of gene action in morphogenesis has been studied in more detail in *Drosophila* than in any other organism, and a picture is beginning to emerge about the nature of the processes that are involved. What has become most evident is that the spatial patterns that develop in embryos are not to be understood in terms of gradients of positional information that are read or interpreted by genes to give spatial patterns of gene transcripts and products which then determine states of cell differentiation and morphogenesis. Such a picture is both far too static and too programmatic to fit the observations. Rather, there is a sequence of spatial ordering processes in which local detail arises gradually as a result of the development of increasingly finer-grained spatial patterns. Furthermore, there need not be any direct relationship between the spatial distributions of gene transcripts observed during this dynamic patterning process and the phenotypes arising from mutations, although in most instances a transcript pattern develops at some stage which correlates with the mutant phenotype. Let us now examine some examples that illustrate these points.

Eleven genes are currently known which are involved in establishing the dorso-ventral (D-V) axis of *Drosophila*. Most of these have mutant alleles that result in varying degrees of dorsalization of the axis, involving reduction or loss of ventral pattern elements (denticle bands) and an increased extent of dorsal cuticle. It was expected that the gene transcripts of at least one of these genes would be distributed in a gradient from dorsal to ventral in the embryo, thus identifying it as a candidate for the role of primary determinant of the D–V axis. One gene in particular, *toll*, had mutant alleles with the properties expected of such a determinant, since not only is there a series of alleles of this gene showing varying degrees of dorsalization in embryos, but there is also a dominant allele in which there is strong ventralization. However, when gene transcripts were studied using radio-labelled cDNA probes to mRNA from this and other genes which affect the D–V axis, no graded distribution was observed. Thus there is as yet no identified primary determinant of this axis. Such a determinant there must be. However, it need not be a gene product. It is perfectly possible that the symmetry-breaking event that initiates the D–V axis is based upon a physiocochemical process such as an electrical gradient involving differential ion distributions and fluxes, and/or some graded physical state of the cytoskeleton. Such processes have been directly implicated as primary axial initiators in a number of other developing systems (Jaffe, 1981; Nuccitelli, 1986; Weisenseel and Kicherer, 1981; Goodwin and Pateroniche-lakis, 1979; Goodwin *et al.*, 1983; Goodwin and Trainor, 1985). This could then interact with uniformly distributed gene products which amplify the initial gradient and consolidate an initially weak and labile axis, leading through a progressive elaboration of local detail to ventral and dorsal pattern elements at the cellular level. The morphogenetic field in such a process consists of the total set of spatial patterning phenomena, involving electrochemical activities, the cytoskeleton and gene products, all acting together in a cascade system to initiate, stabilize, and elaborate in detail the structural complexity of pattern elements along the D–V axis.

SEGMENTATION DYNAMICS

After the establishment of the A–P and D–V axes, the next major spatial patterning process in *Drosophila* is segmentation, the emergence of a periodic sequence of elements along the A–P axis. As a result of the intensive studies of Nüsslein-Volhard and Wieschaus (1980), the unexpected result emerged that there are only three major categories of genetic perturbations of segmentation, and probably no more than 20 genes influencing this process. The three categories are gap (deletions extending continuously over several segments), pair-rule (deletions of pattern elements with double-segment periodicity), and polarity (alterations of polar organization of every segment). Why is there no set of mutations deleting one segment at a time? Why none involving deletions with a periodicity other than double segment, or that give an increased number of segments of normal polarity? Some of these may yet turn up, but the categories observed so far are already very suggestive of the dynamics involved: a sequence of global spatial patterns that progresses from longer to shorter wavelengths, with wave number doubling (as revealed by the pair-rule mutations) being a characteristic feature.

The study of gene transcript distributions has borne this out. The relevant period for segmentation is from early syncytial blastoderm to cellular blastoderm. One of the genes whose activity has been investigated is *paired* (Kilchherr *et al.*, 1986). *Paired* gene transcripts first show up during the 12th nuclear cycle of the syncytial blastoderm stage, when the embryo is still a single multinucleated cell, the nuclei being located in the continuous peripheral periplasm just inside the plasma membrane (Figure 4). Radio-labelled *prd* cDNA can be seen to be located in a band 6 nuclei wide in the cortical part of the periplasm, stronger in the ventral than in the dorsal region and located about 13 nuclei from the anterior pole. During the following nuclear cycle, the band broadens to about 12–16 nuclei on the ventral side and 8–10 dorsally, in a region where the primordial head segments will form, but as yet without any sign of a segmental pattern. During the 14th nuclear cycle (late syncytial blastoderm), 5 additional bands of transcripts light up at approximately equal intervals posterior to the first band, with a width of some 5–7 nuclei. At the same time, the anterior band splits giving 2 bands whose width is about the same as the others. These 7 bands increase in intensity as the nuclei elongate and plasma membranes grow inwards to produce the cellular blastoderm. The 7 corresponding dorsal bands are closer together and weaker than those ventrally, revealing the dorso-ventral polarity of the pattern. The nuclei transcribing the *paired* gene are separated by gaps in which no paired transcripts are detectable.

Shortly before the completion of cellular blastoderm, an 8th band appears posterior to the 7th, while bands 2–7 begin to split into two. This is a result of the disappearance of transcripts in the 2 middle cells of each band, giving 14 transcription domains of about 2 cells in width. This splitting process occurs later dorsally than ventrally. The periodic pattern that occurs by the completion of the cellular blastoderm corresponds to the segmental periodicity of the segmented germ band embryo which emerges some 4 hours later. A further

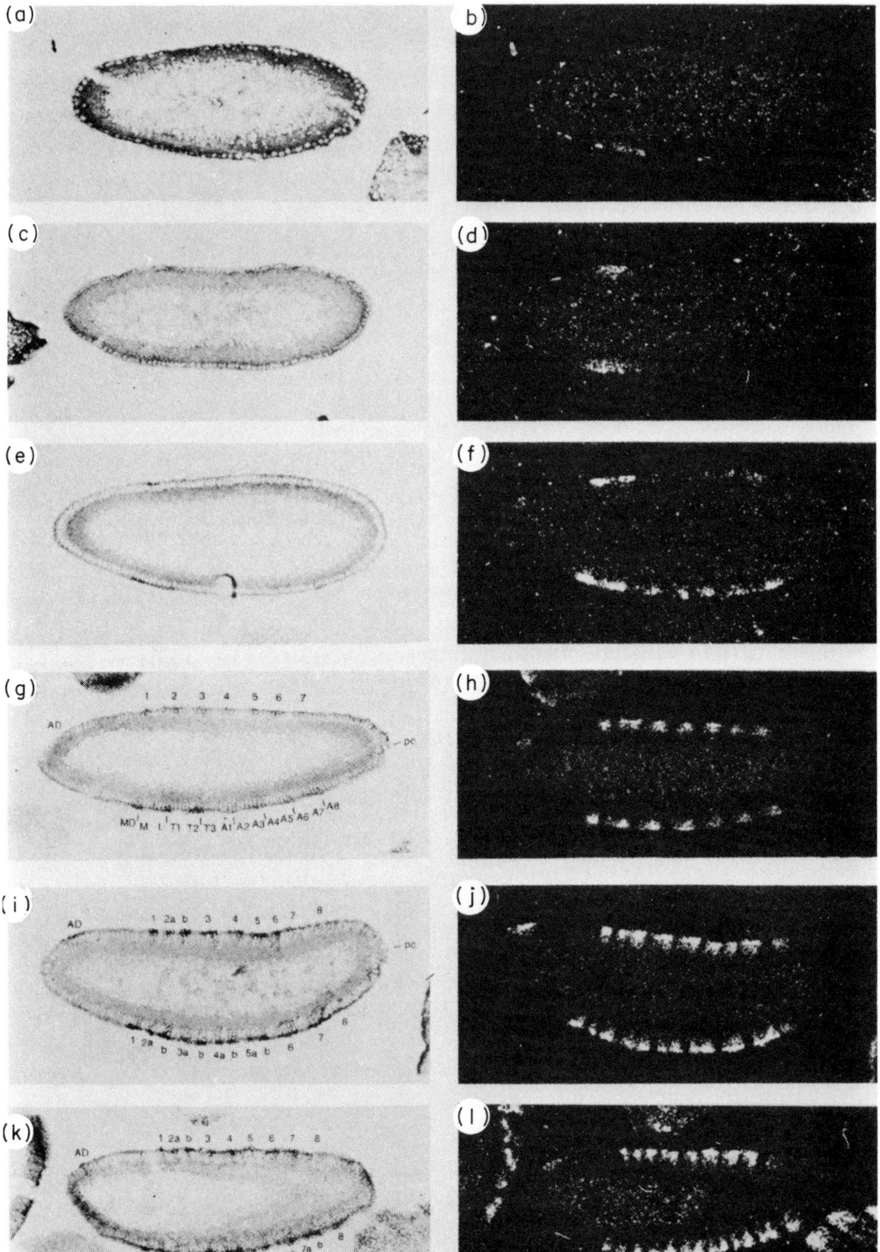

Figure 4 Localization of *paired* gene transcripts in early *Drosophila* embryos; *left* photograph, *right* autoradiograph of the same embryo. The bands are labelled 1–7 at early, and 1–7b and 8 at late, cellular blastoderm

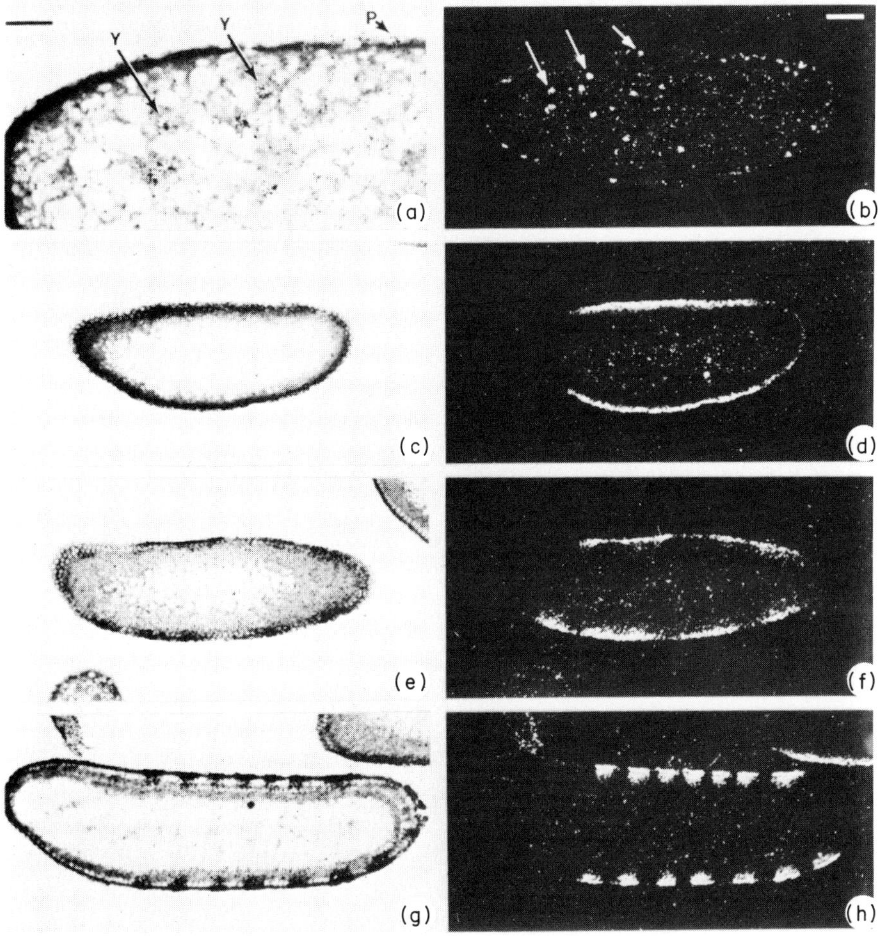

Figure 5 Localization of *fushi tarazu* gene transcripts in early *Drosophila* embryos: *left* photograph, *right* autoradiograph of the same embryo

region of transcription shows up at cellular blastoderm: a band about 6 cells wide located in the dorsal anterior part of the embryo.

Another pair-rule gene, *fushi tarazu*, also shows a changing spatial pattern of transcript distributions, but different in certain respects from *paired* (Weir and Kornberg, 1985). Initially *ftz* RNA is observed in syncytial blastoderm embryos during the 10th nuclear cycle, with autoradiographic grains localized over both yolk and peripheral nuclei along the entire egg length, though not all nuclei are labelled (Figure 5). By the 12th cycle, transcripts light up in the cortical region of the embryo, occupying a continuous band from approximately 15 to 65 per cent of egg length. This spatial pattern is stable until the 14th cycle (late syncytial blastoderm), when 4 broad bands of transcripts can be seen, each of which corresponds to about 4 segments of the larva. As the nuclei elongate and before membrane furrows penetrate into the embryo, the 3 posterior bands split in anterior-to-posterior sequence to give a pattern of 7 discrete regions with 2-segment periodicity, phase-shifted relative to the corresponding *prd* distribution. These 7 bands correspond quite closely to the deletion pattern of the mutant phenotype, with odd-numbered segments missing. The transition to a pattern with single-segment periodicity, as observed in the paired transcripts, does not occur in *ftz* RNA at cellular blastoderm, although it is observed later in every segment of the nervous system.

THE HARMONIC HIERARCHY

What the gene transcript distributions reveal is a hierarchy of patterns proceeding from longer to shorter wavelengths as development proceeds. This is evident already in the categories of segmentation mutants: the gap genes are involved in generating large sections of the pattern; pair-rule genes function in relation to the double-segment stage of the process; while the polarity genes affect single-segment organization. The transcript patterns reveal intermediate levels of the hierarchy, such as the 4-segment periodicity of the *ftz* distribution, which are not reflected in mutants. What we are seeing here are sequences of spatial states in the morphogenetic field of the embryo.

Why do patterns occur that have no obvious relation either to mutant or to normal phenotypes, such as the single *prd* band 13 nuclei wide (4-segment wavelength during the 13th nuclear cycle), or the 4 bands of *ftz* mRNA during the 14th cycle, each one also with a 4-segment wavelength? And why is there a 2-segment periodicity before the emergence of the final segmented pattern, the surprise that accompanied the discovery of the pair-rule mutants? A possible answer is that the morphogenetic field of the embryo has a particular dynamic such that, to reach the single-segment pattern, it must pass through a series of what are called *harmonics*. These are spatial patterns of decreasing wavelength, just like the standing waves in organ pipes or on violin strings, whose fundamental wavelength determines the dominant note and whose harmonics are the overtones. The most basic harmonic is the octave, which is a doubling of the number of waves (halving of wavelength); and this is the dominant harmonic observed in the sequence of segmentation gene transcription patterns, bands splitting into two to give a doubling of wave number, halving the wavelength.

From this viewpoint gene products interact with the global morphogenetic field patterns in the embryo, influencing these in particular ways so that a particular sequence of harmonics occurs. Gene products are thus an integral part of the field, whose spatial characteristics are clearly altered in specific ways by mutant products. The whole space–time pattern may therefore be a single unfolding process, proceeding from larger to smaller scale patterning as the wavelength of the field harmonics decreases in a characteristic manner. The different spatial locations of different gene transcripts, such as the relative positions of *prd* and *fts* transcripts at the 2-segment harmonic, presumably reflect differential responses of genes to different states of the pattern, such as the top and bottom of a stationary wave. And the same holds for the gap genes in the earlier stages of the process, when the long wavelength pattern predominates. The different positions of the gaps reflect the different regions of the primary pattern within which the various gap genes are expressed, as has been confirmed by observations of their transcript patterns. However, what gives spatial coherence and coordination to the whole segmentation process throughout the embryo, and the striking regularity of the periodic patterns of *prd* and *ftz* transcripts, is the harmonic patterns of the morphogenetic field and its transitions, of which gene activity is an integral part.

One way of identifying and characterizing the dynamics of the integrative system that defines the context within which gene activity is coordinated is to study the effects of perturbations to early embryos that phenocopy some of the segmentation mutants, such as gap and pair-rule, and also generate a spectrum of pattern disturbances that reveals further aspects of the dynamics. A paper on ether perturbations is in press (Ho *et al.*, 1987), and others are in preparation. The objective of this work is to complement the molecular genetic data by identifying the physicochemical components of the morphogenetic field with which gene transcripts interact and which they influence, particularly those involving ion fluxes across the plasmalemma and the role of the cytoskeleton. The latter structure is an obvious candidate for generating fields and harmonics with the wavelengths observed in *Drosophila*, as argued elsewhere (Goodwin, 1985) and as revealed in the beautiful preparations by Karr and Alberts (1986). Clearly, what needs to be identified is the precise nature of these electromechanochemical processes, possibly of the type described by Odell *et al.* (1981), Oster, Murray and Harris (1983) and Goodwin and Trainor (1985), which are involved in initiating the spatial patterns with which gene transcripts interact. These interactions stabilize, amplify and impose constraints upon the harmonic sequences that result finally in overt patterns of morphology and cell structure characteristic of the species.

HOMEOLOGICAL TRANSFORMATIONS

The types of morphological transformation observed in *Drosophila* as a result of either genetic or environmental disturbance are more general than homeotic transformations, which refer to a change of segment character into that typical of another segment. Gap mutants and their phenocopies, for example, involve extensive alteration of the pattern, together with characteristic changes in the pattern elements that remain. These morphological changes are not, however, arbitrary: they fall into distinct categories that indicate the presence of development constraints limiting the set of possible forms that are generated. The conclusion drawn above is that it is the set of spatial harmonics which the morphogenetic field can generate that limits these forms; and that, further, the specific mechanisms whereby gene products stabilize and influence these harmonics in a particular species further constrain their transformations, giving rise to such perturbations as the pair-rule mutants and their phenocopies.

The term homeological transformation will be used to describe the dynamic relationship between the members of this broader class of morphological variants (see also Ho, 1987). This reflects both the terms homeotic and homeomorphism, the latter being a similarity transformation or mapping. It also resonates with homology, but is sufficiently different to escape from the unfortunate genealogical implications of this term in relation to its usual meaning of descent from a common ancestor. In order to pursue a systematic study of morphological relationships, whether within or between species, synchronic (structural) or diachronic (genealogical), one needs a definition of similarity that is history free. This perfectly clear in relation to the use of protein homologies in the study of genealogies, for example. In this context, homology is defined in terms of similarity of primary protein structure, a logical definition with no implications about descent from a common ancestor. This purely synchronic relationship between members of the set of possible proteins can then be used to study the taxonomic relations of different species, either in a history-free manner (a rational taxonomy, as in transformed cladistics), or to study genealogical relationships, assuming some functional relation between homological difference and evolutionary time. But if a genealogical assumption is built into the definition of homology, as in the notion of descent from a common ancestor, then contradictions and absurdities arise.

This becomes evident in de Beer's (1971) booklet on homology, where he points out that the fore-limbs and the hind-limbs of the members of a particular species are not homologous because they do not both descend from a common ancestral structure, there being no evidence of a vertebrate ancestor with only one pair of limbs from which tetrapods have evolved. The result is the conclusion that human arms and bat wings, for example, are homologous, but human arms and legs are not, despite the much higher degree of structural similarity between the latter than the former. This bizarre conclusion is discussed at greater length in Goodwin (1984b, 1986), where a resolution is proposed which involves using a history-free definition of homology, as was done in pre-Darwinian rational morphology.

Evidently it is not as easy to find an unambiguous criterion for structural similarity between morphological characters as it is between proteins, whose primary building blocks remain unchanged, only their order varying. No such atomistic decomposition is available for adult morphology. However, in studying the processes underlying the generation of these forms, a degree of decomposition into primary generative elements becomes possible. This emerges from the dynamics of segmentation in *Drosophila*, and the perturbations which this process can undergo. What this reveals is that there is a systematic sequence of harmonic transitions of the morphogenetic field, as visualized by the spatial distributions of gene transcripts, both environmental and genetic influences acting on the states of this sequence to destabilize certain ones and stabilize others, resulting in a particular morphology. Homeological transformations refer to changes in the sequence of harmonic transitions in this potential set, defined by the spatial harmonics which are accessible to the morphogenetic fields as a result of both specific constraints on the harmonic sequence due to gene products influencing the state of the field, and to inherent (generic) properties of morphogenetic fields. The latter are developmental universals, the former particulars that specifically constrain transitions of field state. The harmonics, then, are the elements out of which morphogenetic fields are constructed, a specific sequence during development leading to a particular morphology.

One of the expectations of this dynamic transformational view of morphogenesis is that there are many more possible morphologies available to a species than is suggested by homeotic transformations, in which a segment alters its character to that typical of another in the normal organism. The basic reason is that, if global fields are involved in generating characters by a sequence of harmonics of the type observed in the gene transcript patterns, then characters do not arise from discrete blocks of differentiating tissue whose fate is specified by local genetic interpretation of an initial global pattern of positional information. The whole process is much more dynamic than this, involving a sequence of global fields (including gene products) so that perturbations occurring at different times have complex consequences, both global and local. Thus, under perturbation, embryos are expected to undergo transformations which generate a spectrum of forms many of which will be different from those observed in the normal organism, though still generically related to them. This is precisely what is observed in the *Drosophila* mutants and their phenocopies. Although experimenters label segments in mutants with the labels used for normal segments, it is often the case that the only correspondence between them is relative position, the actual spatial patterns of the denticle belts being quite different. An example is the segments remaining in the gap mutant, *Krüppel*. The phenotype of a larva with a strong allele has what is labelled abdominal segments 8, 7, and two 6s fused in mirror symmetry (see, e.g. Preiss *et al.*, 1985). However, comparison of these denticle belts with those of the normal larva shows little resemblance of detail, and an experimenter presented with an isolated denticle belt would have great difficulty in deciding which normal segment was most similar to it. The whole pattern, global and local, undergoes transformation, as expected. And this, of course, is the basis of evolutionary change.

The final point I wish to make about homeological transformations is that this concept not only does away with the difficulties and contradictions inherent in the notion of homology as used since Darwin; it also replaces the unfortunate term 'phenocopy', which implies that genetic perturbation is primary and environmental perturbation is secondary. However, the phenomenon of phenocopying itself demonstrates the causal equivalence of the two categories of influence *within an ontogenetic perspective* (Goldschmidt, 1938; Goodwin, 1984a). The fact that quite different types of disturbance can generate identically transformed phenotypes means that, in relation to developmental processes, their effects are identical. Hence the developmental states they induce are identical, and in this respect they are causally equivalent; neither copies the other. There is no reason to regard either as primary.

Homeological transformations, then, refer to the full spectrum of morphological changes in organisms due to alterations in morphogenesis, independently of the causes initiating the alterations. They are analogous to amino acid substitutions in proteins, independently of the causes of such substitutions. Homeological equivalence of two forms means that they can be transformed one into the other by a change of the morphogenetic process that generates them.

Within this comprehensive sense of equivalence, there can be as many categories of more restricted relationship as are useful. Homeotic transformations, involving only segmental identity change, constitute a more restricted class, defining homeotic equivalence. Clearly, in order to get a quantitative measure of degree of equivalence, analogous to that used in protein structure, a hierarchy of developmental equivalences is required.

The classification scheme that emerges from these different degrees of equivalence will certainly be different from a Darwinian classification based upon genealogical relationship, because there is no reason to expect that a logical ordering of forms based upon generative principles will be the same as an historical ordering of organisms. Using protein homology as an example, it is clear that the purely structural (logical) relationship of proteins do not correspond to the historical sequence of their realizations. Evolutionary convergence provides many examples of similar morphological types despite divergent histories; while historically distant species such as coelenterates and vertebrates are much more similar in the earliest stages of embryogenesis (cleavage) than either is to insects. Ho (1987) gives other interesting examples of such differences between genealogical and rational taxonomies. So a rationalist programme of the type described here has different goals and serves a different purpose from the historical enterprise of Darwinism, which is concerned with stability, not generation. There is no reason why they should be in conflict. In physics, historical questions are pursued within the context of a subject that has a unified rational structure, and are made intelligible within such a context. The unification and intelligibility result from the existence of a simpler set of generative processes than the set of entities produced by these processes — i.e. they follow from an abstract reduction of manifest diversity to an underlying generative unity. The falling apple and the orbiting moon are specific instances of the same principles. Such principles in no sense limit the potential diversity of forms; they simply make them intelligible.

Similarly, all organismic forms may be realized through hierarchies of harmonic sequences in morphogenetic fields which combine in characteristic ways electrical, mechanical and biochemical processes, defining a type of space–time organization which is distinctive to the living state. In this chapter I have not described the physics and mathematics that underly such a view, nor the electrical and ionic evidence for global dynamic processes that act together with interacting gene products to generate biological form. My objective has been to outline the general characteristics of a research programme that seeks to make biology a rational as well as an historical unity. If such an enterprise can be realized, then development and genetics, morphogenesis and heredity, will have been reunitied in a science of biological form and function that has been remarkably elusive since Goethe introduced the term morphology into biology in the late eighteenth century, and indicated how this could be achieved in terms of transformation, or metamorphosis.

REFERENCES

de Beer, Sir Gavin (1977). *Homology: An Unsolved Problem*, Oxford University Press.

Dunn, L.C. (1963). Genetics, *Encyclopaedia Brittanica*.

Goldschmidt, R.B. (1938). *Physiological Genetics*, McGraw-Hill, New York.

Goodwin, B.C. (1984a). A relational or field theory of reproduction and its evolutionary implications. In M.-W. Ho and P.T. Saunders (eds), *Beyond Neo-Darwinism*, Academic Press, pp. 219–41.

Goodwin, B.C. (1984b). Changing from an evolutionary to a generative paradigm in biology. In J.W. Pollard (ed), *Evolutionary Theory: Paths into the Future*, Wiley, pp. 99–120.

Goodwin, B.C. (1985). What are the causes of morphogenesis? *BioEssays*, **3**, 32–6.

Goodwin, B.C. (1986) Is biology an historical science? In S. Rose and L. Appignanesi (eds), *Science and Beyond*, Blackwell, Oxford.

Goodwin, B.C., and Pateromichelakis, S. (1979). The role of electrical fields, ions, and the cortex in the morphogenesis of *Acetabularia, planta*, **145**, 427–35.

Goodwin, B.C., Skelton, J.L., and Kirk-Bell, S.M. (1983). Control of regeneration and morphogenesis by divalent cations in *Acetabularia mediterranea, Planta*, **157**, 1–7.

Goodwin, B.C., and Trainor, L.E.H. (1985). Tip and whorl morphogenesis in *Acetabularia* by calcium regulated strain fields, *J. Theor. Biol.*, **117**, 79–106.

Ho, M.-W. (1987). How rational can rational morphology be? *Revista di Biologia* (in press).

Ho, M.-W., Matheson, A., Saunders, P.T., Goodwin, B.C., and Smallcombe, A. (1987). Ether-induced segmentation defects in *Drosophila melanogaster, Roux Arch. Dev. Biol.*, **196**, 511–521.

Jaffe, L.F. (1981). The role of ionic currents in establishing developmental pattern, *Phil. Trans. Roy. Soc. (B)*, **295**, 553–66.

Karr, T.L., and Alberts, B.M. (1986). Organization of the cytoskeleton in early *Drosophila* embryos, *J. Cell Biol.*, **102**, 1494–509.

Kilchherr, F., Baumgartner, S., Bopp, D., Frei, E., and Noll, M. (1986). Isolation of the *paired* gene of *Drosophila* and its spatial expression during early embryogenesis, *Nature*, **321**, 494–9.

Lovelock, J.E. (1979). *Gaia: A New Look at Life on Earth*, Oxford University Press, Oxford.

Nuccitelli, R. (ed) (1986). *Ionic Currents in Development*, Alan R. Liss Inc.

Nüsslein-Volhard, C., and Wieschaus, E. (1980). Mutations affecting segment number and polarity in *Drosophila, Nature*, **287**, 795–801.

Odell, G., Oster, G.F., Burnside, B., and Alberch, P. (1981). The mechanical basis of morphogenesis, *Devel. Biol.*, **85**, 446–62.

Oster, G.F., Murray, J.D., and Harris, A. (1983). Mechanical aspects of mesenchymal morphogenesis. *J.Embryol. Exp.Morph.*, **78**, 83–125.

Preiss, A., Rosenberg, U.B., Kienlin, A., Seifert, E., and Jäckle, H. (1985). Molecular genetics of Krüppel, a gene required for segmentation of the *Drosophila* embryo, *Nature*, **313**, 27–32.

Sachs, T. (1978). Patterned differentiation in plants, *Differentiation*, **11**, 65–73.

Webster, G.C. (1984). The relations of natural forms. In M.-W. Ho and P.T. Saunders (eds), *Beyond Neo-Darwinism*, Academic Press, pp. 193–218.

Webster, G.C., and Goodwin, B.C. (1982). The origin of species: a structuralist approach, *J.Soc. Biol. Struct.*, **5**, 15–47.

Weir, M.P., and Kornberg, T. (1985). Patterns of *engrailed* and *fushi tarazu* transcripts reveal novel intermediate stages in *Drosophila* segmentation, *Nature*, **318**, 433–5.

Weisenseel, M.H., and Kicherer, R.M. (1981). Ionic currents as control mechanisms in cytomorphogenesis. In O. Kiermayer (ed), *Cytomorphogenesis in Plants*. Cell Biology Monographs, vol. 8, pp. 379–99.

Evolutionary Processes and Metaphors
Edited by M.-W. Ho and S.W. Fox

CHAPTER 9

Panbiogeography: method and metaphor in the new biogeography

R. Craw and R. Page

ABSTRACT

The importance of biogeography to Darwin's discovery of evolution by natural selection is well known. Any attempt to develop an alternative to the Neo-Darwinian evolutionary synthesis must address the question of what such an alternative means in the discipline of biogeography. Croizat's panbiogeography offers the foundations for the development of a viable alternative to Darwinian biogeography by providing a new methodology for analysing geographic distribution, a means by which biogeography and geology can be integrated and a new system of world biotic regions. The latest New Zealand developments in clarifying and quantifying Croizat's method are reviewed in detail. The role of metaphor in Darwin's and Croizat's work is discussed with particular reference to concepts of absolute and relative space/time.

> When we are dead and gone what a noble subject will be Geographical Distribution! (Darwin, 1858, in Darwin, 1903:117)

It is widely recognized that biogeography, the study of animal and plant geographic distribution, and the development of Darwin's evolutionary theory are intimately connected. An enormous amount of empirical biogeographic information was published in the first half of the nineteenth century and biogeography was seen by a number of workers, especially English biologists and naturalists, as a subject area that might hold the key to the question of the origin of species (Kinch, 1980).

Recently, Darwinian scholars have recognized the dynamic interplay between Darwin's own empirical studies in biogeography, his laborious compilations of biogeographic statistics from taxonomic manuals, and the genesis of his theory of evolution by natural selection (Browne, 1980; Richardson, 1981; Parshall, 1982). During the development of Darwin's theory (1837–58) biogeography supplied the only sufficiently detailed data base on species in relation to their environments against which Darwin could test his new ideas. It is clear that Darwin made the decision to adopt natural selection as the explanation for evolutionary change on the basis of his biogeographic studies. The Darwinian revolution emerged from the discipline of biogeography (Ghiselin, 1969; Richardson, 1981; Desmond, 1982).

Once he had rejected the concept of species fixity Darwin had to demonstrate that animal and plant geographic distribution could be explained by species origin in single centres of origin, descent with modification, subsequent migration across barriers and colonization:

> When we can feel assured that all the individuals of the same species, and all the closely allied species of most genera, have within a not very remote period descended from one parent, and have migrated from some one birth-place; and when we better know the many means of migration, then, by the light which geology now throws and will continue to throw, on former changes of climate and of the level of the land, we shall surely be enabled to trace in an admirable manner the former migrations of the inhabitants of the whole world.
>
> Darwin, (1911:666–7)

Darwin devoted much time and effort to attempts at understanding how the means of dispersal of different species and the effects of various physical barriers on such dispersal acted as determinants of geographic distribution. Natural selection operating on a species' means of dispersal was the explanation underlying geographic distribution:

> Widely-ranging species, abounding in individuals, which have already triumphed over many competitors in their own widely-extended homes will have the best chance of seizing on new places, when they spread into new countries. In their new homes they will be exposed to new conditions, and will frequently undergo further modification and improvement; and thus they will become still further victorious, and will produce groups of modified descendants. (Darwin, 1860:350)

Darwin's theory on how organisms achieve their geographic distributions was elaborated upon by a number of workers whose books and papers have had an immense influence on the development of biogeography since 1859 (e.g. Wallace, 1876, 1880; Matthew, 1915). More recently supporters of the Darwinian dispersal model have included major architects of the Neo-Darwinian synthesis (e.g. Simpson, 1940, 1953, 1965, 1980; Darlington, 1957; Mayr, 1976).

Any attempt to construct an alternative to the Neo-Darwinian evolutionary synthesis must involve a consideration of what such an alternative might mean for the discipline of biogeography. In facing this question one is drawn irresistibly to the flame of Leon Croizat's panbiogeography.

CROIZAT'S PANBIOGEOGRAPHY: JUST ANOTHER CRITIQUE OF DARWINISM?

Leon Croizat's early scientific work was primarily concerned with the systematics and phylogeny of the plant families Euphorbiaceae and Cactaceae (Croizat, 1934, 1940; reviewed in Heads, 1984). During the course of these studies Croizat came to be dissatisfied with the Darwinian explanation that animal and plant geographic distribution is a result of their means of dispersal (Croizat, 1952). Rather than just criticizing Darwinism, Croizat chose instead to attempt to test it empirically by posing the question: 'Does the geographic distribution of organisms in space/time support Darwin's theory [i.e. that natural selection operating on a species' means of dispersal followed by migrations by those means out of a specific centre of origin was the process responsible for geographic distribution?]' (Craw, 1984a). In order to attempt to answer this question Croizat began to develop a method by which means geographic distribution patterns could be analysed and compared. Geographic distribution patterns of literally hundreds of different plant and animal taxa were analysed in a series of books and papers (full listing in Heads and Craw, 1984). Through this series of analyses Croizat discovered that organisms with completely differing dispersal propensities and colonizing abilities exhibited very similar repetitious distribution patterns that he termed generalized or standard tracks (Craw, 1984a, 1988). The standard tracks bore little relation to the present-day geography of the world, but rather joined areas of the globe now widely separated by major sea and ocean basins.

Croizat's findings through the use of track analyses (which he termed the panbiogeographic method) were completely opposed to what one might expect on the basis of Darwin's theory. It would be no exaggeration to claim that Croizat's work constitutes a unique critique of Darwinism and the synthetic theory in that core ideas concerning the relation between natural selection, means of dispersal and geographic distribution have been falsified by the application of the panbiogeographic method to one of the richest empirical data bases known to biologists, the immense body of facts concerning the geographic distribution of organisms that has steadily accumulated since the Enlightenment.

Critics of Darwinism and the synthetic theory have been frequently chided for being just that, critics, and offering little in the way of viable alternatives to the theories that they condemn. This is not true of Croizat's panbiogeography which offers a practical methodology through which evolutionary change can be tied to particular space/time events (Gray, this volume), a means by which biogeography can be integrated with geology and a new system of world biogeographic regions. Recent work in panbiogeography has focused on clarifying and quantifying the track method (Craw 1983a, 1985; Page, 1987), extending Croizat's insights into the relationship between biogeography and geology (Craw and Weston, 1984), applying Croizat's new system of regions to the biogeographic classification of New Zealand (Craw, 1983b, 1988) and incorporating the metaphor of reciprocally constrained construction into panbiogeography (Gray, 1987 and this volume). These developments are reviewed and extended below. Detailed reviews of Croizat's work have appeared elsewhere (Craw and Gibbs, 1984; Grehan and Henderson, 1988).

PANBIOGEOGRAPHIC METHOD: GRAPH ANALYSIS OF PLANT AND ANIMAL GEOGRAPHIC DISTRIBUTION

Croizat was primarily interested in the similarity of distribution patterns rather than the similarity between disjunct biotas. His approach to analysis of this problem was to develop a graphical technique, vocabulary and a set of symbols (Croizat, 1968, 1982) that could be applied to maps of geographic distributions. The initial step in panbiogeographic analysis of a particular taxon is to construct the track or graph of its geographic distribution. Tracks are line graphs drawn on maps of the geographic distributions of particular taxa connecting the disjunct localities of those taxa. The simplest method of constructing such a graph is to form a minimal spanning tree, i.e. an acyclic graph that connects all the localities occupied by a taxon such that the sum of the lengths of the links connecting each locality is the smallest possible (Figure 1). For n localities the line graph connecting them will consist of $n-1$ links. The choice of a minimal spanning tree over other types of graph is made because this type of graph can be calculated exactly and efficiently. When taxa are distributed over a large area straight lines between their disjunct collection localities are not accurate estimates of the true distance between those localities because of the curvature of the earth's surface. The formula of Phipps (1975) can be used to correct for this distortion. For taxa that have differentiated into distinct entities a minimal spanning tree can be constructed for each entity and then these separate subgraphs are all linked together to form a single minimal spanning tree.

Next an hypothesis of the baseline (diagnostic character) is proposed for each track depending on the particular ocean or sea basin that the track crosses or circumscribes (Figure 1). The baseline is interpreted as a primary diagnostic biogeographic component for the taxon being analysed (Craw, 1983, 1988). Baselines provide the initial basis on which tracks are oriented, i.e. converted from non-directed line graphs into directed graphs. Finer resolution of track orientation and baselines can be obtained through the employment of additional biogeographic and phylogenetic criteria.

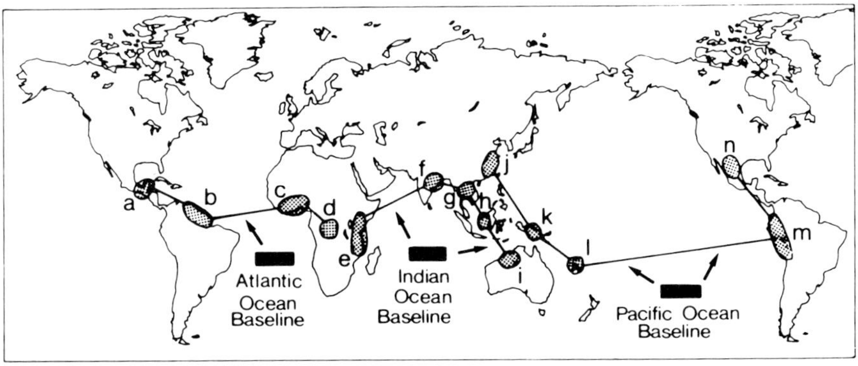

Figure 1 The concepts of a track and its baseline. Tracks are minimal spanning graphs linking distributional localities/areas of a particular taxon or group of taxa. For instance, the taxa distributed in the four areas a to d are linked by a three link minimal spanning graph. As the graph for these taxa crosses the Atlantic Ocean their track is interpreted as having an Atlantic Ocean baseline. Similarly the taxa distributed in five areas e to i are linked by a four link graph and have an Indian Ocean baseline while those distributed in the five areas j to n are linked by a four link graph and have a Pacific Ocean baseline

Consider the taxa belonging to a more inclusive taxon that are distributed in the Australasian area (Figure 2a). This group of taxa is described as having a main massing around the Coral Sea because four out of the six taxa are distributed directly around the shores of that sea basin. The track for the group is constructed and, on the basis of the group's main massing, a Coral Sea baseline is proposed for the group's track (Figure 2b). Phylogenetic information for the group is limited to a single known sister group relationship (Figure 2f). It is not necessary to have a fully resolved phylogenetic hypothesis for the group before the biogeographic analysis can proceed further. The single known sister group relationship is sufficient to confer direction to the remainder of the graph. Taxa in New Caledonia (area 4) and New Zealand (area 5) are nearest phylogenetic neighbours so a line is drawn to link these two areas (Figure 2c). This link is bidirectional but this ambiguity is removed by adding Australian taxa (areas 2 and 3) to the track. Since the Australian areas are closer geographically to New Caledonia then they are to New Zealand, Australia is linked to New Caledonia (Figure 2d). Similarly, New Guinea is added to the track while the section of the track for the taxon occurring in Fiji (area 6) derives its direction from its nearest geographic neighbour, New Caledonia (Figure 2e). The biogeographic analysis can now be presented as a directed minimal spanning graph with baselines identified as a basis for further research (Figure 2g).

Graphs can be represented in matrix form and this has allowed for the development of track analysis into a quantitative biogeographic method (Page, 1987). The connection matrix and the incidence matrix are the two types of

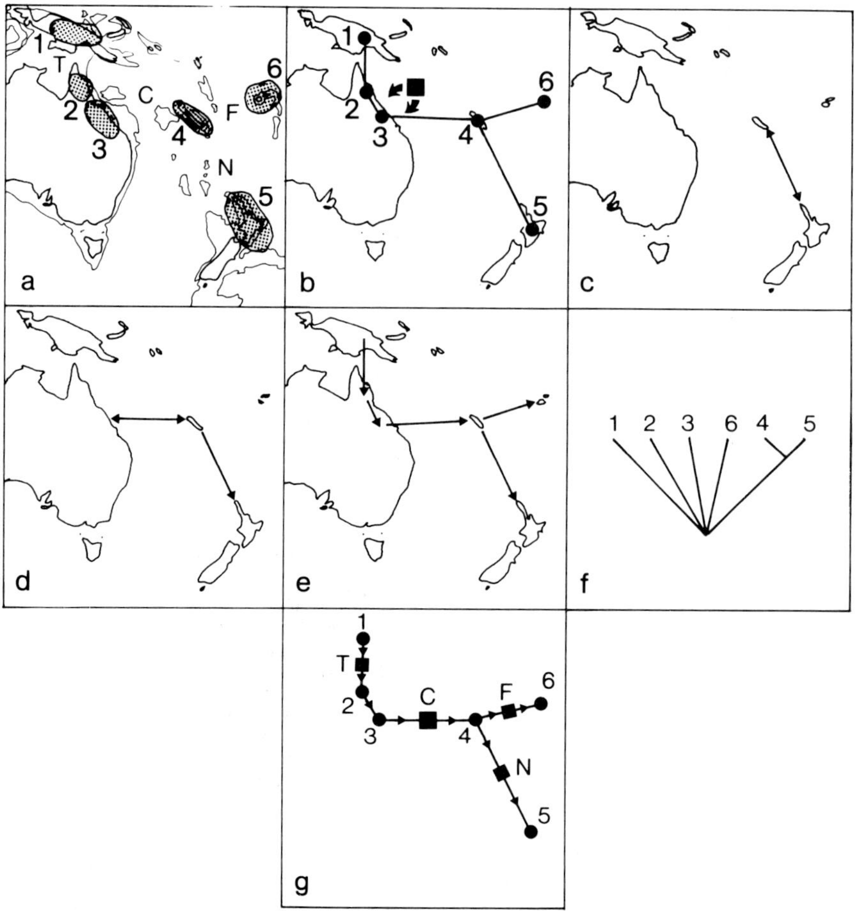

Figure 2 Track construction, orientation and baseline identification:
(a) Taxa 1 to 6 (which could be populations or subspecies of a species, species of a genus, genera of a family, etc.) are geographically distributed in various parts of Australasia. Abbreviations: C—Coral Sea, F—Fiji Basin, N—Norfolk Ridge, T—Torres Strait.
(b) The track for taxa 1 to 6 with the main baseline (Coral Sea) identified (see text).
(c) The initial step in orienting the track by linking the sister taxa in areas 4 and 5.
(d) The second link is formed between areas 3 and 4 which then serves to orient the link 4 to 5.
(e) The links 1 to 2 to 3 are formed with area 6 being linked to the nearest neighbour in area 4.
(f) The known unresolved phylogeny of taxa 1 to 6.
(g) The fully oriented track for taxa 1 to 6 with baselines identified

matrices applicable to track analysis. Connection matrices are $n \times n$ matrices where n is the number of points (that can correspond to localities or distri-

	1	2	3	4	5	6
1	–	1	0	0	0	0
2	1	–	1	0	0	0
3	0	1	–	1	0	0
4	0	0	1	–	1	1
5	0	0	0	1	–	0
6	0	0	0	1	0	–

(a)

to (columns), from (rows)

	1	2	3	4	5	6
1	–	1	0	0	0	0
2	0	–	1	0	0	0
3	0	0	–	1	0	0
4	0	0	0	–	1	1
5	0	0	0	0	–	0
6	0	0	0	0	0	–

(b)

Figure 3 Matrices of the tracks in Figure 2. The symbol '0' indicates no link, '1' indicates a link:
(a) Connection matrix for the track in Figure 2(b).
(b) Connection matrix for the oriented track in Figure 2(g)

bution areas) in the graphs. For any two points that are connected by a link in the graph the matching element in the matrix is a '1', and for any two points that are not connected the matching matrix element is a '0' (e.g. Figure 3a represents the resulting matrix for the track graph of Figure 2b). For tracks that have been completely oriented and thus converted into directed graphs, the corresponding connection matrix is constructed so that the matching matrix element is a '1' only if the direction of the connection link is from the first to the second point (e.g. Figure 3b represents the directed connection matrix for the directed graph of Figure 2g). The incidence matrix that can be constructed to represent a track is an $i \times j$ matrix where i is the number of points, localities or areas and j is the number of links connecting those points, localities or areas. In a directed track the incidence matrix element is '1' if the jth link is connected to the ith point and is directed away from it. If the link is directed towards the ith point then the matrix element is '-1' and if there is no connecting link then the matrix element is '0'.

Conversion of track graphs into connection matrices allows for a quantitative comparison of tracks for a number of different taxonomic groups. Connection matrices are a type of distance matrix which means that they are open to statistical hypothesis testing to determine whether a statistical association exists between corresponding elements in the matrices for different tracks (Page, 1987). It is now possible to test statistically Croizat's claim that organisms with widely differing means of dispersal and ecologies have similar distribution patterns. This is not the only possibility for biogeographic analysis that is presented by the construction of track matrices. Page (1987) has noted how track connection matrices can be used to identify incorrect phylogenetic hypotheses and composite geographical and geological areas. Incidence matrices can be used to recognize shared diagnostic biogeographic components or baselines.

THE PANBIOGEOGRAPHIC METHOD SUMMARIZED

The method outlined above and by Page (1987) can be summarized as follows:

Step 1 For the plant or animal group(s) under biogeographic analysis construct a (the) track(s) by either:

(a) connecting the collection localities using a minimal spanning tree graph. The links that join main massings or that are coincident with previously recognized baselines are possible baselines and serve to orient tracks; or

(b) using the minimum distance criterion along with available phylogenetic information to construct a spanning tree where localities are linked to the geographically nearest sister group.

Step 2 Construct connection matrices for the individual tracks and a summary connection matrix for all tracks combined.

Step 3 Use the Mantel test to evaluate statistically track congruence between the individual matrices. Search the summary connection matrix for circuits indicating incongruent tracks.

Step 4 Construct incidence matrices for individual tracks. Compare individual incidence matrices in order to recognize shared track elements. The presence of these indicate that the tracks are homologous and that the taxa have had a common history.

Step 5 Present the results of the analysis on a map with tracks, track orientations, baselines and main massings identified as a basis for further research.

INTEGRATING BIOGEOGRAPHY, GEOLOGY AND TECTONICS

Until the advent of panbiogeography a successful synthesis between biogeography and geology was not possible (Nelson, 1978). Earlier biogeographers either slavishly followed the fashionable geological ideas of their times, trimming their biogeographic data to fit the geology, or branched out on their own erecting outrageous palaeogeographical hypotheses that were limited only by the constraints imposed by a spherical earth. The lack of a method by which biogeography and geology could be efficiently integrated is best illustrated by the ease with which former opponents of continental drift rapidly transferred their original biogeographic models, developed on the basis of the permanence of continents and oceans, to the mobile models of earth history developed once the concepts of plate tectonics became widely accepted in the early 1970s (Craw, 1978). It is important to note that the emphasis in panbiogeography is on the correlation of biogeographic tracks and major tectonic features rather than on support for particular palaeogeographical models to which the biogeographic data are fitted. One of the main purposes of Croizat's develop-

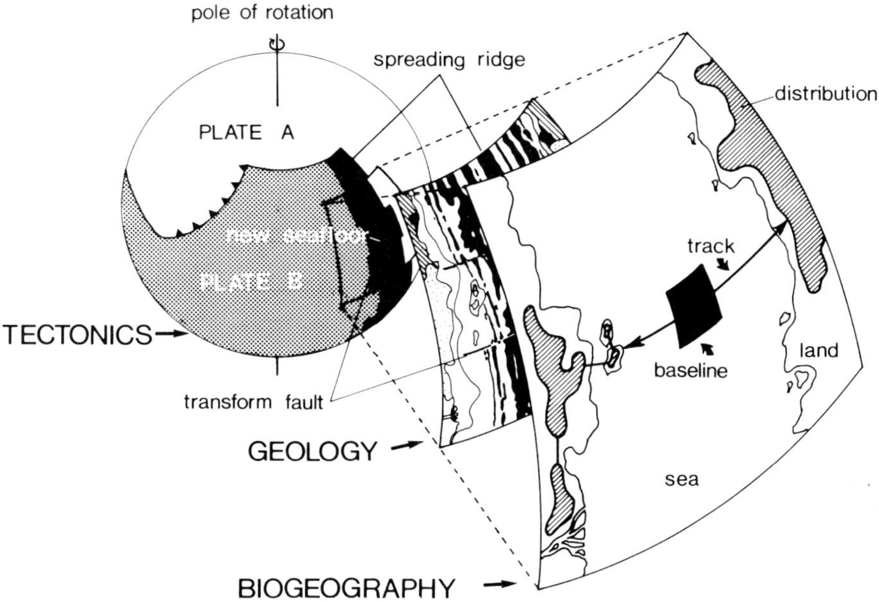

Figure 4 Integrating biogeography with tectonics. The baseline of a track reinterpreted in terms of the crossing of a major tectonic feature

ment of panbiogeography was not only to establish biogeography as an independent science free from a prior commitment to any one geological or geophysical theory (Craw and Weston, 1984) but also to ensure that a synthesis between biogeography and geology could be achieved.

Panbiogeographers attempt to integrate track geometry with the geometry of the major tectonic features on the earth's surface (e.g. mid-oceanic ridges, rift lines and transform faults) rather than with particular palaeogeographic models (Croizat, 1961, 1964). The concepts of track and baseline can be integrated with tectonic features allowing for the conversion of a biogeographic analysis into a dynamic process explanation of geographic distribution in terms of a biogeographic time/space. The initial idea of a track baseline as an ocean or sea basin is transformed so that the baseline for a track is where that track crosses a major tectonic feature (Figure 4).

Mid-ocean spreading ridges between tectonic plates can be visualized as resulting from the relative rotation between two plates. This relative motion is a rotation around a pole. By drawing tracks for taxa with trans-Pacific biogeographic relations on maps reprojected using poles of rotation for the Pacific plate it can be shown that the biogeographic tracks are approximately parallel to one another and to the transform faults, and orthogonal to the spreading ridge (Figure 5). This symmetry between trans-Pacific biogeographic tracks and major tectonic features in the Pacific seafloor is entirely unexpected

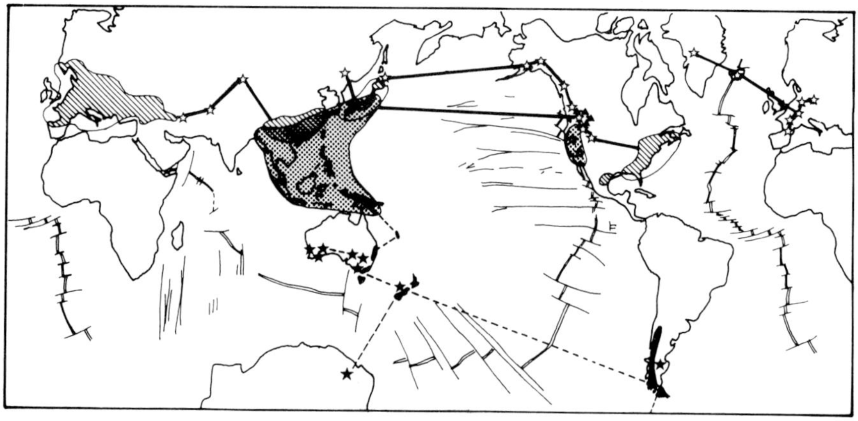

Figure 5 Symmetry between the tracks of *Lithocarpus* (living — dotted area, fossil — black triangles), *Fagus* (living — diagonal areas, fossil — white stars) and *Nothofagus* (living — black areas, fossils — black stars) and major tectonic features in the floor of the Pacific Ocean (single lines — transform faults, double lines — spreading ridge)

from the perspective of neo-Darwinian biogeography with its emphasis on centres of origin and random dispersal across a world geography essentially similar to that of today or, in its modern post-plate tectonic version, across a Mesozoic supercontinent Pangaea (e.g. Schuster, 1976). In fact, leading architects of the Neo-Darwinian synthesis such as Simpson (1976) and Mayr (1982) maintain that acceptance of the new concepts of a mobile earth history do not necessitate much revision of Darwinian dispersal biogeography despite the fact that palaeontologists (e.g. Hallam, 1981) and philosophers of science (e.g. Frankel, 1984) have acknowledged that Croizat's approach is more compatible with the new geology.

From his earliest major biogeographic work (1952) onwards Croizat argued that there must be substantial agreement between biogeographic analysis and geological and geophysical theories. It was for this reason that he was opposed to Wegener's version of continental drift where all the continents are originally bunched around Africa, closing the Atlantic and Indian Oceans and greatly increasing the width of the Pacific Ocean (Figure 6). Croizat discovered that trans-Pacific biogeographic relationships in terrestrial organisms were just as commonplace as trans-Indian and trans-Atlantic Ocean relationships. From this basis he argued that the Pacific Ocean must have widened through time rather than narrowed as in Wegener's theory. He proposed a number of 'potential lines of tectonic stresses' in the world's major sea and ocean basins by which means modern geography might have developed (Figure 7).

Croizat's recognition of the importance of trans-Pacific relationships is by no means original, though his documentation of the phenomenon is by far the most comprehensive. As early as the 1830s Darwin, in his first 'Transmutation of Species' notebook (de Beer, 1960), had drawn attention to these relations

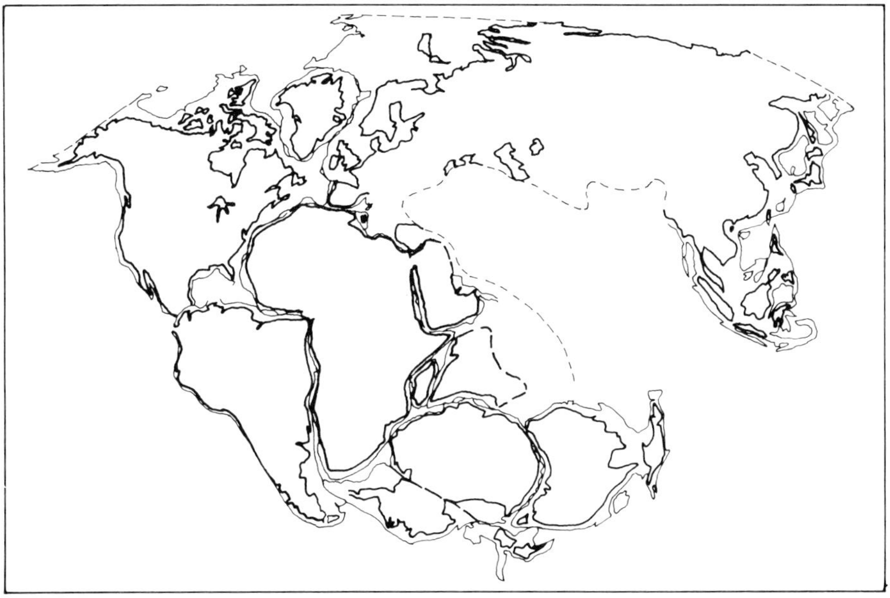

Figure 6 The commonly accepted pre-drift palaeogeographical reconstruction of the supercontinent Pangaea

and suggested an explanation in terms of a novel continental drift model which is the opposite of Wegener's in that the continents are grouped together in the middle of the Pacific (Craw, 1984b; Grinnell, 1974). Since then a number of biologists (e.g. Murray, 1873; Hallier, 1912; Germain, 1928; Guillaumin, 1928; Skottsberg, 1925, 1928; Steenis, 1962) have discussed trans-Pacific relations and postulated various forms of continental connections in the Pacific Ocean. Some biogeographers have even drawn attention to the anomalies such relationships provide for Wegener's reconstruction that became widely accepted in its broad outlines by most geologists and biogeographers in the 1970s (e.g. Wulff, 1943; Melville, 1966, 1981; Skottsberg, 1956). Though Croizat's recognition of this problem has no originality his attempted solution was completely novel (Craw and Weston, 1984).

Croizat, in his classic study of the panbiogeography of the New World (1958, Vol. 1), identified two areas of North and Central America as of major importance because they contained numerous organisms with trans-Pacific biogeographic affinities (Figure 8). North and Central America were also shown to have biogeographic relations with Europe and Africa across the Atlantic Ocean (e.g. Croizat, 1958, Vol. 2b: Fig. 259). In a later work Croizat (1961) compared these biogeographic conclusions with Wegener's Pangaea theory, where North America is a single unit, and proposed instead that the Americas were composite continents (Figure 9). Large portions of western North and South America were postulated to have drifted from out of the

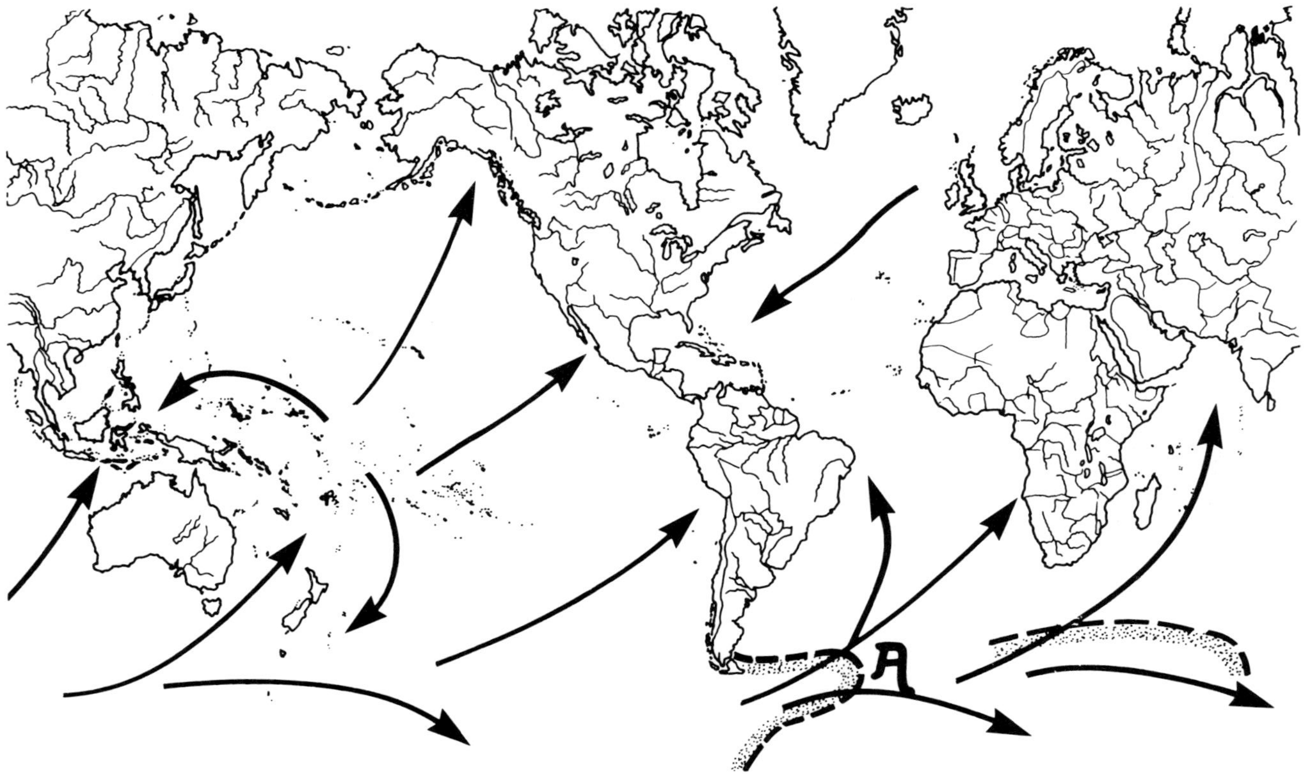

Figure 7 The tectonic stresses (arrows) that have made modern geography. Note their location in the world's major ocean basins. (From Croizat, 1952: Fig. 99. Reproduced by permission of Dr W. Junk, Publishers, Dordrecht, The Netherlands.)

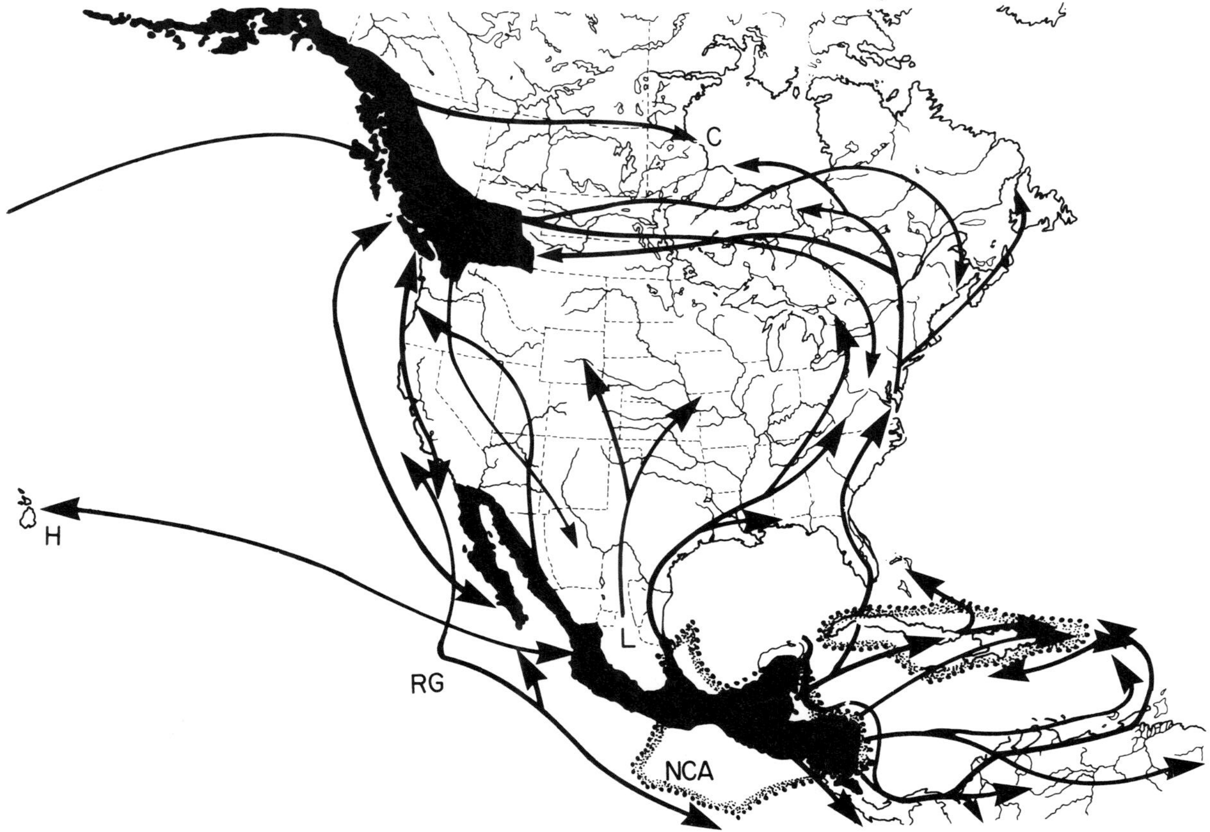

Figure 8 The two main areas of western North America involved in trans-Pacific biogeographic relations with western Asia. Note the close correspondence between the northern area and the Cordilleria micro-continent (Figure 10). (From Croizat, 1958: Fig. 56. Reproduced by permission of Catalina Croizat, Coro, Venezuela.)

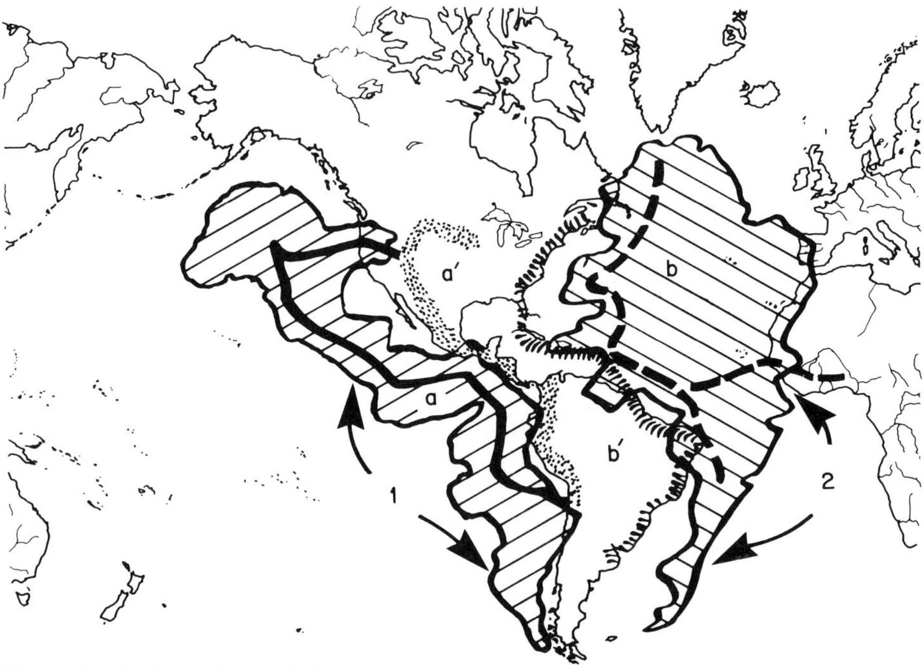

Figure 9 An hypothesis of the composite nature of the North and South American continents as proposed by Croizat in 1961. Part of the original caption to the figure reads: 'We may suppose that the New World was originally split into two halves (1 and 2) which later 'floated' to yield the current geography. *Halenia* being generally dispersed on the western half to agree with main range *a* (continuous line); and *Drosera* on the eastern to agree with main range *b* (broken line); the ranges of the two genera would still easily agree with the distribution now ruling (stippled line *a'* for *Halenia*; hatched line *b'* for *Drosera*) after junction of the continental halves.' Lines *a*/*a'* and *b*/*b'* are components of trans-Pacific and trans-Atlantic Ocean standard tracks respectively. (From Croizat, 1961: Fig. 8. Reproduced by permission of Catalina Croizat, Coro, Venezuela)

Pacific Ocean along with their associated biota to compact with eastern North and South America which had been connected with Africa and Europe before separation through the formation of the Atlantic Ocean.

This was geological heresy in 1961 when most geologists and biogeographers had not even accepted continental drift but recently it has become accepted that parts of western North America are not North American but exotic in origin (e.g. Coney, Jones and Monger, 1980; Kerr, 1980; McWilliams and Howell, 1982; Saleeby, 1983; Schermer, Howell and Jones, 1984; Tozer, 1982; Tarduno *et al.*, 1986) and may once have been part of a Pacific continent (Nur and Ben-Avraham, 1977, 1982; Melville, 1981; Pal, 1983) or united with East Asia (Hughes, 1975). Of particular interest is the recent recognition of a microcontinent Cordilleria which moved from somewhere in the Pacific to dock with cratonic North America in the late Mesozoic (Chamberlain and

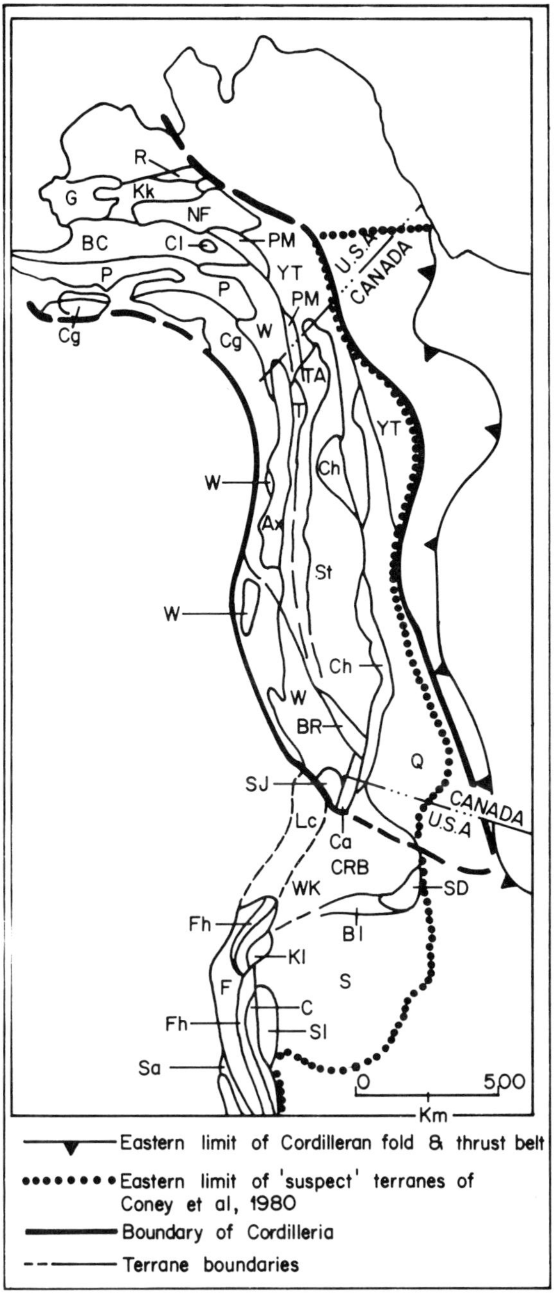

Figure 10 The recently recognized microcontinent of Cordilleria. (From Chamberlain and Lambert, 1985. Reproduced by permission of the authors and MacMillan Journals Ltd, London, England)

Lambert, 1985) (Figure 10). This newly defined microcontinent bears a striking similarity to Croizat's north western North American region involved in trans-Pacific affinities (cf. Figures 8 and 10).

This capacity of panbiogeography to generate novel predictions about palaeogeography testifies to the success of Croizat's suggestions for the integration of biogeography and geology. This is in contrast to the work of Darwinian dispersalist biogeographers who have consistently explained geographic distribution in terms of current geological knowledge rather than attempting to proceed beyond what is already accepted.

TOWARDS A NEW GLOBAL SYNTHESIS IN BIOGEOGRAPHY

Taxonomies not only summarize and order our experience; they also mirror differing ideas about how the world is organized. Biogeographic taxonomies are no exception. Croizat's work heralds the possibility for radical revision in the field of biogeographic classification.

The beginnings of modern phytogeographic and zoogeographic classifications can be traced to the work of Candolle (1820, 1838) and Wallace (1876) respectively. Candolle established 40 botanical regions in total, and these have remained the basis for subsequent work by phytogeographers. For instance, the modern classification of world floristic regions of Good (1974) is clearly based on Candolle's work (Nelson, 1983). Wallace (1876) building on earlier work by Sclater (1858), proposed a system of six major zoogeographic regions for animal life (Figure 11). These major regions were each divided into four sub-regions and there is a close correspondence between Candolle's regions and Wallace's sub-regions (Nelson, 1983). The relationship between Candolle's and Wallace's regions and the biogeographic classifications of later workers, has been reviewed elsewhere (Craw, 1988).

Wallace (1894:613) specifically developed his system of regions 'solely for the purpose of facilitating the study of the existing geographical distribution of animals in its bearing on the theory of evolution'. The larger landmasses of the Northern Hemisphere were regarded by Darwin and Wallace as the major centres of evolution. Matthew (1915) adopted Wallace's ideas and extended them in terms of Darwin's theory of natural selection by proposing the concept of faunal dominance. He shifted the major centres of life into the northern parts of the Holarctic and argued that the alterations of arid and cold climatic zones in the northern Northern Hemisphere resulted in the evolution of more competitive organisms. This occurred because inhospitable climates increase the force of natural selection allowing for the evolution of more dominant, competitive life forms, in such areas.

Matthew was a vertebrate palaeontologist at the American Museum of Natural History in New York. Although he died in 1930 before the evolutionary synthesis was developed he had a great influence upon G.C. Simpson who

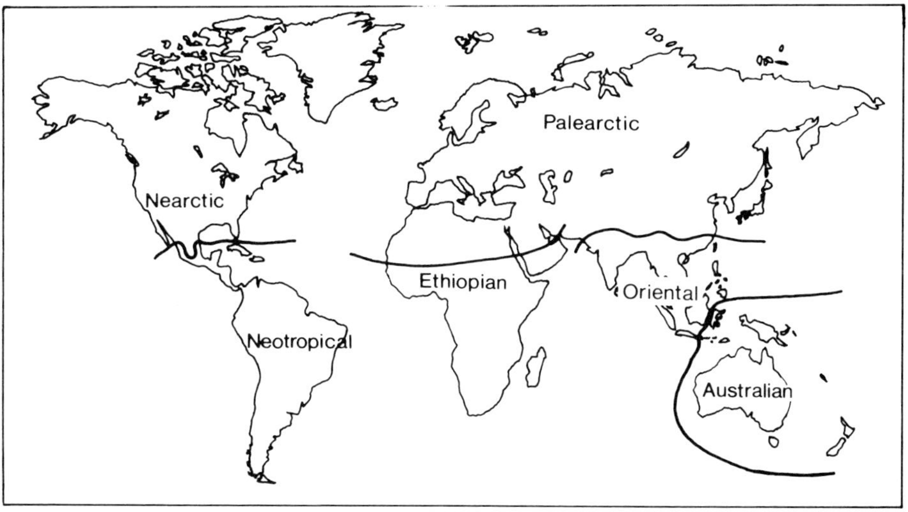

Figure 11 The boundaries of Wallace's six zoogeographic region. Note their close relation to present-day land areas

was the leading palaeontologist and zoogeographer involved in the birth of the synthesis. Indeed, Simpson (1980) has written in his autobiography about how Matthew both 'awed' and 'instructed' him, while Schmidt (1955) has commented that Matthew's papers became 'a Holy Writ to his disciples'. Botanists were also influenced by Matthew's ideas. Axelrod, an important American palaeobotanist and phytogeographer, who had opposed continental drift in the early 1960s (e.g. 1963) but later changed his mind (e.g. 1970), has written 'I was indoctrinated under the Matthewian dicta' (*in litt.*, 1975).

Darlington (1957) modified Matthew's views by shifting the location of the major evolutionary centre to the Old World tropics. The reason for this was that he rejected Matthew's idea of competitive, dominant forms evolving under adverse climatic conditions arguing that these conditions would lead to cold adapted forms. Darlington still accepted the earlier Darwinian view that organisms disperse from large to small areas but added the alternative idea that they disperse from favourable to unfavourable climatic areas. From their centre of origin in the Old World tropics the major groups of the vertebrates dispersed along three independent routes into the Southern Hemisphere.

The above views depend on the idea of the permanence of the continents and oceans since the time of origin of modern life. While some commentators have seen the recent acceptance of a mobile earth history in the form of the concepts of plate tectonics and continental drift as indicating that the Wallace/Matthew/Simpson synthesis 'was based on some fundamental error' (Nelson, 1981: 524), others have argued the exact opposite:

> the Mathew–Darlington mode of global analysis, based on a relatively static continental framework, is still valid. (Wilson, 1985: 491)

Similarly, Mayr (1982:451–2) has argued that continental drift theory supports the Wallacean position: 'Although the major land masses (plates) are still considered permanent their positions and connections are changing in the course of time . . . '. There is much truth in these statements by Mayr and Wilson since Wallace's regions and sub-regions can be readily fitted on to reconstructions of Gondwana and Laurasia (Craw, 1982:Fig. 3).

Application of the panbiogeographic method to geographic distribution leads to a totally new system of world biogeographic regions (Craw, 1984a, 1988). The standard tracks recognized by Croizat (1958) and others (e.g. Craw, 1983b, 1985; Henderson, 1985) bear no relationship at all to the present-day configuration of land and sea. They join areas now widely separated by ocean and sea basins. Portions of continents, larger islands and island groups are related to one another by standard tracks that span ocean and sea basins. These standard tracks overlap in these land areas indicating that these areas are composite biogeographically rather than being unit regions, sub-regions or areas of endemism. These regions and areas of endemism of de Candollean phytogeography and Wallacean zoogeography are not parts of the real natural world (Craw, 1983a). They are biogeographic and geological boundaries where fragments of two or more ancestral biological and geological worlds meet and combine in space/time (Craw, 1983a, 1984b; Craw and Weston, 1984). The natural biogeographic regions for terrestrial and freshwater organisms are not present-day land areas but the world's major ocean basins (Figure 12).

BIOGEOGRAPHIC METAPHORS: OLD AND NEW

Scientific writing is rich in metaphors. These metaphors serve more than a decorative purpose by providing a 'rich soil' within which theories are rooted (Haraway, 1976). The uses of metaphors by scientists are ways of relating theories to the empirical world (Leatherdale, 1974; MacCormac, 1976). Metaphors are an important part of a scientific paradigm performing not only an explanatory function bridging the gap between an abstract system and the real world (Hesse, 1966) but also serving as the 'basic organizing relation of a paradigm (Haraway, 1976).

Darwin's writing is rich in metaphors which capture succinctly the relation between his root metaphor of 'natural selection' and the concepts of means of dispersal, biogeographic regions and faunal dominance. For instance, all these concepts are captured marvellously within his chapters in *The Origin* on geographical distribution where 'the more dominant forms, generated in the larger areas and more efficient workshops of the North' (1860:380) become 'the living waters [that] have flowed' (1860:382) southwards into the rest of the world. These descriptive biogeographic metaphors reflect not only the intimate connection between natural selection and Darwin's biogeography but are indicative also of the social milieu of the time. Both Darwin and Wallace

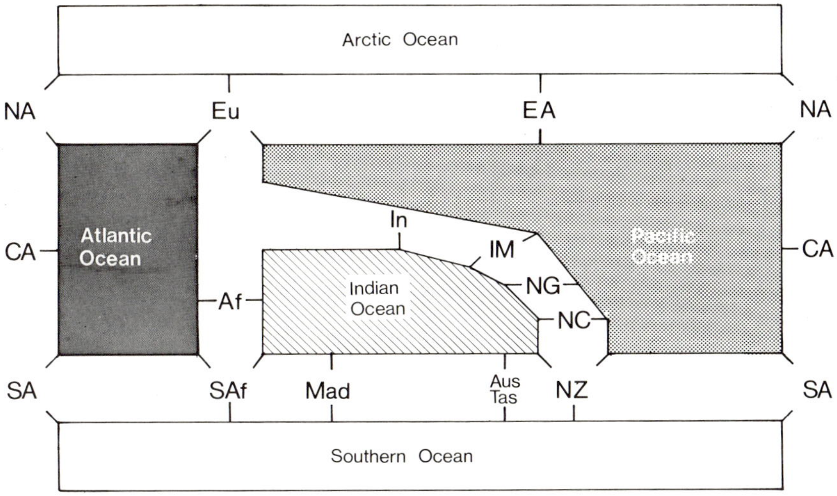

Figure 12 A new world system of biotic regions based on panbiogeographic analyses. Abbreviations: Af — Africa; Aus/Tas — Australia/Tasmania; CA — Central America; EA — East Asia; EU — Europe; In —India; IM — Indo-Malaya; Mad — Madagascar; NA — North America; NC — New Caledonia; NG — New Guinea; NZ — New Zealand; SA — South America; SAf — South Africa.

formulated their ideas during an age of colonialism (Nelson, 1978) and in retrospect the notion of white racial superiority can be clearly seen to have influenced their biogeography. Consider the following passage from an early disciple of Darwinian biogeography:

> If we accept the general configuration of the earth's surface as permanent a continuous and progressive dispersal of species from the center to the circumference, i.e. southwards, seems inevitable. If an observer were placed above a point in St George's Channel from which one half of the globe was visible he would see the greater quantity of land spread out in a sort of stellate figure. The maritime supremacy of the English race has perhaps flowed from the central position of its home. That such a disposition would facilitate a centrifugal migration of land organisms is at any rate obvious and fluctuating conditions of climate operating from the pole would supply an effective means of propulsion. (Thiselton-Dyer, 1909: 308)

Darwinian biogeographic metaphor has continued to be developed. Wallace's regions have recently been described as 'the major theaters of evolution by natural selection' (Wilson, 1985:490). In Darlington's synthesis the concepts of faunal dominance, chance dispersal and dispersal pathways are expressed in terms of the descriptive metaphors of 'road blocking fauna' (1957:575) 'the gigantic whirlpool of dispersal' (1965:215) and the 'zoogeographic Kaleidoscope' (1971:216).

Croizat's panbiogeography is also structured around a root metaphor —

'nature forever repeats' (e.g. 1964:423). This expresses concisely his important finding that organisms of diverse dispersal propensities, colonizing abilities and ecologies exhibit homologous distribution patterns which are related to the tectonic structure of the earth rather than to the force of natural selection operating on means of dispersal in centres of origin. Croizat frequently used two descriptive metaphors — 'earth and life evolve together' and 'life is a geological layer' — to express his view that a symmetry exists between the distribution patterns of organisms and the tectonic structure of the earth. There is a danger that through a literal reading of these metaphors Croizat's work could be interpreted as a crude type of geological determinism where completely passive organisms have their geographic distributions and evolution governed completely by geological events. Such an interpretation would deny the revolutionary implications of Croizat's work and could only be understood as an attempt to assimilate his work within the traditional organism/environment dichotomy. To the contrary Croizat's metaphors reflect his rejection of this dichotomy in favour of the three-fold parallelism of space/time/form (Gray, this volume). In contrast to Darwinian dispersalism where mobile organisms flow across a stable geographical environment, and cladistic vicariism (e.g. Nelson and Platnick, 1984) where populations of passive organisms are fragmented by a highly mobile geographical environment, panbiogeography proposes instead that earth and life have moved in unison (Croizat, pers. comm.). Croizat's metaphor of 'earth and life evolve together' is compatible with the Gaian hypothesis of ecology. This proposes that the earth's atmosphere is to be understood not as having originated as a separate single physical and chemical system but rather as having developed through a dynamic equilibrium between biological life that is maintained within it and that in turn maintains it (Lovelock, 1979; Sagan and Margulis, 1983).

The traditional organism/environment dichotomy is related to the use in Darwinian dispersalism of an absolute concept of space (Craw, 1988). Two conceptions of space have been identified as relevant to spatial analysis: absolute and relative space, associated with the philosophers Newton and Leibniz respectively (Friedman, 1983; Gatrell, 1983). The Newtonian concept of absolute space involves the idea that space forms a fixed, immovable three-dimensional container within which objects/bodies are embedded. Within this container forces are impressed upon and external to these bodies. Centres of origin/dispersal biogeography can be directly equated with this absolutist approach since organisms originate in fixed centres of origin (containers) from which they flow outwards to other areas (containers) driven by the external force of natural selection. The centres of origin are located in Wallace's regions, the 'theaters of evolution' (larger containers), and the taxa that disperse out of them constitute distinct faunal elements in the regions they migrate to (e.g. Wallace's Nearctic region contains, apart from endemic

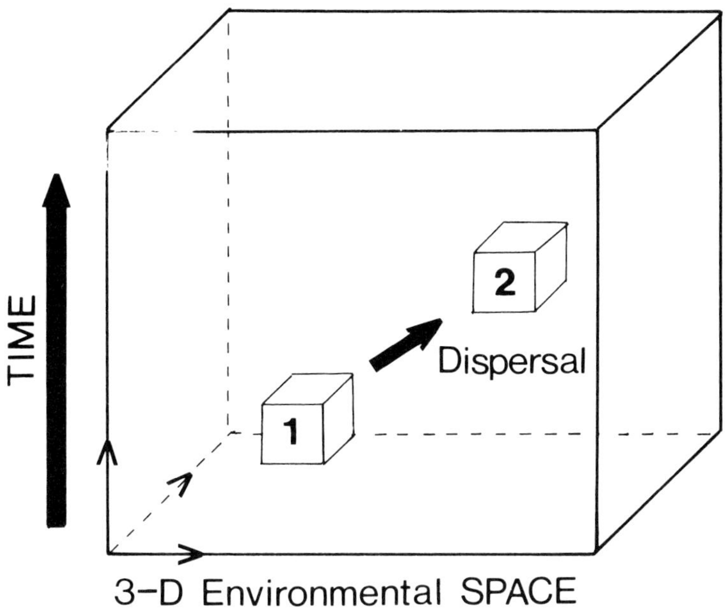

Figure 13 The concept of absolute space with reference to the organism/environment dichotomy. The boxes numbered 1 and 2 represent an organism or taxon at locations 1 and 2 respectively

elements, a whole series of elements from elsewhere such as a Neotropical element from South America — see Mayr, 1976). Dispersal biogeography postulates a separate universe of taxa constantly moving through a separate category of absolute geological time and extended in a three-dimensional environmental space (Figure 13). The Newtonian nature of Darwinian and modern dispersalist biogeography meshes well with recent interpretations of nineteenth- and twentieth-century Darwinism in terms of this analogy. Darwin's theory is clearly an explicit extension of the Newtonian paradigm (Depew and Weber, 1985). For instance, Sober (1986) has demonstrated the conceptual relationship between Darwin's natural selection and Newton's conception of force.

In contrast, panbiogeography is based around a relational concept of space/time (Craw, 1988) and is a relativistic spatio-temporal approach to biogeographic analysis and explanation (Figure 14). Using the panbiogeographic method the spatio-temporal background of the development of many different plants and animals can be reconstructed. Panbiogeography lays down a means for integrating the biogeographic, ecologic, phylogenetic and morphogenetic attributes of taxa within a coherent framework. This is illustrated by the novel panbiogeographic system of world biotic regions where the elements (the organisms) and the regions (the environments) are synthesized into one (Craw,

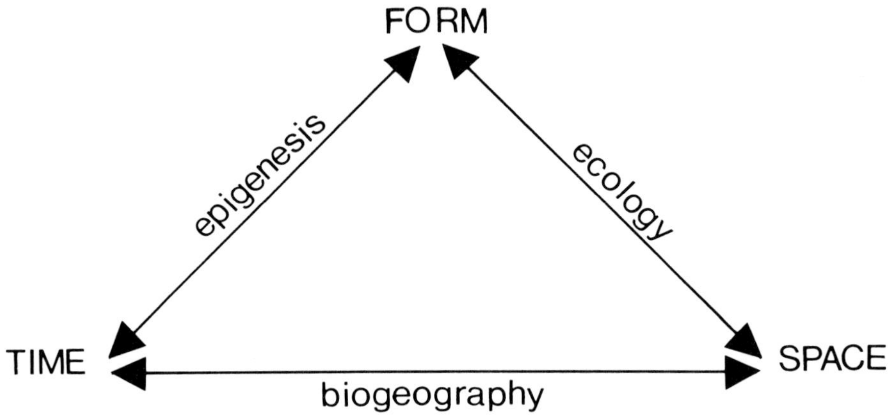

Figure 14 The concept of relational space with reference to the collapse of the organism/environment dichotomy into the three-fold parallelism of space/time/form. Space and time are no longer absolute categories but relations on form (understood here to refer to the structural, functional and behavioural aspects of organisms). Likewise form is no longer an absolute category (e.g. the 'organism') but a relation on space/time. Along the edges of the triangle are indicated the biological sub-disciplines particularly associated with each relation

1988). Here the organism/environment dichotomy is collapsed into a dynamic process metaphysics where 'geographic distribution [of plants and animals] . . . is by far not a pack of chance records, but the coherent byproduct of a constant interplay of space, time, form the world over' (Croizat, 1965: 702–3).

CONCLUSION

The '*Origin of Species*' was described by Darwin as 'one long argument from beginning to end'. Darwin was forced by circumstances to adopt the centres of origin/dispersal model. This was because Darwin's arguments about biogeography were constructed as a refutation of the views of idealist biologists like Louis Agassiz who postulated independent multiple creations as an explanation for discontinuities in the distribution patterns of taxa (Croizat, 1958; Mayr, 1982). Seen from this perspective the model of dispersalist biogeography has no foundations in the real world but was rather developed as a rhetorical device against idealist biogeography. What reason is there to continue to utilize this model and its attendant metaphors as a basis for analysing and explaining geographic distribution? Ho (1986 and this volume) has called for a search for new metaphors for a new biology. What better place to begin that exploration than the rich mine provided by the writings of Leon Croizat?

REFERENCES

Axelrod, D.I. (1963). Fossil floras suggest stable not drifting continents, *J. Geophys. Res.*, **68**, 3257–63.

Axelrod, D. (1970). Mesozoic paleogeography and early angiosperm history, *Bot. Rev,*, **36**, 271–319.

Browne, J. (1980). Darwin's botanical arithmetic and the 'Principle of Divergence', 1854–1858, *J. Hist. Biol.*, **13**, 53–89.

Chamberlain, V.E., and Lambert, R. St J. (1985). Cordilleria, a newly defined Canadian microcontinent, *Nature*, **314**, 707–13.

Coney, P.J., Jones, D.L., and Monger, J.W.H. (1980). Cordilleran suspect terranes, *Nature*, **288**, 328–33.

Craw, R.C. (1978). Two biogeographical frameworks: implications for the biogeography of New Zealand, *Tuatara*, **23**, 81–114.

Craw, R.C. (1982). Phylogenetics, areas, geology and the biogeography of Croizat: a radical view, *Syst. Zool.*, **31**, 304–16.

Craw, R.C. (1983a). Panbiogeography and vicariance cladistics: are they truly different? *Syst. Zool.*, **32**, 431–8.

Craw, R.C. (1983b). *Biogeography of New Zealand: a methodological and conceptual approach*. PhD thesis in Zoology, Victoria University of Wellington, New Zealand.

Craw, R.C. (1984a). Leon Croizat's biogeographic work: a personal appreciation. In R.C. Craw and G.W. Gibbs (eds), *Croizat's Panbiogeography and Principia Botanica*, Victoria University Press, Wellington as *Tuatara*, **27** (1), pp. 8–13.

Craw, R.C. (1984b). Biogeography and biogeographical principles, *NZ Entomologist*, **8**, 49–52.

Craw, R.C. (1985). Classic problems of Southern Hemisphere biogeography reexamined: panbiogeographic analysis of the New Zealand frog *Leiopelma*, the ratite birds and *Nothofagus, Z.f. zool. Systematik u. Evolutionsforschung*, **23**, 1–10.

Craw, R.C. (1988). Panbiogeography: method and synthesis in biogeography. In A. Myers and P.A. Giller (eds), *Analytical Biogeography: an Integrated Approach to the Study of Animal and Plant Distribution*, Chapman and Hall, London.

Craw, R., and Gibbs, G.W. (eds) (1984). *Croizat's Panbiogeography and Principia Botanica: Search for a Novel Biological Synthesis*, Victoria University of Wellington Press, Wellington as *Tuatara* **27** (1).

Craw, R., and Weston, P. (1984). Panbiogeography: a progressive research program? *Syst. Zool.*, **33**, 1–13.

Croizat, L. (1934). De Euphorbio Antiquorum atque Officinarum, a study of succulent Euphorbiae long known in cultivation. New York.

Croizat, L. (1940). On the phylogeny of the Euphorbiaceae and some of their presumed allies. *Rev. Univ. (Univ. Catól. Chile)*, **25**, 205–20.

Croizat, L. (1952). *Manual of Phytogeography or an Account of Plant Dispersal Throughout the World*, Junk, The Hague, Netherlands.

Croizat, L. (1958). *Panbiogeography*. Vols 1, 2a and 2b. Published by the author, Caracas, Venezuela.

Croizat, L. (1961). *Principia Botanica*. Published by the author, Caracas, Venezuela.

Croizat, L. (1964). *Space, Time, Form: The Biological Synthesis*, Published by the author Caracas, Venezuela.

Croizat, L. (1965). An introduction to the subgeneric classification of *Euphorbia* L. with stress on the South African and Malagasy species. I. *Webbia*, **20**, 573–706.

Croizat, L. (1968). The biogeography of the tropical lands and islands east of Suez–

Madagascar: with particular reference to the dispersal and form-making of *Ficus* L., and different other vegetal and animal groups. *Atti Ist. Univ. Lab. Crittogam Pavia*, **(6)**4, 1–400.

Croizat, L. (1982). Vicariance/vicariism, panbiogeography, 'vicariance biogeography', etc., a clarification, *Syst. Zool.*, **31**, 291–304.

Darlington, P.J. (Jr) (1957). *Zoogeography: The Geographical Distribution of Animals*, John Wiley, New York.

Darlington, P.J. (1959). Area, climate and evolution, *Evolution* **13**, 488–510.

Darlington, P.J. (1965). *Biogeography of the Southern End of the World*, Harvard University Press, Cambridge, Mass.

Darlington, P.J. (1971). The carabid beetles of New Guinea. Part IV. General considerations; analysis and history of fauna; Taxonomic supplement, *Bull. Mus. Comp. Zool.*, **142**(2), 129–337.

Darwin, C. (1860). *On the Origin of Species by Means of Natural Selection or the Preservation of Favoured Races in the Struggle for Life*, 2nd edn, John Murray, London.

Darwin, C. (1911). *On the Origin of Species by Means of Natural Selection or the Preservation of Favoured Races in the Struggle for Life*, 6th edn, John Murray, London.

Darwin, F. (ed) (1903). *More Letters of Charles Darwin*, Vol. 1, John Murray, London.

De Beer, G. (ed) (1960). Darwin's note books on transmutation of species. Part 1. First notebook (July 1837 – February 1838), *Bull. Brit. Mus. (Nat. Hist) Hist.* **Ser. 2** (2), 27–73.

De Candolle, A.P. (1820). Géographique botanique. In *Dictionaire des sciences naturelles*, **18**, 359–422. Strasbourg.

De Candolle, A.P. (1838). *Statistique de la famille des Composées*. Treuttel and Würtz, Paris and Strasbourg.

Depew, D.J., and Weber, B.H. (1985). Innovation and tradition in evolutionary theory: an interpretative afterword. In D.J. Depew and B.H. Weber (eds), *Evolution at a Crossroads: The New Biology and the New Philosophy of Science*, MIT Press, Cambridge, Mass. pp. 227–60.

Desmond, A. (1982). *Archetypes and Ancestors: Palaeontology in Victorian London 1850–1875*. Blond and Briggs, London.

Frankel, H. (1984). Biogeography, before and after the rise of sea floor spreading. *Stud. Hist. Phili Sci.*, **15**, 141–68.

Friedman, M. (1983). *Foundation of Space–Time Theories: Relativistic Physics and Philosophy of Science*, Princeton University Press, Princeton, NJ.

Gatrell, A. (1983). *Distance and Space: A Geographical Perspective*, Clarendon Press, Oxford, England.

Germain, L. (1928). L'Origine et l'évolution de la faune de Hawaii, *Proc. 3rd. Pacif. Sci. Congr.* Vol. 1.

Ghiselin, M.J. (1969). *The Triumph of the Darwinian Method*, University of California Press, Berkeley and Los Angeles.

Good, R. (1974). *The Geography of Flowering Plants*, 4th edn, Longman, London.

Gray, R. (1986). Faith and foraging: a critique of the paradigm argument from design. In A.C. Kamil, J.R. Krebs and H.R. Pulliam (eds), *Foraging Behaviour*, Plenum Press, New York, pp. 69–140.

Grehan, J.R., and Henderson, I. (1988). Panbiogeography. In M. Archer, G. Clayton and J. Long (eds), *Vertebrate Zoogeography and Evolution in Australasia*, Hesperian Press, Marrickville, Australia (in press).

Grinnell, G. (1974). The rise and fall of Darwin's first theory of transmutation, *J. Hist. Biol.*, **7**, 259–73.

Guillaumin, M.A. (1928). Les régions floristiques du Pacifique d'après leurs endémismes et la répartition de quelques plantes phanérogames, *Proc. 3rd Pacif. Sci. Congr.* Vol. 1, pp. 920–38.

Hallam, A. (1981). Relative importance of plate movements, eustasy, and climate in controlling major biogeographical changes since the early Mesozoic, In G. Nelson and D. Rosen (eds), *Vicariance Biogeography: A Critique*, Columbia University Press, New York, pp. 303–30.

Hallier, H. (1912). Über fruhere landbrucken Pflanzen und Völkerwanderungen zwischen Australasien und Amerika. *Mededeelingen van Rijk Herbarium, Leiden*, pp.1–32.

Haraway, D.J. (1976). *Crystals, Fabrics and Fields: Metaphors of Organicism in Twentieth-Century Developmental Biology*, Yale University Press, New Haven and London.

Heads, M. (1984). Principia Botanica: Croizat's contribution to botany. In R.C. Craw and G.W. Gibbs (eds) *Croizat's Panbiogeography and Principia Botanica*, Victoria University Press, Wellington, as *Tuatara*, **27**(1), pp. 26–48.

Heads, M. and Craw, R.C. (1984). Bibliography of the scientific work of Leon Croizat, 1932–1982. In R.C. Craw and G.W. Gibbs (eds), *Croizat's Panbiogeography and Principia Botanica*, Victoria University Press, Wellington, as *Tuatara*, **27** (1), pp. 67–75.

Henderson, I. (1985). *Systematic studies of New Zealand Trichoptera and critical analysis of systematic methods.* PhD thesis in Zoology, Victoria University of Wellington, New Zealand.

Hesse, M. (1966). *Models and Analogies in Science*, University of Notre Dame Press, South Bend, Indiana.

Ho, M.-W. (1986). Genetic fitness and natural selection: myth or metaphor (ms. in press).

Hughes, T. (1975). The case for the creation of the North Pacific Ocean during the Mesozoic era, *Palaeogeogr. Palaeoclimatol. Palaeoecol.*, **18**, 1–43.

Kerr, R.A. (1980). The bits and pieces of plate tectonics, *Science*, **207**, 1059–61.

Kinch, M.P. (1980). Geographical distribution and the origin of life: the development of early nineteenth century British explanations, *J. Hist. Biol.*, **13**, 91–119.

Leatherdale, W.H. (1974). *The Role of Analogy, Model and Metaphor in Science*, North Holland, Amsterdam and Oxford.

Lovelock, J.E. (1979). *Gaia: A New Look at Life on Earth*, Oxford University Press, Oxford.

MacCormac, E.R. (1976). *Metaphor and Myth in Science and Religion*, Duke University Press, Durham.

McWilliams, M.O., and Howell, D.G. (1982). Exotic terranes of western California, *Nature*, **297**, 215–17.

Matthew, W.D. (1915). Climate and evolution, *Ann. NY Acad. Sci.*, **24**, 171–318.

Mayr, E. (1976). *Evolution and the Diversity of Life: Selected Essays*, Harvard University Press, Cambridge, Mass.

Mayr, E. (1982). *The Growth of Biological Thought: Diversity, Evolution and Inheritance*. Harvard University Press, Cambridge, Mass.

Melville, R. (1966). Continental drift, Mesozoic continents and the migrations of angiosperms, *Nature*, **211**, 116–20.

Melville, R. (1981). Vicarious plant distributions and paleogeography of the Pacific region. In G. Nelson and D. Rosen (eds), *Vicariance Biogeography: A Critique*, Columbia University Press, New York, pp. 238–74, 298–302.

Murray, A. (1873). On the geographical relations of the chief Coleopterous Faunae, *J. Linn. Soc. (Zool.)*, **11**, 1–89.

Nelson, G. (1978). From Candolle to Croizat: comments on the history of biogeography, *J. Hist. Biol.*, **11**, 269–305.

Nelson, G. (1981) Summary. In G. Nelson and D. Rosen (eds), *Vicariance Biogeography: A Critique*, Columbia University Press, New York, pp. 524–37.

Nelson, G. (1983). Vicariance and cladistics. Historical perspectives with implications for the future. In R.W. Sims, J.H. Price, and P.E.S. Whalley (eds) *Evolution, Time and Space: The Emergence of the Biosphere*, Academic Press, London, pp. 469–92.

Nelson, G., and Platnick, N. (1984). *Biogeography*, Carolina Biology Reader No. 114.

Nur, A., and Ben-Avraham, Z. (1982). Oceanic plateaux, the fragmentation of continents and mountain building, *J. Geophys. Res.*, **87**, (B5), 3644–61.

Nur, A. (1977). Lost Pacifica continent, *Nature*, **270**, 41–3.

Page, R. (1987). Graphs and generalized tracks: Quantifying Croizat's panbiogeography, *Syst. Zool.*, **36** (in press).

Pal, P.C. (1983). Palaeocontinental configurations and geoid anomalies, *Nature*, **303**, 513–16.

Parshall, K.H. (1982). Varieties as incipient species: Darwin's numerical analysis, *J. Hist. Biol.*, **15**, 191–214.

Phipps, J.B. (1975). Kilometric distance, *Can. J. Bot.*, **53**, 1116–19.

Richardson, R.A. (1981). Biogeography and the genesis of Darwin's ideas on transmutation, *J. Hist. Biol.*, **14**, 1–41.

Sagan, D., and Margulis, L. (1983). The Gaian perspective of ecology, *The Ecologist*, **13**, 160–7.

Saleeby, J.B. (1983). Accretionary tectonics of the North American cordillera, *Ann. Rev. Earth Planet. Sci.*, **11**, 45–73.

Schermer, E.R., Howell, D.G., and Jones, D.L. (1984). The origin of allochthonous terranes, *Ann. Rev. Earth Planet. Sci.*, **12**, 107–31.

Schmidt, K.P. (1955). Animal geography. In *A Century of Progress in the Natural Sciences*, pp. 767–94. California Academy of Sciences, San Francisco.

Schuster, R.M. (1976). Plate tectonics and its bearing on the geographical origin and dispersal of the angiosperms. In C.B. Beck (ed), *Origin and Early Evolution of Angiosperms*, Columbia University Press, New York, pp.48–138.

Sclater, P.L. (1858). On the general geographical distribution of the members of the class Aves, *J. Proc. Linn. Soc. Lond. (Zool.)*, **2**, 130–45.

Simpson, G.G. (1940). Mammals and land bridges, *J. Wash. Acad. Sci.*, **30**, 137–63.

Simpson, G.G. (1953). *Evolution and Geography*, Oregon State System of Education, Eugene, Oregon.

Simpson, G.G. (1965). *The Geography of Evolution*, Chilton Books, Philadelphia and New York.

Simpson, G.G. (1976). The compleat paleontologist? *A. Rev. Earth Plan. Sci.*, **4**, 1–13.

Simpson, G.G. (1980). *Why and How: Some Problems and Methods in Historical Biology*, Pergamon Press, Oxford.

Skottsberg, C. (1925). Juan Fernandez and Hawaii: A phytogeographical discussion, *Bernice P. Bishop. Mus. Bull.*, **16**.

Skottsberg, C. (1928). Remarks on the relative independency of Pacific floras, *Proc. 3rd Pan-Pacif. Sci. Congr.*, **1**, 914–20.

Skottsberg, C. (1956). Derivation of the flora and fauna of Juan Fernandez and Easter Island, In C. Skottsberg (ed), *The Natural History of Juan Fernandez and Easter Island*, Vol. 1, Almquist and Wiksell, Uppsala, pp. 193–438.

Sober, E. (1986). Darwin on natural selection: a philosophical perspective. In D. Kohn (ed), *The Darwinian Heritage*, Princeton University Press, Princeton, NJ, pp. 867–99.

Steenis, G.G.G.J. van (1962). The land-bridge theory in botany, *Blumea II*: 235–542.

Tarduno, J.A., McWilliams, M., Sliter, M.V., Cook, H.E., Blake, M.C., and Premoli-Silva, I. (1986). Southern hemisphere origin of the cretaceous Laytonville limestone of California, *Science*, **231**, 1425–8.

Thiselton-Dyer, W. (1909). Geographical distribution of plants. In A.C. Seward (ed), *Darwin and Modern Science*, University Press, Cambridge, pp. 298–318.

Tozer, E.T. (1982). Marine Triassic faunas from North America: their significance for assessing plate and terrane movements, *Geol. Rund.*, **71**, 1077–104.

Wallace, A.R. (1876). *The Geographical Distribution of Animals*, Vols 1 and 2, MacMillan, London.

Wallace, A.R. (1880). *Island Life*, Macmillan, London.

Wallace, A.R. (1894). What are zoological regions? *Nature*, **49**, 610–13.

Wilson, E.O. (1985). The search for faunal dominance. In G.E. Ball (ed), *Taxonomy, Phylogeny and Zoogeography of Beetles and Ants*, Dr W. Junk, Dordrecht, Netherlands, pp.489–93.

Wulff, E.V. (1943). *An Introduction to Historical Plant Geography*, Waltham, Mass.

Evolutionary Processes and Metaphors
Edited by M.-W. Ho and S.W. Fox

CHAPTER 10

The active role of behaviour in evolution

Patrick Bateson

ABSTRACT

The behaviour of animals may play a major role in introducing evolutionary change. Such changes may arise from active choices or from alterations by the animal of its physical or social environment. Evolutionary change may also be initiated when animals are able to modify their behaviour in response to changed conditions or when they expose themselves to new conditions which reveal heritable variation that was not previously expressed. Heritable variation does not mean that the characters involved are developmentally stable or even that the variation in the characters depends on genetic variation. How an individual animal develops during its life is conditional, and also dependent, on the reciprocal character of the transactions with the physical and social world about it. When this knowledge about development is recoupled to questions about evolution, it becomes much easier to perceive how an animal's behaviour can initiate and shape long-term changes.

INTRODUCTION

Many biologists (including myself) have unthinkingly accepted the Darwinian image of selection, with nature picking those organisms that fitted best into the environments in which they lived. The picture of an external hand doing all the work is so vivid that it is easy to treat organisms as if they were entirely passive in the evolutionary process. That is not, of course, to suggest that any biologist would deny that organisms, and animals especially, are active. But the notion of 'selection pressure' does subtly downplay the organisms' part in the processes of change. In this chapter I shall argue that when the role of active organisms is brought into evolutionary theory, the selection metaphor becomes an encumbrance. I shall then consider a feature of Neo-Darwinist thought, of which I have already been critical in my previous writing about evolutionary theory (e.g. Bateson, 1982, 1986), that is the tendency to place so much emphasis on genes as the units of evolution and so little emphasis on how they are expressed. When developmental issues are recoupled to questions about evolution, it becomes much easier to perceive how an organism's behaviour can initiate and direct lines of evolution. Developmental processes do not merely act as constraints, as much current thinking would seem to suggest (e.g. Maynard Smith *et al.*, 1985), they can make certain types of evolutionary change more likely.

Before proceeding any further I must emphasize that, unlike some other contributors to this volume, I still think that the evolutionary mechanism proposed by Darwin is an exceedingly powerful historical explanation for the existence of current adaptations. The explanation proposes heritable variation of a character and unequal representation of those variants in subsequent generations. It would be empty if all this meant was 'the survival of the survivors', but it doesn't. The circularity is removed when the explanation is made specific. For example, if the behaviour of a long-tailed tit that lined its nest with feathers were to be explained in Darwinian terms, the cause of differential survival of nestlings in the past might have been in terms of the thermal insulation that was provided for them. Up to some limit, those parents that provided better insulation produced more offspring that survived to do the same for their own young than those parents that did not weave so many feathers into the lining of their nest. The assumption that insulation matters to the young can be tested directly. The other crucial assumption that differences in the amount of insulation provided by the birds were heritable in the past could only be examined if sufficient variation in behaviour remains in the existing population of long-tailed tits. Nevertheless, when the proposal is expressed in terms of particular consequences of expressing a character and differential survival, it is not tautological (nor is it teleological).

A totally different type of historical explanation involves inheritance of acquired characters. Such inheritance of individually acquired characteristics through social modes, such as imitation by offspring of parent, poses few problems for us in terms of mechanism (see Bonner, 1980). However, in recent years Piaget (1979) has favoured an explanation involving biological inheritance, as did Ho and Saunders in their contributions to the book they edited together (Ho and Saunders, 1984). When applied to the nest example, it might take the following form. At some time in the past birds were able to detect that nest lining benefited their offspring, they modified their behaviour accordingly and after some generations, their physiology or genomes may become permanently changed so that an offspring subsequently behaved like its parents without modification of its own nest-building behaviour. This form of explanation assumes that mechanisms for detecting benefits to offspring and for translating acquired adaptations possibly back into the genome were already present at the time of the evolutionary change. How these mechanisms evolved needs to be explained. Furthermore, a coherent account of how a change in behaviour could lead to an appropriate change in the genome remains to be suggested. That is not to say that such a mechanism does not exist, but merely that, in proposals of this type, considerable gaps remain.

Undoubtedly the case for Darwinism is frequently overstated. I think that Dawkins (1983:403) did this when he wrote that the Darwinian theory of evolution '. . . is probably the only theory that *can* adequately account for the phenomena that we associate with life' (his emphasis). Indeed, I agree with Hailman (1982) that many examples of phylogeny and speciation can be explained plausibly and coherently without recourse to this theory. Certainly, Darwin's theory never explained how variation in phenotypes is first generated and much of the current ferment of ideas is precisely about this issue (see, for example, Dover, 1986). The enthusiastic proponents of universal Darwinism claim too much and, as a consequence, so do their critics. The way forward, I shall suggest, is to examine in considerable detail the dynamics of what happens in evolutionary change. In order to make this argument more concrete, I shall focus on ways in which the behaviour of animals might have changed the course of evolution.

BEHAVIOUR AND EVOLUTION

As far as I can tell four major proposals have been made for the ways in which an animal's behaviour could affect subsequent evolution. These have often been jumbled up, partly because they do indeed grade into each other. Also the various processes that have been proposed could interact. However, I believe that it is helpful to start by keeping them clearly distinguished. The proposals are briefly listed and then considered in greater detail.

1. Animals make active choices and the results of their choices have consequences for subsequent evolution.
2. By their behaviour, animals change the physical or the social conditions with which they and their descendants have to cope and thereby affect the subsequent course of evolution.
3. Animals are able to modify their behaviour in response to changed conditions. This allows evolutionary change that otherwise would probably have been prevented by the death of the animals exposed to those conditions.
4. By their behaviour animals often expose themselves to new conditions which may reveal heritable variability. This opens up possibilities for evolutionary changes which would not otherwise have taken place.

Active choice

The role of choice in evolution was clearly recognized by Darwin (1871) in his principle of sexual selection. Fisher (1930) noted that Darwin's postulated process of sexual selection could have a runaway character to it. These positive feedback effects and other complexities have been the subject of intense theoretical activity among orthodox Neo-Darwinians in recent years (e.g. Kirkpatrick, 1982; Lande, 1981; O'Donald, 1983). Since these dynamic consequences of behaviour on evolution are now widely accepted, they do not demand extensive comment here. Nevertheless, it is worth noting that the extension of the selection metaphor to such evolutionary processes has caused difficulties. When editing a book on mate choice, I found that in the writings of many of my contributors 'sexual selection' had acquired two meanings: (a) what animals do; and (b) what the consequences are in terms of evolutionary change. It helped when 'mate choice' was used for what animals do when they respond actively and differentially to members of the opposite sex prior to mating (see Bateson, 1983a). 'Sexual selection' was then to be used exclusively for the postulated evolutionary change that results from such behaviour. Even so, the possibilities for continuing confusion remained and I now feel that 'sexual selection' causes so much trouble that it ought to be abandoned.

The role of active choice in other aspects of evolutionary changes has been favoured by many biologists over the years. The great pioneering ecologist, Elton (1930), pointed out that the choice of the environment by the animal was important in evolution. Such a sentiment was congenial to ethologists who consistently placed emphasis on the active animal which engages its environment. For example, in a review paper that deserves to be more widely cited, Hinde (1959) drew attention to the variety of ways that the choices of animals could influence the subsequent course of evolution. As might have been expected, Waddington (1959) was also sympathetic to such a view since he, more than anybody, attempted to forge links between developmental and evolutionary processes when it was unfashionable to do so. He wrote as follows:

> Animals — and the following considerations do not apply so directly to plants — are usually surrounded by a much wider range of environmental conditions than they are willing to inhabit. They live in a highly heterogeneous 'ambience', from which they themselves select the particular habitat in which their life will be passed. Thus the animal by its behaviour contributes in a most important way to determining the nature and intensity of selective pressures which will be exerted on it. Natural selection is very far from being as external a force as the conventional picture might lead us to believe. (Waddington, 1959)

More recently, Dover (1986) has suggested that when the internal dynamics at the molecular level lead to changes in the genetic composition of multigene families, alterations may occur at the level of the whole organism. If, as a result of such 'molecular drive', the cohesively evolving population of organisms develops a feature that would have been maladaptive in the environment in which its ancestors lived, its members can move to another environment where the feature works or, at least, is not a handicap. Although the initiating event in such a postulated process is molecular, behavioural choice could play an important role in maintaining the course of the evolutionary change.

Active alteration of conditions

The environment does not simply set a problem to which the organism has to find a solution. The organism can do a great deal to create an environment to which it is best suited. This should make us hesitate if we consider evolution purely in terms of selection by external forces (see Lewontin, 1983). By leaving an impact on their physical and social environment, organisms may affect the evolution of their own descendants, quite apart from changing the conditions for themselves. Some of the impact is subtle, such as when a plant sheds its leaves which fall to the ground and change the characteristics of the soil in which its own roots and those of its descendants grow. Some of the impact is conspicuous and active such as when beavers dam a river, flood a valley and create a private lake for themselves.

The effect of behavioural control on evolutionary change could be especially great when a major component of the environmental conditions with which animals have to cope is provided by their social environment. A similar type of positive feedback to that flowing from the effects of mate choice could operate in such circumstances (Jolly, 1966). If individuals compete with each other within a social group and the outcome of the competition depends in part on each individual's capacity to predict what the other will do, the evolutionary outcome might easily acquire a runaway property. In discussing the social function of intelligence, Humphrey expressed the idea as follows:

> An evolutionary 'ratchet' has been set up, acting like a self-winding watch to increase the general intellectual standing of the species. In principle the process might be expected to continue until either the physiological mainspring of intelligence is full-wound or else intelligence itself becomes a burden. The latter seems most likely to be the limiting factor; there must surely come a point where the time required to resolve a social 'argument' becomes insupportable. (Humphrey, 1976: 311)

As Humphrey notes, such an explanation makes sense of the astonishing rate of increase in the cranial capacity of humans, if it is assumed (reasonably) that cranial capacity and intellectual ability are correlated.

Adaptability

Modern thinking about the importance of behavioural plasticity in evolution is usually thought to stem from Baldwin (1896), but Lloyd Morgan (1896) and Osborn (1896) independently developed ideas about 'organic selection', as the subject was called at the time. Lloyd Morgan's (1896) account of the process was particularly clear and may be paraphrased as follows:

1. Suppose that a group of organisms that are capable of change in their own lifetimes are exposed to new environmental conditions.
2. Those whose ability to change is equal to the occasion survive. They are modified. Those whose ability is not equal to the occasion are eliminated.
3. The modification takes place generation after generation in the changed environmental conditions, but the modification is not inherited. The effects of modification are not transmitted through the genes.
4. Any variation in the ease of expression of the modified character which is due to genetic differences is liable to act in favour of those individuals that express the character most readily.
5. As a consequence, an inherited predisposition to express the modifications in question will tend to evolve. The longer the evolutionary process continues, the more marked will be such a predisposition.
6 Thus plastic modification within individuals might lead the process and a change in genes that influence the character would follow; the one paves the way for the other.

It is obvious from this outline of the proposed process, that Lloyd Morgan was not suggesting genetic inheritance of acquired characters as a mechanism. Nor, less obviously, was Osborn or Baldwin. It is equally obvious that the process does not depend on unlikely temporal coincidence. The crucial postulate is a cost of operating the original process of phenotypic adaptation, a cost which can be subsequently reduced by genotypic change. Adaptability to new conditions might be physiological, such as coping with high altitudes by enhancing the oxygen carrying capacity of the blood. Nevertheless, an important component would be the ability of the animal to adapt to new circumstances by behavioural means, particularly by learning.

While these ideas about the role of the animal in shaping the course of evolution have been discussed by the major writers on evolutionary theory, the postulated process was not generally regarded as being important. Hardy (1965) has done more than most in recent times to stress that, on the contrary, the process could be of great significance. He put it as follows:

> If a population of animals should change their habits (no doubt often on account of changes in their surroundings such as food supply, breeding sites, etc., but also sometimes due to their exploratory curiosity discovering new ways of life, such as new sources of food or new methods of exploitation) then, sooner or later, variations in the gene complex will turn up in the population to produce small alterations in the animal's structure which will make them more efficient in relation to their new behaviour pattern; these more efficient individuals will tend to survive rather than the less efficient, and so the composition of the population

will gradually change. This evolutionary change is one caused initially by a change in behaviour. (Hardy, 1965:170)

Hardy envisaged a cascade of changes flowing from the initial behavioural event. Even without structural change, control of behavioural development might alter over time. A requirement for this to happen is that adapting to the new conditions be more costly than doing it easily without active modification. One instance might involve differential responsiveness to particular types of food. Many cases of choice of an unusual food for a given species are probably not due to genetic changes, but to the functioning of normal mechanisms in unusual circumstances. A group of animals might be forced into living in an unusual place after losing their way, but they cope by changing their preferences to suitable foods that are locally abundant. Later, those descendants that didn't need to learn so much might be more likely to survive than those that could only show a fully functional phenotype by learning. A cost was incurred in the time taken to learn. As a consequence, what started as a purely phenotypic difference between animals of the same species living in different habitats becomes a genotypic difference.

Alteration in gene expression

Even within the confines of Neo-Darwinian theory, it is obvious that anything that markedly changes the character of gene expression during development could be of great importance in evolution. One candidate for generating such changes is an alteration in environmental conditions. Conditions might be imposed by, say, a glaciation or a catastrophe resulting from earth's collision with a comet or asteroid. However, a change in the environment of a given animal might be brought about because it can move.

When discussing his results of experiments on what he called 'genetic assimilation', Waddington (1953) suggested that the heat shock, applied to the larvae of fruitflies, led to the expression of genes that were carried in only a part of the population. Waddington bred from the flies that had developed a particular character (lack of a cross-vein in the wings) as a result of their larval experience. He continued to apply heat shock in each generation and to breed selectively from the flies with cross-veinless wings. After many generations of heat shock and selective breeding, cross-veinless wings developed spontaneously in the absence of the external triggering condition of heat shock. Of course, some link must be found between his procedures and the eventual spontaneous expression of cross-veinless wings. Many suggestions which do not involve the inheritance of acquired characters have been proposed. For instance, only genotypes that are homozygous at particular loci lead to the spontaneous expression of cross-veinless wings; the much commoner genotypes which are heterozygous at those loci express the character only when exposed to heat shock in larval life (see Bateson, 1982). Consequently, selective breeding of the heterozygotes after heat shock increases their frequency and eventually leads to the breeding of the homozygotes that express the character spontaneously without experiencing heat shock. As Ho (1984) notes, Waddington's experiment lacked a control, namely flies that were exposed to heat shock but not subjected to selective breeding. If such a population had eventually developed cross-veinless wings spontaneously, it would have been necessary to look for more complex explanations.

Genetic assimilation, as defined by Waddington, is often regarded as an example of the Baldwin effect. Hardy (1965) certainly believed it and, as a consequence, probably did a disservice to his own promotion of adaptability as a potent force in evolution. Waddington (1957), by contrast, attempted to distance his own thinking from the proposals of Baldwin, Lloyd Morgan and Osborn. I believe that Waddington was right and that there is, indeed, an important and interesting difference. The Baldwin effect involves necessary compensation for the effects of a new set of conditions and immediate adaptation to the challenge. The compensation is not inherited and differential survival of different genotypes may arise from subsequent differences in the ease with which the new adaptation is expressed.

Waddington's proposal also involves expression of a novel character in a new environment, but the character is not necessarily an adaptation to the triggering condition, even though it may confer some advantage on its possessor. Cross-veinless wings do not bear any functional relation to the environment which supplied a heat shock when the flies were larvae. Nor need there be such a relationship under natural conditions. All that is required initially is that the environmental conditions trigger the expression of phenotypic variation that is heritable. This would be enought to speed up the rate of evolution.

It should be noted that, if changes analogous to those in Waddington's experiments occurred under natural conditions, rather more is required. Animals that express a character that is initially only seen after environmental disturbance must have an advantage in terms of survival of reproductive success. Finally, after many generations the character must be expressed in the absence of the disturbance. Genetic fixation could occur in the way that I have already suggested.

The major point that I wish to establish here is that alteration of the pattern of development by external influences uncovers an important source of heritable variation. That being the case, one crucial way in which animals can bring themselves into contact with such potent influences is through their own behaviour.

BEHAVIOURALLY INDUCED EVOLUTIONARY CHANGE

High mobility by animals, such as that involved in active exploration or migration, would have frequently placed them in conditions that revealed heritable variation not previously apparent in the population. For that reason, behaviour, along with other forms of dispersion, was likely to be important in initiating evolutionary change. In addition, behavioural adaptability of the animals would have helped buffer them against extinction in new conditions. Consequently, interplay between this process and the expression of new variation could have readily occurred. Similarly, both processes might have been enhanced if the animals were capable of choosing the local conditions to which they were most suited and if they could actively alter their environment.

Hardy (1965) believed that such processes have played an important role in evolution. It is worth quoting him again.

> It is adaptations which are due to the animal's behaviour, to its restless exploration of its surroundings, to its initiative in seeking new sources of food when its normal supply fails or becomes scarce through competition, that distinguish the main diverging lines of evolution; it is these dynamic qualities which lead to the different roles of life that open up to a newly emerging group of animals in that phase of their expansion technically known as adaptive radiation. What are the main features that do in fact distinguish the different diverging lines of evolution to be found both in the outbursts of the reptiles in the mesozoic age and of the mammals in the tertiary period? They are behavioural differences associated with their newly exploited environments; the development of new habits giving us the lines of runners, climbers, burrowers, swimmers and conquerors of the air. It cannot have been a new mutation or reassortment of genes that made an animal long used to terrestrial life begin to take to the water for its food. Again and again in the long history of the terrestrial vertebrates we have seen different forms, particularly among reptiles and mammals, but also, of course, among birds (and the more terrestrial lines of amphibia) turning to the water for their food. Some, of course, have become completely aquatic like the extinct Icthyosaurus and modern whales which show such a wonderful convergence in evolution between reptiles and mammals, towards a fish-like form. (Hardy, 1965:192–3)

All the forms of behaviour involved in such interplay require complex neural processing. So it would not be surprising if the rate of behaviourally induced evolutionary change were related to neural capacity and a special feature of complex animals such as birds and mammals. The size of the brain relative to the rest of the body is not a perfect measure of neural capacity, but it is a surprisingly good one (e.g. Jerison, 1973). Presumably *relative* brain size rather than absolute brain size is the better measure because animals with larger bodies have more sense organs to monitor than those with smaller bodies, they have more muscles to control and the general housekeeping jobs of maintaining the larger body are greater.

Given that relative brain size is a measure of the animal's ability to change its behaviour in ways that might increase the rate of evolution, is there any evidence for such a relationship? A striking and intriguing correlation discovered by Wyles, Kunkel and Wilson (1983) suggests that there is. Taking an admittedly controversial measure of anatomical variability within a taxonomic group as a unitary measure of the rate of evolution within that group, they found the groups that were evolving most rapidly by this measure were also the groups with the biggest brains relative to body size (see Figure 1). They attributed the correlation to enhanced ability to imitate on the part of the animals with larger relative brain sizes. My own view is that all the four ways in which I have suggested that behaviour might affect evolutionary rate could be tied into the measure of brain size. Of course, people will argue about the way in which Wyles *et al.*, obtained their measures, what the measures mean, and whether the correlation, strong though it is, is due to a third and unmeasured variable. Nonetheless, I think that we should welcome the appearance of such evidence. It shows that we are not necessarily stuck with merely producing plausible scenarios. We can start to test them by imaginative comparative studies.

WHAT IS THE ROLE OF DEVELOPMENT?

The proposals about the active involvement of animals in evolution places great emphasis on their phenotypes and how they develop. By contrast, a pronounced feature of a certain style of Neo-Darwinism has been to place all the emphasis on changes in gene frequencies in the course of evolution and remove the organism from consideration. The justification for doing this has undoubtedly seemed compelling to many since genes generally survive generation after generation, whereas individual organisms never do. The consequences of such an approach was that, when the effect of a gene on an organism was considered, the gene was supposed to *determine* the outcome.

The meaning of 'determine' is not wholly clear. Many people would use it in the mildest possible way to mean 'influence'. Others would harden it up a bit and mean 'necessary but not sufficient' for the expression of a character. However, a common practice has been to refer to a gene as *coding* for a character. The clear implication of such language was that a one-to-one correspondence existed between that gene and that character. Here is an example of such language from a justifiably popular textbook:

> In a sense all behaviour must be coded for by genes; reduced to its simplest form behaviour is nothing more than a series of nervous impulses and muscle contractions and the protein structure of nerve and muscle is coded for by genetic instruction. (Krebs and Davies, 1981: 12)

When the biologists, who were primarily interested in evolution, employed such phrases, they made a strong statement about how the mediating developmental processes are supposed to work. They had plainly trespassed on to the domain of the developmentalists, even though it was an area in which they

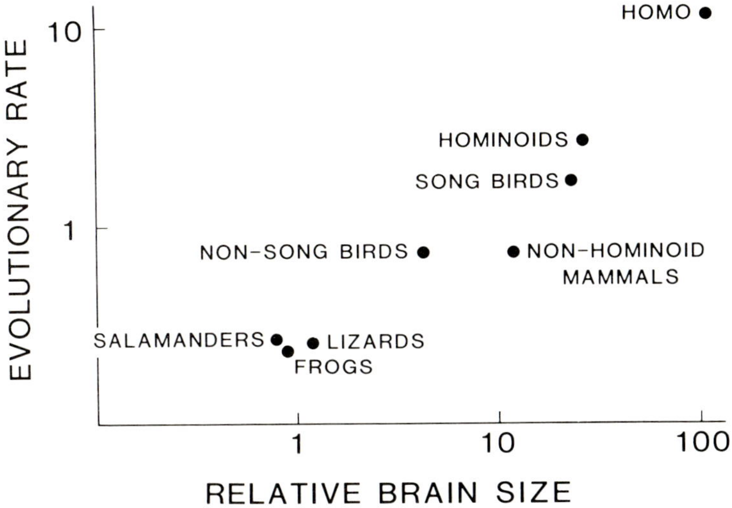

Figure 1 The rate of evolution in different groups of terrestrial vertebrates plotted against relative brain size. The rate of evolution within a given taxonomic group is obtained by assessing the differences between the relative lengths of eight morphological characters of members of that group. Relative brain size = $yx^{-0.67}$ where y = brain weight and x = body weight. Both scales on the figures are logarithmic. (Replotted from data in Wyles, Kunkel and Wilson, 1983)

often professed not to be interested. Since the phrase of a gene coding for a character occurs so frequently, it is worth examining how plausible is the belief that genes can do that.

It is certainly possible now to measure the DNA content of an individual organism accurately. However, when the amount of DNA is compared to the relative size of the nervous system and the complexity of behaviour it generates, the lack of correlation is surprising (see Changeux, 1983). Mice have 6000 times as much DNA as bacteria which makes sense if the differences in the genes are responsible for the differences in complexity. However, humans have no more DNA than mice. In other words, whatever else happened in evolution, the gradual emergence of behavioural complexity within the mammals was not achieved by accumulating genes. Some of the lack of association may be attributed to 'junk' DNA which has no effect on the organism's phenotype. Even so, the implication for studies of development is profound. At the very least, it means that the growth of nervous systems and the emergence of behaviour are critically dependent on regulation and the combinatorial action of genes. This means that the correspondence between genes and behaviour is never likely to be simple.

Genes really do code for polypeptides, of course, but they *represent* nothing else since the process of development is generative. A small subset of genes and cytoplasmic conditions start the whole process off after fertilization of the egg.

These starting conditions create products that switch off some active genes, switch on others and bring the developing components into contact with new influences from outside. And so the whole process continues until death.

The attempts to modernize the old preformationist belief that simple correspondence would be found between the starting points and the end points of development have turned out to be a damaging waste of time. We have to deal with dynamic systems that develop particular characteristics only under certain sets of conditions (Bateson, 1986). However, if developmental processes were such that genetic influences failed to lead to the same phenotypic outcome generation after generation then Darwinian evolution could not work. Since behaviour is plainly changed by environmental conditions, it follows from the Darwinian premise that either the environment has been stable or the behaviour is buffered against environmental change. Since the environment does not seem particularly stable, the naive evolutionary biologist tends to suppose that behaviour patterns which are likely to be the products of Darwinian evolution must be 'unlearned'.

Those who have given rather more thought to the matter will accept that learning may be involved in an individual's development even though the outcome of the developmental process was almost certainly the product of a Darwinian process of evolution. The ability to recognize close kin as a consequence of an imprinting-type process in early life would be a good case in point (e.g. Lorenz, 1965).

When we have to deal with examples of behaviour that change in response to alterations in the environment, it seems highly likely that the specific ways in which animals tune their behaviour to local conditions are themselves the product of Darwinian evolution. If we postulate rules for learning, which fit the animal's information-gathering equipment to particular problems and which may have been subject to Darwinian evolution, then the conditions necessary for their development must be transmitted in some way from one generation to the next. While the mode of transmission may commonly involve genes, the rules for modifying behaviour are hardly likely to spring fully armed out of the genome. The criticism of genes coding for unlearnt behaviour applies just as forcibly to the rules which are involved in the development of behaviour. Such rules represent the workings of an already functional nervous system and body. They themselves have to develop and depend on structures that require for their development a complex interplay between the products of many genes and many conditions external to the genome.

Many features of the inferred rules for learning seem to be profoundly modified by experience. For instance, whether or not initially neutral cues are treated as potentially relevant or ignored is greatly affected by the animal's prior history (e.g. Dickinson, 1980). Such selectivity in responsiveness to external conditions can be of great use to the animal. To take another example, monkeys used in many experiments on associative learning are rewarded with food. The machines which dispense the food are commonly designed to drop peanuts into a cup when the monkey is to be rewarded. However, many monkeys do not like the peanuts at first. They have to be deprived of their regular food and accustomed to the peanuts for weeks before they will take them with any readiness, let alone treat the nuts as rewards for appropriate behaviour (Weiskrantz and Cowey, 1963). In such cases, which are not exceptional, experience expands what the monkeys regard as acceptable food. However, at earlier stages in development, experience usually narrows the range by imprinting-like processes (for a review see Immelmann, 1984).

It might be argued that an unlearnt programme could still be detected at work behind the scenes since the general category of food, and its effectiveness as a reward, were in some sense built in. In other cases though, it becomes more difficult to pinpoint what might or might not act as a reward without extensive knowledge of the animal's previous experience. For instance, the condition in which it becomes possible for an animal to perform an act that would bring it food becomes rewarding in itself. So the animal will work in order to provide itself with those conditions. In this way lengthy chains of behaviour can be developed with any one event providing the terminating condition for one action and the enabling condition for the next (Kelleher, 1966). This is the basis for many complex circus acts performed by animals.

It must be stressed that we are not faced with the possibility of an infinite regress with learning rules affected by other learning rules and so on back to conception. Learning does not have to be involved in the development of the rules for learning. For instance, even very young rats selectively associate taste with poison and texture with electric shock (Domjan, 1980). Possibly some of the selectivity was acquired during embryonic life. However, it seems plausible in this case that the rules for forming some associations and not others are not dependent on prior learning for their development. Even so, that should not make anybody complacement about what generates them. The classic mistake has been to confuse experiences involved in learning with all other kinds of experience which the animal can have during its development (see Lehrman, 1970). All sorts of environmental conditions can have non-specific but profound influences on behavioural development without involving learning (see Bateson, 1983b). Change the social or physical conditions in which the animal is growing up and you may find it ends up with a different set of rules for learning.

A final point, often made but equally often ignored, is that correlations between the behaviour patterns of the parent and those of the offspring do not necessarily arise because they share genes in common. They may arise because other conditions that are necessary for the peculiarities of their behaviour are shared. Common odours and preferences for familiar smells might arise from the particular combination of bacteria that break down the fats secreted into the body surface (e.g. Gorman, Nedwell and Smith, 1974). When the bacteria pass from mother to offspring, so does the source of her special smell. In such cases the Darwinian process of evolution can work but the mode of inheritance does not depend on variation in the genes of the host. Such examples emphasize the need for good knowledge of the biology and life history of the organism in question.

The sophisticated response by the evolutionary biologists to such developmental arguments is to reply that none of this matters (e.g. Dawkins, 1982). The notion of a gene coding for a character or a developmental rule was wrong, but also irrelevant. So long as developmental systems and the conditions surrounding the animal produce beneficial outcomes by one means or another, it does not matter how they do so. It is only the end product that is 'seen'. The question, though, is within what limits will the developmental systems, dynamic as they are, produce the same result. It would not be in the least surprising if a developmental system produced markedly different products when sufficiently perturbed.

Sometimes the perturbations produced by the new set of conditions may be such that developmental processes generate maladaptive outcomes — such as flippers instead of arms when the human embryo had been exposed to thalidomide. A more subtle and relevant example might be the reward mechanisms which are highly sensitive to sugar intake. Behaviour patterns that lead to increased sugar intake are reinforced and the individual rapidly acquires a repertoire of patterns that increase the likelihood that sugar will be found and eaten. This mechanism would have worked well in the conditions in which humans evolved, where doubtless the sugar was in short supply. However, it becomes disastrous when the supplies of the stuff are virtually unlimited.

I have argued that a rule for learning, or for any other kind of developmental process, is not a gene written large. We have no reason to suppose that any simple correspondence will be found between gene and rule for changing behaviour any more than there is an isomorphic relationship between gene and behaviour. The same point applies with equal force to all the other epigenetic rules which bring order to development. Presumably, if the rules have any universality in natural conditions, the experience which affects them must be a common feature of all the animals having the rules. Alternatively, they must be well buffered against change by the particular conditions in which an individual finds itself. However, if conditions are changed enough, developmental stasis is no longer maintained. The biological equivalent of an earthquake occurs and the system suddenly adopts a new configuration. If such a bifurcation occurs early in development, the effects may ramify and generate a radically different phenotype. The general point is that even unlearned behaviour is only buffered from environmental conditions within certain limits. Changes in conditions may kill the animal, but they can also open up important new avenues for subsequent evolutionary change. That is why knowledge of development impinges on studies of evolution and why genetic determinism has seriously stultified thought about the nature of evolution.

CONCLUSION

Darwin's theory of evolution has plenty of life in it and remains extraordinarily fertile in stimulating further thought. As Hailman (1982) pointed out, that does not mean that it explains everything. Moreover, it was unfortunate that the metaphor of selection, so strongly associated with the theory, has portrayed the living organism in such a passive way. The major thrust of this chapter has been to suggest that the organism should be treated as an active agent, particularly if it is an animal with a complicated nervous system. The argument is not original, but it is worth a restatement and an amplification, because it is still so widely ignored.

I suggest that an animal's behaviour might affect the course of evolution in at least four ways. First, animals make active choices and the consequences of their choices are often important. Second, animals change the conditions in which they live by altering the physical or the social environment and again the consequences are likely to be important. Third, animals are able to modify their behaviour in response to changed conditions and thereby make further genetic change possible. Finally, active animals often expose themselves to new conditions which may reveal variability, with some variants more likely to survive than others. In general terms, it seems highly likely that the capacities to change and cope with new environments and the induction of unsuspected variation by new combinations of conditions have been of profound importance in evolution. Also, as Dover (1986) argued, behaviour may be important if the dynamics of molecular events have, as side effects, produced phenotypic changes which are shared by most members of a population. As a result, while the organisms may no longer be suited to the environment in which their ancestors lived, they are now especially well suited to another environment—if they can get there.

The notion of genetic determination, so firmly embedded in evolutionary theory, has seriously interfered with attempts to understand the dynamics of behaviour and its development. If anything has been learned in recent years, it is that what an individual animal does in its life is conditional, and also dependent, on the reciprocal character of the transactions with the world about it. This knowledge also points to ways in which an animal's own behaviour can provide the variation which influences the subsequent course of evolution. When developmental issues are recoupled to questions about evolution, it becomes much easier to perceive how an organism's behaviour can initiate and shape evolutionary change. The interaction between ontogeny and phylogeny is commonly expressed in terms of developmental *constraints* (e.g. Maynard

Smith *et al.*, 1985). However, the active features of behaviour and the developmental capacities to modify them could have been the *instigators* of evolutionary change. If that is correct the metaphor of selection by an external agent has outlived its usefulness.

ACKNOWLEDGEMENTS

I am very grateful to George Barlow, Tim Caro, Tim Clutton-Brock, Gabby Dover, Jack Hailman, Robert Hinde, Mae-Wan Ho, Susan Oyama and James Serpell for their comments on an earlier version of this chapter.

REFERENCES

Baldwin, J.M. (1896). A new factor in evolution, *Amer. Natur.*, **30**, 441–51, 536–53.

Bateson, P.P.G. (1982). Behavioural development and evolutionary processes. In *Current Problems in Sociobiology* (eds King's College Sociobiology Group), Cambridge University Press, Cambridge, pp. 133–51.

Bateson, P.P.G. (ed) (1983a). *Mate Choice*, Cambridge University Press, Cambridge.

Bateson, P.P.G. (1983b). Genes, environment and the development of behaviour. In T.R. Halliday and P.J.B. Slater (eds), *Animal Behaviour*, Vol. 3. *Genes, Development and Learning*, Blackwell, Oxford, pp. 52–81.

Bateson, P.P.G. (1986). Sociobiology and human politics. In S. Rose and L. Appignanesi (eds), *Science and Beyond*, Blackwell, Oxford, pp. 79–99.

Bonner, J.T. (1980). *The Evolution of Culture in Animals*, Princeton University Press, Princeton, NJ.

Changeux, J.-P. (1983). *L'homme neuronal*, Fayard, Paris.

Darwin, C. (1871). *The Descent of Man, and Selection in Relation to Sex*, Murray, London.

Dawkins, R. (1982). *The Extended Phenotype*, Freeman, Oxford.

Dawkins, R. (1983). Universal Darwinism. In D.S. Bendall (ed.), *Evolution from Molecules to Men*, Cambridge University Press, Cambridge, pp. 403–25.

Dickinson, A. (1980). *Contemporary Animal Learning Theory*, Cambridge University Press, Cambridge.

Domjan, M. (1980). Ingestional aversion learning: unique and general processes. *Adv. Study Behav.* **11,** 275–336.

Dover, G.A. (1986). Molecular drive in multigene families: how biological novelties arise, spread and are assimilated, *Trends Genet.*, **2**, 159–65.

Elton, C. (1930). *Animal Ecology and Evolution*, Oxford University Press, Oxford.

Fisher, R.A. (1930) *The Genetical Theory of Natural Selection*, Oxford University Press, Oxford.

Gorman, M.L., Nedwell, D.B., and Smith, R.M. (1974). An analysis of the anal scent pockets of *Herpestes auropunctuatus* (Carnivora: Viverridae), *J. Zool.*, **172**, 389–99.

Hailman, J.P. (1982) Evolution and behavior: an iconoclastic view. In H.C. Plotkin (ed), *Learning, Development, and Culture*, Wiley, Chichester, pp. 205–54.

Hardy, A. (1965). *The Living Stream*, Collins, London.

Hinde, R.A. (1959). Behaviour and speciation in birds and lower vertebrates, *Biol. Revs.,* **34**, 85–128.

Ho, M.–W. (1984) Environment and heredity in development and evolution. In M.–W Ho and P.T. Saunders (eds), *Beyond Neo-Darwinism*, Academic Press, London, pp. 267–89.

Ho, M.–W., and Saunders, P.T. (1984). *Beyond Neo-Darwinism*, Academic Press, London.

Humphrey, N.K. (1976). The social function of intellect. In P.P.G. Bateson and R.A. Hinde (eds) and *Growing Points in Ethology*, Cambridge University Press, Cambridge, pp. 303–17.

Immelmann, K. (1984). The natural history of bird learning. In P. Marler and H.S. Terrace (eds), *The Biology of Learning*, Dahlem Konferenzen, Springer-Verlag, Berlin, pp. 271–88.

Jerison, J.H. (1973). *Evolution of the Brain and Intelligence*, Academic Press, New York.

Jolly, A. (1966). Lemur social behavior and primate intelligence, *Science*, **153**, 501–6.

Kelleher, R.T. (1966). Chaining and conditioned reinforcement. In W.K. Honig (ed.), Operant Behavior, Appleton-Century-Crofts, New York, pp. 160–212.

Kirkpatrick, M. (1982). Sexual selection and the evolution of female choice, *Evolution*, **36**, 1–12.

Krebs, J.R., and Davies, N.B. (1981). *An Introduction to Behavioural Ecology*, Blackwell, Oxford.

Lande, R. (1981). Models of speciation by sexual selection on polygenic traits, *Proc. Nat. Acad. Sci. USA*, **78**, 3721–5.

Lehrman, D.S. (1970). Semantic and conceptual issues in the nature–nurture problem. In L.R. Aronson, E. Tobach, D.S. Lehrman, and J.S. Rosenblatt (eds), *Development and Evolution of Behavior*, Freeman, San Francisco, pp. 17–52.

Lewontin, R.C. (1983). Gene, organism and environment. In D.S. Bendall (ed.), *Evolution from Molecules to Men*, Cambridge University Press, Cambridge, pp. 273–85.

Lloyd Morgan, C. (1896). On modification and variation, *Science*, **4**, 733–40.

Lorenz, K. (1965). *Evolution and Modification of Behavior*, University of Chicago Press, Chicago.

Maynard Smith, J., Burian, R., Kauffman, S., Alberch, P., Campbell, J., Goodwin, B., Lande, R., Raup, D., and Wolpert, L. (1985). Developmental constraints and evolution, *Quart. Rev. Biol.*, **60**, 265–87.

O'Donald, P. (1983). Sexual selection by female choice. In P. Bateson (ed.), *Mate Choice*, Cambridge University Press, Cambridge, pp. 53–66.

Osborn, H.F. (1896). Ontogenic and phylogenic variation, *Science*, **4**, 786–9.

Piaget, J. (1979). *Behaviour and Evolution*, Routledge & Kegan Paul, London.

Waddington, C.H. (1953). Genetic assimilation of an acquired character, *Evolution*, **7**, 118–26.

Waddington, C.H. (1957). *The Strategy of the Genes*, Allen & Unwin, London.

Waddington, C.H. (1959). Evolutionary systems — animal and human, *Nature*, **183**, 1634–8.

Weiskrantz, L., and Cowey, A. (1963). The aetiology of food reward in monkeys, *Animal Behaviour*, **11**, 225–34.

Wyles, J.S., Kunkel, J.G., and Wilson, A.C. (1983). Birds, behavior, and anatomical evolution, *Proc. Nat. Acad. Sci. USA*, **80**, 4394–7.

Evolutionary Processes and Metaphors
Edited by M.-W. Ho and S.W. Fox

CHAPTER 11

Metaphors and methods: behavioural ecology, panbiogeography and the evolving synthesis

RUSSELL D. GRAY

ABSTRACT

Recent debates in evolutionary theory have been characterized by rhetoric, dogma and a lack of communication between defenders and critics of the modern synthesis. Two causes are suggested to account for this situation. First, the belief that adherence to Darwinism involves an adherence to certain immutable positions, and second, the belief that the problems with the functional approach are primarily theoretical rather than methodological. The methodological problems of the functional approach, and their connection to the metaphor of adaptation, are outlined using optimal foraging theory as the paradigm case. An alternative metaphor, reciprocally constrained construction (RCC), is presented as a more appropriate model of organism–environment relations in development and evolution. The implications of RCC for our thinking about constraints on foraging behaviour, natural selection, heredity and the nature of species are discussed. The metaphor is used to guide the development of a new methodology for evolutionary studies. This methodology involves a process of reciprocal illumination between investigations of cascading ecological, developmental and evolutionary changes, and panbiogeographic analyses. The combination of ecological, behavioural, developmental and biogeographical studies is an attempt to create a truly modern synthesis that is both Darwinian and yet beyond Neo-Darwinism.

DARWINISM HAS NO ESSENCE

> Darwinian theory will not resolve to a single significance nor yield a single pattern. It is essentially multivalent. It renounces Descartian clarity, or univocality. Darwin's method of argument and the generative metaphors of *The Origin* lead . . . into profusion and extension. (Beer, 1985:9)

Traditionally, evolutionary biologists have set themselves at least two related tasks — to explain the diversity of life and to account for the apparent fit between organisms and their environments (Maynard Smith, 1978; Lewontin, 1983a). After a period of relative consensus since the advent and hardening of the modern synthesis (Gould, 1983), evolutionary biologists are currently engaged in a variety of intense debates on the nature, validity and sufficiency of current evolutionary theory (Craw and Gibbs, 1984; Ho and Saunders, 1984; Pollard, 1984; Brooks and Wiley, 1986). Unfortunately, during many of these debates defenders of the modern synthesis accuse their critics of attacking extreme, unrepresentative positions, while the critics respond that the defenders are dogmatic and myopic. The arguments seem to fall easily into old mutually reinforcing oppositions such as chance v. necessity, adaptation v. non-adaptation and selection v. constraint (see Gray, 1987a). This crude oppositional thinking leads to caricature arguments in which 'adaptationists' fight 'anti-adaptationists' and 'Darwinists' battle against 'anti-Darwinists'. More importantly it means that, while taking different sides, both defenders and critics accept the terms of the debate. In particular, it seems that both sides act as if Darwinism itself were at stake and thus if certain positions 'essential' to it are undermined then Darwinism falls. These essential positions are often posited to include a belief in the fundamental importance of natural selection, the ubiquity of adaptation, the central role of population genetics in evolutionary explanations, or perhaps the absence of 'Lamarckian', orthogenetic and saltational processes in evolution. And yet almost invariably criticism of the modern synthesis for some unpardonable sin (adaptationism, atomism, gradualism, reductionism, genetic determinism) leads to the familiar homily, 'You are attacking a strawman. We have always acknowledged non-adaptive phenomena / the unity of the organism / the possibility of rapid change / levels above the genome / the role of experience in development . . .'. Sadly it seems that beneath the rhetoric and the labels there is little communication and no synthesis.

I think that an alternative, more positive, attitude to the Darwinian research programme is possible. Darwinism, like a species, is an evolving entity — it has no essence. Thus the meaning of Darwinism can change and at any one time can be read in a variety of ways. A position that at one stage may seem essential can later be peripheral. (For a discussion of the idea that research programmes, like Darwinism, are what philosophers would term individuals rather than classes see Hull, 1984, 1985.) This evolutionary, anti-essentialist view of research programmes may seem to make the whole of science impossible. (Hull, 1984, notes that most biologists go white at the knuckles when this idea is suggested.) After all, if research programmes have no essence what is there to define, defend and criticize? Actually I think it might help to free us from the sterility of current disputes if biologists thought of research programmes in this way. If the modern synthesis is considered as an evolving entity then its defenders would not need to brand any critic as a crank or creationist — they could admit instead that Darwinism today might contain many contradictory elements and unresolved problems. It would not be the essence of evolutionary biology that was under attack but rather particular positions and theories. Those who see faults with current evolutionary theory could then focus on these specific problems, and perhaps explore new readings of Darwinism, rather than claiming the entire enterprise is fundamentally flawed. If Darwinism is an individual then it can evolve without risking extinction.

It is in this spirit of an evolving synthesis that I wish to examine some problems associated with the functional approach to evolutionary biology and then attempt to resolve these problems by outlining some alternative biological metaphors and methods.

FUNCTIONALISM AS A METHOD AND ADAPTATION AS A METAPHOR

Both defenders and critics of Darwinism frequently seem to treat the functional approach as if it were purely a theory. In contrast I think that evolutionary functionalism, like structuralism (see Piaget, 1968; Barthes, 1964; Broekman, 1974; Hammen, 1985), is best understood as an activity or method guided by certain metaphors. This abstract distinction between a theory and a method may seem to be of little practical relevance to the current debates in evolutionary biology. On the contrary I will argue that this distinction can be quite helpful in enabling us to understand the lack of communication between defenders and critics of the functional method. To make the discussion more specific I will use Optimal Foraging Theory (OFT) as an exemplar of the functional method.

The name optimal foraging theory is really a bit misleading. There is no one theory that *is* OFT. Instead OFT is a collection of theories that have been generated by a common approach, or set of procedures. OFT is better understood as a method rather than a theory (Heinrich, 1983).[1] The procedures for this method could be informally represented as follows:

1. Think of a 'problem' an animal might face while foraging (e.g. the need to maximize energy uptake). This is the maximization criterion.
2. Define the 'strategies' available to the forager (e.g. in a two prey type situation catch one or both prey types).
3. Define the 'constraints' in this system (e.g. handling time).
4. Use the above to formulate an explicit optimality model (e.g. the classical diet model — specialize on prey type 1 when

$$\frac{1}{\lambda_1} < \frac{E_1}{E_2} (h_2 - h_1)$$

where $\frac{1}{\lambda_1}$ is the mean time taken to search for the next type 1 prey item, E_1 and E_2 are the respective energetic returns for each prey type and h_1 and h_2 are the respective handling times).

5. Derive some predictions from this model.
6. Test these predictions.
7. Interpret these results and plan further research. If the predictions are confirmed the researcher typically presumes that he or she has correctly understood the selective 'pressures' that have shaped the study animal. If the predictions are not confirmed then more work and refinement of the model are called for. The maximization criterion could be inappropriate, the strategy set may be unrealistic, or perhaps further constraints should be added to the model. This approach is simple and elegant. Its procedures have given unity and order to behavioural ecologists' thoughts and actions.

The OFT methodology is nothing fundamentally new. It is simply a more explicit, quantitative version of the functional method. The functional method is, in turn, based on metaphor with very deep historical roots in Western culture. Adaptation, or the metaphorical fit of organism to environment, is a legacy from pre-Darwinian thought. Prior to the Darwinian revolution this fit was used as an argument for the existence of a creator. In today's version of Paley's 'argument from design' natural selection, rather than God, is seen as the designer. (It is with some justification therefore that Dawkins (1983) has suggested that biologists who use the functional method could be termed 'transformed Paleyists'.) The notion of adaptation has become so central and so ingrained in the thinking of current evolutionary biologists that its metaphorical nature is generally overlooked. And yet it is undoubtedly metaphorical to talk of a species adapting to a 'problem' posed by its environment. Central to the metaphor of adaptation is this image of a fixed external environment *to* which species are fitted (Lewontin, 1983b). Just as we design a key to fit a specific lock so we speak of natural selection adapting species *to* their niches. Metaphors such as adaptation can be quite helpful in enabling us to deal with complex situations. They can also become stale, overworked and unproductive. While the metaphor of adaptation has proved reasonably successful in the past, its limitations and negative consequences have recently been the subject of considerable debate. In the next section I will review some of these negative consequences and use the distinction between a theory and a method to clarify some of the miscommunication that has characterized these debates.

NEGATIVE CONSEQUENCES OF THE FUNCTIONAL METHOD

Panselectionism (in practice but not in theory)

It is common for critics of the functional approach to label its proponents as 'panselectionists'. The proponents frequently reply that they have always acknowledged the existence of non-selective phenomena (see Mayr, 1983). The proponents then feel quite happy to continue using the functional approach which, of course, makes the critics feel quite frustrated. Why is there this miscommunication? I think it may be because both sides are interpreting the problem as a theoretical question (do non-selective phenomena exist?) when it is, at root, a methodological issue. Since both sides agree on the existence of selective and non-selective phenomena the real question is, can the non-selective phenomena be incorporated into the functional method, in anything other than a trivial manner, without undermining the whole approach? My answer to this question is simply no. Rosenberg (1985) argues that extremal approaches, like the functional method, are committed to explaining everything in their domain. There is no scope for treating the functional approach as only a partial account of the behaviour of objects in its domain as its power derives from the fact that it can, in principle, be used to explain any biological phenomenon. As Krebs, Charnov and Houston (1981:4) note in a review of OFT, 'The widespread occurrence of maladaptive traits . . . would seriously undermine the rationale for optimality models . . . '. The problem is that there is no rigorous method currently available for distinguishing between the effects of developmental constraints, molecular drive, evolutionary lag and chance from those of natural selection. Current utility does not enable us to disentangle the direct products of natural selection from all these other phenomena. Because it is not possible to limit applications of the functional method to a domain of known validity, and because anything from the evolution of molecules to the evolution of human sexual habits (Weinrich, 1977) can be given a functional explanation, the continued use of the functional method will inevitably lead us to construct an inaccurate panselectionist picture of the world. The functional method forces researchers to be panselectionist in practice even if they are not in theory.

Atomism

Proponents of the functional approach are also frequently criticized for being excessively atomistic (see Gould and Lewontin, 1979; Webster and Goodwin, 1982; Taylor, 1987). Again, defenders of this approach reply that they have always acknowledged the interdependence of parts that characterize organisms. In an effort to prove this they point to a lengthy history of well-known 'functionalists' who have emphasized 'laws of growth' and 'correlation of parts'

(Darwin — see Ghiselin, 1969; Craw, 1984), developmental integration (Dobzhansky, 1956), and the 'unity of the genotype' (Mayr, 1975). Again the proponents of the functional method cannot understand what all the fuss is about. However, as the previous argument about panselectionism suggested, it is one thing to pay lipservice to a phenomenon, it is quite a different matter to incorporate the phenomenon rigorously into your method. Hailman (this volume) argues that the functional method requires this atomistic, piece by piece approach to be operationally tractable. For example, most behavioural ecologists would agree that the hierarchical and interactive nature of foraging choices is a general phenomenon (see Gray, 1987b), yet in OFT there exist separate theories for diet choice, patch choice, patch time allocation, patterns of movement and habitat selection. While occasionally a few problems may be pieced together they are normally considered separately and it is hoped that 'all other things will be equal'. This means that the integration and spatio-temporal organization of behaviour is virtually ignored within OFT.[2] OFT is just one example of the general manner in which atomism and the functional method go hand in hand. One does not speak of the function or functioning of a whole but rather we attribute functions to parts. As a consequence of using the functional method, an image of organisms is constructed as merely mechanical aggregates of parts, each of which is a solution to a particular environmental 'problem'.

The separation of organism and environment

From the functionalist perspective, the environment is separate and ontologically prior to organisms (Lewontin, 1983b, c). It poses 'problems' to which populations of organisms must adapt. Thus the environment is pictured as the driving force of evolution and organisms are rendered passive *objects* shaped by these autonomous environmental *forces*. Several commentators have remarked on the similarity between this view of evolution and the Newtonian view of motion (Sober, 1984a 1985; Brooks, Le Blond and Cummings, 1984; Ho, 1987). Just as in Newtonian physics bodies are considered to maintain their state of motion unless an external force acts on them, so in current biology populations are seen as being at equilibrium unless perturbed by an external agent like selection.

The separation of organism and environment is not just an abstract problem of little consequence for our empirical endeavours. It leads to major difficulties when we attempt to construct and test realistic theories. In classical optimal diet theory prey items are given fixed values and ranked in a linear order independently of the context in which they occur. Thus the properties of the environment (e.g. the 'value' of a diet type) do not vary with the properties of the organism (a set of 'strategies') and vice versa. Furthermore, both are ripped from their ecological and developmental context despite the wide range of evidence from the inter-ramification of foraging organisms and their environment. Prey items cannot be realistically assigned 'intrinsic values'. Their value is contingent upon the familiarity of the forager with its prey. The discrimination of prey types, the success of the forager at capturing prey and their subsequent handling time and digestibility may all vary with experience (see Gray, 1987b). The value of a prey item is also contingent upon other prey in the diet (Rapport, 1981). The developmental and ecological embeddedness of foraging choices cannot be easily incorporated into OFT as it causes violation of the transitivity assumption that is an essential requirement for the description of any system in optimization terms (McCleery, 1978).

Behavioural ecology could perhaps advance beyond these problems by mining some relatively unexplored regions of Darwin's texts. Despite assertions to the contrary (e.g. Patten, 1982; Lewontin, 1983b), Darwinian biology is not necessarily based on the separation of organism and environment. For Darwin the organism took an active role in its interactions with its environment and in determining the course of its species future evolution (Ghiselin, 1986). Darwin's work on the formation of coral reefs (1842) and his brilliant experimental studies on earthworms (1881) are paradigm examples of the reciprocal interaction between organisms and their environments. The metaphor of the 'entangled bank' with which Darwin closed *The Origin* superbly evokes the flux, mutual control and interdependency that characterize organisms' webs of changing ecological interactions. I shall return to this theme later.

Do development and evolution stop?

The idea that organisms fit their environment is a very static conception of organism–environment relations. For pre-evolutionary biologists this was no problem, but for modern evolutionists this causes some difficulty, because it is very difficult to allow for evolutionary lag within the functional method. Thus a picture of organisms as adapted rather than as adapting is created (Lewontin, 1981). The functional method also tames changes in a deeper manner. To deem something an adaptation it is necessary to make a comparative judgement (Hinde, 1975). A one-legged person may be deemed maladapted in comparison to a two-legged person, but adapted when compared to a person without legs. Thus implicit in the idea of adaptation is a fixed set of possibilities within which comparisons can be made. In OFT this is explicitly recognized as the 'strategy set'. Clearly, however, the strategy set can change in both development and evolution. Even under tightly controlled laboratory conditions, where researchers attempt to dictate the available 'strategies', animals can create an existence in the most unpredictable ways (see Gray, 1987b). The essentially innovative nature of development and evolution frustrates our efforts to specify a fixed point for comparative judgements of adaptation.

'Whig biology' and evolutionary determinism

The fixed strategy set assumption and the emphasis on current utility lead to a simple deterministic view of history. History is seen as a linear unfolding of events inexorably to the current optimal state. In the study of human history, explanations of this kind are pejoratively labelled as 'Whig history'. Lewontin (1980) and Taylor (1987) suggest that functional explanations produce a similar kind of 'Whig biology'.

Understanding the Whig nature of the functional method sheds some light on the highly charged arguments about the alleged deterministic nature of sociobiology. In response to the allegation of genetic determinism defenders of sociobiology (e.g. Dawkins, 1982; Bateson, 1982) have argued that sociobiology does not require rigid determination of the phenotype by the genotype. All that is required is the possibility of heritable differences in behaviour. However, both the initial charge and the responses are wide of the mark. It is not developmental determinism that is the fundamental problem but rather evolutionary determinism. While individual development may vary, such *deviations* from the optimal state will, presumably, lead to decreased reproductive success. Animals that are less efficient in choosing a diet than their competitors will be selected against and so the population will evolve towards optimal patterns of diet selection. Thus the present state of affairs is seen as inevitable, not because it is genetically determined but because it is optimal in an evolutionary sense. The crude, deterministic nature of the functional method creates an impoverished and distorted view of history. The ecological and developmental contingency of evolutionary transformations disappears as the dynamics of evolutionary change are reduced to the change of gene frequencies under the unitary force of natural selection (Taylor, 1987). Evolutionary history is far more subtle and capricious than current utility would suggest.

Genes, individuals and groups — is there a fundamental level?

A further consequence of the functional method that I wish briefly to discuss concerns a form of methodological reductionism implicit in the functional approach. In selectionist explanations phenomena are explained in terms of the level on which selection is presumed to act. In OFT, where the concern is generally with fitness rather than inclusive fitness arguments, foraging behaviour is explained in terms of individual advantages as the current consensus suggests that group selection is unlikely to be common. Thus although foraging is a social activity for many species, group behaviour is explained in terms of advantages occurring to individuals. For instance Pulliam and Caraco (1984) model group size and habitat choice as an epiphenomenon of individual decisions. Similarly, genic selectionism renders the gene as the fundamental level of explanation (Dawkins, 1982). The functional method, at least as it is currently formulated, requires a commitment to one level of analysis as ontologically prior to others.

Maladaptive canalization of the functional method

Perhaps the biggest practical problem with the functional method is its ability to buffer itself against change. This can be seen very clearly with OFT. Most reviews of OFT create the impression that its predictions have been overwhelmingly confirmed (see Schoener, 1987). OFT is presented as *the* success story of behavioural ecology. However, when the tests of specific optimal foraging models are examined in a detailed, systematic manner the mass of supposedly confirmatory evidence evaporates (see Pyke, 1984; Gray, 1987b). Not only have few of the predictions of optimal foraging models been consistently confirmed, the predictions that have been confirmed are often rather trivial and previously well known. The failure of OFT has not led its proponents to search for other ways of gaining insight into the diversity of life. When things go wrong the prescription is simply to keep doing more of the same kind of research (Pyke, 1984) and so the approach is protected by enumerable *ad hoc* hypotheses. While it is true that ' . . . today's *ad hoc* hypothesis may well be tomorrow's law of nature' (Hull, 1985: 806), this is no justification for permanently immunizing the approach from contradictory evidence. Ultimately, whether such a research programme is seen as degenerate or progressive depends on whether one thinks other approaches might be at all possible. So ingrained is the functionalist habit, and so seductive are adaptive stories, that for most biologists there is no alternative. Barash epitomizes this religious conviction; ' . . . the central principle of sociobiology [optimality] is better than nothing' (Barash 1982:55). As long as the functional method can only be contrasted with random research it will remain a popular and highly canalized approach (see Jamieson, 1986). The challenge is, therefore, to generate new metaphors and methods.

NEW METAPHORS FOR DEVELOPMENT AND EVOLUTION

Biologists who have spent years patiently observing animals in the field tend to be very impatient with the 'theoretical' arguments over adaptation. They retort that one only has to observe the wonders of mimicry to be convinced of the reality of adaptation. However, what is self-evident to some may, with subsequent investigation, turn out to be more complicated than previously thought. In the following sections I will outline an alternative to the metaphor of adaptation.

Reciprocity and constraint in development

There was a major advance in our thinking about development when Kuo, Schneirla, Lehrman, Hailman, Gottlieb and others questioned what had previously seemed obvious — the idea of instinct. They helped us to reformulate our thinking so that many (although not all) behavioural biologists now realize that it is impossible to attribute phenotypic characters to either the genome or the environment. Instead, internal (genetic) and external factors are co-determined and co-act. Thus current approaches to behavioural development stress what Bateson (1986) has termed the 'truly conditional' character of development. By this he means that the effects of a genetic or environmental factor will be conditional on the context in which they occur and the time at which they act. The reciprocal, conditional nature of development is only crudely conveyed by the static and passive 'norms of reaction' concept (Ho, 1987; cf. Lewontin, 1983b). A more realistic view of development would incorporate the manner in which organism and environment are co-implicative, co-defining and co-constructing.

It is impossible for an organism to exist without an environment. Similarly, the word environment implies that there must be something to be environed (Shaw and Turvey, 1981). In this sense organism and environment are co-implicative. It could be claimed that while organism and environment are indeed co-implicative they can still be defined separately. For example, an organism's environment could be specified in terms of its geographical position and the physicochemical and biological properties of that locality. Several researchers have, however, argued that such a Newtonian, organism-independent description is not as biologically meaningful as an organism-referent description (Shaw and Turvey, 1981; Lewontin, 1983b). Shaw and Turvey argue that one problem with an organism-independent description of the environment is that the environment can be described at an infinite number of space/time scales. In contrast, from the organism-referent point of view, the organism defines the scale at which its environment can be most meaningfully specified. Thus for a bacterium in a pond Brownian motion is a major environmental feature and gravity is not (Lewontin, 1983b), while for a heron in the same pond quite the opposite is the case. Similarly, any meaningful description of environmental resources and information must be implicitly organism-referent. What are resources to one organism may be irrelevant to another. Any statement about resources limiting a particular population must be contingent upon a given range of actions by that population. The information contained in the flow of ambient light or in the vibration of air molecules will also be contingent upon the organism perceiving that information. Not only are different organisms sensitive to different frequencies, they pick up different information from the same physical stimulus. Light reflected from a tree may afford a potential food source to one organism, shelter to another, and a good position to sing from for another. (For a fuller argument for a molar, organism-referent description of environments in terms of their affordance structure see Gibson, 1979; Shaw and Turvey, 1981.)

If environments are best defined by reference to an organism then reciprocally an organism's activities should be described in relation to their environmental context. Students of animal behaviour are increasingly recognizing that descriptions of behaviour which incorporate the particular environmental context are more informative than acontextual descriptions (see Gordon, 1985).

So far I have sketched the co-implicative, co-defining nature of organisms and their environments. From a developmental perspective the most important aspect of organisms and their environments is their mutual construction. Neither organism nor environment pre-exist their relationship. Instead they are constructed in a mutually constraining fashion through ontogeny. Students of development are familiar with the fact that the environment experienced by an organism plays an essential role in the construction of its phenotype (Gottlieb, 1976; Bateson, 1983). What is equally obvious, but perhaps less frequently emphasized, is that the organism selects and modifies (constructs) its own environment based on the materials that are available. This construction may range from the dramatic building of nests, hives and dams (Collias and Collias, 1976; Hansell, 1984) to more subtle forms of external construction such as the release of allelochemicals.

A useful metaphor to summarize this view of development, where internal and external factors are co-defined and co-constructing, is 'reciprocally constrained construction' (Gray, 1987b). I should emphasize that the kind of construction intended in this metaphor is not that which is guided by either a genetic blueprint or programme. That type of construction cannot be truly reciprocally constrained as it gives genes prior ontological status (see Oyama, 1985). The kind of construction I envisage is a process of self-assembly or self-organization where the term 'self' refers to the organism–environment system. Stent (1981) makes a similar point when he notes that development is analogous to an idealized process of ecological succession. Ecological succession proceeds in an orderly manner due to the regular sequence of interactions between organisms and their environments. Ecological succession does not require an ecological blueprint or programme. The absence of an underlying programme does not mean that the process of reciprocal construction is free to proceed in any direction. It is constrained by the current state of the organism (which is a product of past organism–environment transactions) and its current environmental context.

Reciprocity and constraint in evolution

I guess that many biologists who study behavioural development might accept that reciprocally constrained construction is an appropriate metaphor for the ontogeny of organism–environment relations. What I wish to suggest is that this metaphor can be applied to evolution as well as to development. Implicit in the metaphor of reciprocally constrained construction is a rather different view of the evolutionary process from that of adaptation. Instead of seeing populations of organisms as climbing pre-existing adaptive peaks, their evolutionary dynamics are more akin to life on an adaptive trampoline (Lewontin, 1983c). Organisms participate as active subjects in their own development and, therefore, in the evolution of their population. They define and construct their evolutionary landscape. I should hasten to add that it would be quite inaccurate to interpret this idea as 'anti-adaptationist'. There *is* a relationship between organism and environment but this relationship is more reciprocal and dynamic than that implicit in the metaphor of adaptation.

In reaction to the excessive environmentalism of functional explanations it has become fashionable to talk of 'internal' developmental or phylogenetic constraints on evolution. Regardless of whether these constraints are seen as structural (Webster and Goodwin, 1982), or historical (Lauder, 1982; Grehan and Ainsworth, 1985; Maynard Smith *et al.*, 1985; Brooks and Wiley, 1986), they are not what I intend by suggesting that development and evolution can be viewed as a process of reciprocally constrained construction. If development and evolution really are processes of reciprocally constrained construction then the locus of constraint is not fixed in the organism, nor in the environment, but rather it develops in the fluid, contingent relationship between the two (see Gray, 1987a and Oyama, this volume). From this perspective constraints do not pre-exist development processes: they are constructed by them. Constraints are not transmitted preformed from generation to generation but must be reconstructed each life cycle. Furthermore, as these constraints are not fixed in any entity but rather are contingent on a network of relationships in the organism–environment system they are not universally binding (cf. Webster and Goodwin's, 1982, resurrection of 'laws of form'). To paraphrase Ho (1987), organisms are no more manacled by internal constraints than they are pushed and pulled by environmental forces.

IMPLICATIONS OF RECIPROCALLY CONSTRAINED CONSTRUCTION

If proponents of new theories, metaphors or research programmes wish to communicate their views to a wide audience they must illustrate the manner in which these views shed new light on old problems and open up new questions. In the following sections I will discuss the implications of the metaphor of reciprocally constrained construction for our thinking about constraints on foraging behaviour, natural selection, heredity and the nature of species.

TABLE 1 Foraging and adaptive trampolines—examples of the dynamic interactions between environmental, morphological and physiological variables in feeding

Variables	Authors
Foot movements improved prey availability	Rand (1956), Meyerriecks (1959), Tinbergen (1962), Heather (1977 and refs therein), Pienkowski (1983), Moore (1984)
Search rate varied with the crypticity of the prey	Gendron and Staddon (1983)
Prey capture success varied with experience	Lawton, Beddington and Bonser (1974), Polsky (1977), Caro (1980a,b), Brandt (1984 and refs. therein)
Handling time changed as a function of deprivation	Smith and Dawkins (1971), Werner (1974), Zach and Falls (1978), Krebs (1980)
Digestive enzymes changed in response to different nutrients or toxins in the diet	Grossman, Greengard and Ivy (1943), Brattsten, Wilkinson and Eisner (1977), Ahmad (1983), Terrier (1984 and refs. therein)
Micro-organisims in the gut changed in response to dietary changes	Smith (1965), Abe and Iriki (1978)
Gut length varied with food quality	Miller (1975), Kenwood and Sibly (1977), Al-Jaborae (1979)
Cranial and jaw musculoskeletal morphology changed as a consequence of different diets	Greenwood (1965), Moore (1965), Bouvier and Hylander (1981), Beecher, Corruccini and Freeman (1983)
Predators induced predator-resistant forms of their prey	Gilbert (1966), Haukioja and Niemela (1977), Grant and Bailey (1981), Krueger and Dodson (1981), Strong, Lawton and Southwood (1984), Edwards and Wratten (1985), Fowler and Lawton (1985), Lively (1986)

Reciprocity and constraint in foraging behaviour

One of the questions posed at the conference from which this book is derived was, 'Do morphology, physiology or experience constrain behaviour?' If morphology, physiology and experience are considered separately and taken as given then they could be viewed as constraining behaviour. Such an approach would reinforce the separation of organism and environment. Instead the metaphor of reciprocally constrained construction implies that each of these factors must actually be constructed through ontogeny and evolution in a manner contingent upon the other variables in the organism–environment

Figure 1 Examples of the way in which herons can modify their local feeding environment. (a) White-faced heron foot raking. (Redrawn from S. Morris and G. Moon, 1985, *The White-faced Heron*, Heinemann, Auckland.) (b) The reddish egret using the 'underwing' feeding method. (Redrawn from an illustration by Guy Tudor, *Natural History*, vol. 71, p. 59)

(c) The black heron foraging beneath a full canopy. Redrawn from J. Hancock and H. Elliot, 1978, *The Herons of the World*, London Editions, London

system. Let me illustrate this reciprocal construction by examining the morphology, physiology and experience associated with foraging behaviour.

The size, shape and locomotory method of an animal will considerably limit its relations with the environment (Kuo, 1967). The size of an organism will determine its perception of the grain size of the environment. Size and the forager's locomotory method will also affect the rate at which it can move through the environment, and hence its perception of prey density. There may also be qualitative as well as these quantitative changes in the forager's perception of its environment. At slower search speeds a forager may be more likely to detect cryptic prey (Gendron and Staddon, 1983). Thus a heron moving rapidly through an estuarine pond may only detect crabs and 'large' fish, whereas by stalking slowly through the pond it could detect stationary flounder and shrimps. Morphology can also function to determine the mechanical limits of what can be eaten. In fish the mouth position, size and protrusibility, the form of the teeth and gill raker spacing place an upper and lower limit on the size of prey that can be consumed (Chao and Musick, 1977; Wankowski, 1979; Stoner and Livingston, 1984). However, the essential point to recognize is that morphology itself must be constructed. Cranial and jaw musculoskeletal morphology and gut length can be changed considerably as a consequence of different diets (see Table 1). Similarly the environment is not given but must be selected and modified (see Waddington, 1975:55–8). One example of this construction is the manner in which some bird species use foot and wing movements to improve prey availability (see Table 1 and Figures 1a, 1b, 1c). Experience can affect habitat selection (see Gluck, 1984 and refs therein), digestive physiology (see Table 1), diet choice (Rabinowitch, 1965; Kuo, 1967; Jermy, Hanson and Dethier, 1968; Burghardt, 1970; Arnold and Maller, 1977; Cassidy, 1978; Luthardt-Laimer and Roth, 1983), feeding methods (Padilla, 1935; Ewer, 1968; Norton-Griffiths, 1968), and the degree to which a predator will be a specialist or generalist (L. Gray, 1981).

Organism and environment and, hence, morphology, physiology and experience are all part of the same *dynamic* system of feedback, facilitation and constraint, where behaviour at any one level is contingent on both the levels below and the levels above. Foraging morphology, physiology and experience are not isolated, idiosyncratic examples of such reciprocal effects. Similar phenomena can be seen in the classic studies by Lehrman and his coworkers on the reciprocal effects of nest building, social interactions and endocrine physiology in the reproductive cycle of the ring dove (summarized in Lehrman, 1965).

Does nature *select*?

The importance of the analogy between artificial and natural selection in both the development of Darwin's thought and in the presentation of his argument for natural selection, has been emphasized by many Darwinian scholars (Ghiselin, 1969; Young, 1971; Ruse, 1975; Evans, 1984). Young's paper is particularly interesting because it examines the debate among Darwin's contemporaries over whether selection was really an appropriate metaphor. (Both Darwin and Wallace realized that selection was a metaphorical term — see Young, 1971.) While this debate was primarily centred on the anthropomorphic and volitional connotations of Darwin's metaphor, I think it can still be useful today to pose the question that forms the title of Young's paper; 'Does nature select?'

The idea of selection conjures up an image of a choice (which does not need to be conscious or teleological) made from a set of possibilities. The factors generating the set of possibilities and the agent of selection are quite distinct. Thus, implicit in the metaphor of natural selection is a clear separation between the generation of variation and the selection of that variation, that is between organism and environment (see Ho, 1987). Darwin was quite explicit on this point; 'Every breeder knows that he does not produce the modification he selects' (in a letter to Hooker, cited in Young, 1971: 476). While this clear separation between the generation and selection of variation may apply to artificial selection, is it a valid way to view the natural situation? If the environment helps construct organisms through ontogeny and if organisms help construct their environment then this dichotomy does not seem realistic.

Imposing the metaphor of selection on organism–environment relations renders organisms passive *objects* whose *properties* are shaped by autonomous environmental *forces* (for two very clear accounts of the important role that the concepts of property and force have in selectionist explanations see Sober, 1984a, b). In selectionist accounts of evolution organisms do not push back. Of course, I am not suggesting that differential survival and reproduction of variants does not take place, but rather, that the dynamics of these changes are different from those implicit in the metaphor of natural selection. A more adequate view of evolution might attempt to accommodate what Oyama (1985) terms the essence of evolution — characters, genes and environment all changing together. In other words a synthesis of transformational and variational accounts of evolution (see Lewontin, 1983a; Oyama, this volume).

Heredity as an ecological process: synthesizing development and heredity

In his fascinating account of anti-Darwinian evolutionary theories around the turn of the century Bowler (1983) argues that one of the reasons why the Larmarckians were defeated by the Mendelians was their adherence to embryology and embryological explanations of evolution. The Mendelians' belief in autonomous hereditary units and their complete separation of developmental and evolutionary explanations were necessary, at this point in the development of evolutionary theory, to free biology from the transformational (developmental) model of evolution (Lewontin, 1983b). Abandoning developmental explanations and viewing evolution as simply a change in gene frequencies has been a productive research strategy since this time. However, eventually such simplifications become barriers to further progress. The view that hereditary information pre-exists developmental processes and is tied to particular entities (e.g. genes, chromosomes, memes) has recently been challenged as implicitly preformationist (Oyama, 1985). In contrast, Oyama argues that from a relational, constructionist perspective these entities do not possess any semantically meaningful information independent of the developmental processes in which they are embedded. Instead, information has an ontogeny — it must be constructed in a reciprocally constrained fashion from the set of developmental interactants. The constructionist view of heredity shifts our attention from hereditary products (genes, etc.) to the processes responsible for the generation of similar forms, and thus has been labelled 'process view of heredity' (see Ho, 1987 and this volume; Oyama, this volume).

The process view of heredity does not respect the traditional divide between organism and environment. As development is a process of reciprocal construction between organism and environment it is the relative constancy of a wide range of factors in the organism–environment system that explains the repeated construction of similar forms. A lot more is 'passed on' than just genes (see Table 2; cf. Williams, 1985). Of course, just like genetic information such 'extra-genetic' information is conditional on the activities of the organism and

TABLE 2 'Extra-genetic' inheritance

Type of inheritance	Reference
Physical properties of the environment, e.g. temperature, humidity, turbulence and photoperiod	
Cytoplasmic factors	Cohen (1979), Ho (1984)
Chemical traces from parental foraging	Galef and Henderson (1972), Corbet (1985)
Gut micro-organisms	Mattson (1980), Jones, Aldrich and Blum (1981), Jones (1983)
Social traditions	
Feeding methods	Fisher and Hinde (1949), Norton-Griffiths (1968), Galef (1985)
Migration routes and schooling sites	Van Denburgh (1914), Helfman and Schultz (1984)
Home ranges and territories	Neal (1948), Carrick (1963), Jolly (1972), Woolfenden and Fitzpatrick (1978)
Reproductive sites	Harris and Murie (1984)

its eco-development context. A change in any of the factors listed in Table 2 in space/time can cause a change in the distribution of phenotypes in a population (i.e. evolution will take place). To regard only genetic changes as evolution is to give genes prior ontological status over other sources of information, and so return to the distinction that is the root cause of the nature–nurture dispute (see Oyama, 1985). Furthermore, it is to ignore what evolutionary biologists are really attempting to explain. As Lewontin remarked,

> Population geneticists, in their enthusiasm to deal with the changes in genotype frequencies that underlie evolutionary changes, have often forgotten that what are ultimately to be explained are the myriad and subtle changes in size, shape, behavior, and interactions with other species that constitute the real stuff of evolution. (Lewontin, 1974, p.19)

No doubt some biologists would claim that as these extra-genetic modifications are easily reversible they must ultimately become fixed in the genome to constitute a major factor in evolution. There is, however, no *a priori* reason to assume that certain extra-genetic changes are more reversible than genetic changes (Johnston and Gottlieb, ms.). Back-mutations, the random elimination of alleles, a shift in developmental canalization and a change in 'selection pressures' could all result in the reversal of the effects of genetic changes. Extra-genetic changes such as the shift from a marine to a terrestrial environment or a change in host plant type may persist for millions of years. The effects of 'sensitive periods' to environmental stimulation, frequently found in early ontogeny, could establish relatively irreversible changes in habit or habitat without the necessity of genetic changes. In reality, of course, it is most likely that genetic and extra-genetic changes will occur together. However, the point remains that extra-genetic changes can persist for relatively long periods of evolutionary time. This is important as it frees proponents of developmentally based accounts of evolution from having to demonstrate that developmental changes are 'genetically assimilated' or that Weismann's barrier is violated. This was the trap that the Lamarckians fell into when they were confronted by the Mendelians (see Bowler, 1983). It was a trap not only because the Lamarckians were largely unsuccessful in convincing biologists that they had demonstrated such effects, but because it endorses the orthodox premise that evolution really is just a change in genotype frequencies. The ecological, process view of heredity frees evolutionary biologists from this trap and enables us to begin synthesizing developmental and evolutionary explanations.

Species as individuals: the importance of ecology and biogeography

Two contrasting views of biological species have recently been the subject of considerable debate among systematists and philosophers. Proponents of the first view argue that biological species are classes or natural kinds like the element gold. (For a defence of this position see Webster, 1984). Just as gold has a set of unique properties (such as its atomic number), regardless of the specific history of a particular piece of gold, so it is argued that each biological species has unique properties that can be used to define that species regardless of its actual evolutionary history or current context. Proponents of the second view argue that as species evolve through space/time they cannot have universal defining properties or essences (see Ghiselin, 1974, 1981; Hull, 1976, 1978). Instead they are spatio-temporally bounded historical entities or individuals.[3] The species-as-individuals thesis is more in keeping with the arguments I have outlined for the reciprocal construction of organism and environment. Viewing species as individuals does not presuppose that there is a fixed essence or natural phenotype that is produced independently of the environment (see Sober, 1980, 1984b). It is consistent with the context-dependent nature of development.

The arguments over the ontological status of species may seem to make little difference to the actual research methods of systematists. In fact I think that, in combination with the view of development and evolution I have outlined above, the species-as-individuals thesis implies an important reorientation of systematic methods. While many systematists have used biogeographic and ecological data for species and higher level systematics (see Hennig, 1966; Kruckeberg, 1969), such 'extrinsic' data (termed documents by Dupuis, 1984) are often regarded as less informative than the so-called 'intrinsic' attributes of a species. In many ways this is just a hangover from the essentialist concept of species where some of the 'intrinsic' attributes of species were seen as fixed, and were therefore necessary and sufficient to define the species, while its external relations were fluid and hence less informative. The species-as-individuals thesis denies the validity of this distinction. No characters, be they biochemical, morphological, ecological, or geographical are necessary and sufficient to define a species, as species are by nature variable and varying. Moreover, the allegedly 'intrinsic' attributes of a species are not derived independently of the species' ecological context. If development and evolution are really reciprocally constrained both 'intrinsic' and 'extrinsic' characters should be given equal *a priori* diagnostic weight in systematics.

Recent developments in systematics seem in keeping with this argument. Not only is it being discovered that ecological and biogeographic data are frequently congruent with morphological analysis (see Croizat, 1964; Schuh and Polhemus, 1980; Mitter and Brooks, 1983; Brooks, 1985) but, by using panbiogeographic methods, it has also been demonstrated that biogeography can generate novel phylogenetic hypotheses and tests between established ones (Page, 1986; Craw, 1987). Craw (1987) even goes so far as to argue that a natural classification can only be erected *after* a process of reciprocal illumination between biogeographic and form analyses. In the future ecological data may be increasingly incorporated into this process.

NEW METHODS?

Biologists are, on the whole, very practical people. In the following sections I will propose that investigations based on the idea of 'ecological and evolutionary cascades', when tied to biogeographic and systematic studies, can help us translate the abstract metaphor of reciprocally constrained construction into productive empirical research.

Ecological and evolutionary cascades: evolution as an autocatalytic process

A recurrent problem *within* Darwinism is the contradiction between the atomistic view of organisms constructed by the functional method and the abundant evidence that they are actually integrated systems. Emphasizing the integration of organisms makes it difficult to see how evolution can take place (everything is bound together too tightly). Emphasizing the independence of parts is at odds with our experience. Lewontin (1978), while recognizing this dilemma, argues that since adaptation is 'obvious', quasi-independent variation just must exist. There is, however, an alternative possibility — integrated, self-organizing change.

The idea that small changes in developmental processes can be amplified to produce large phenotypic effects is reasonably familiar to students of development (Lovtrup, 1976; Bateson, 1984; Hinde and Bateson, 1984). Analogous effects could occur in the evolution of populations. Small changes, be they behavioural, physiological, or morphological could trigger large changes in the ecological and internal relations of a population which, in turn, could feed back to cause further changes. Alternatively, environmental changes could cause organismal changes which lead to more changes in the population's ecological relations. For example, a change in geology might produce a change in climate which could cause a change in prey availability which could lead to habitat and diet shifts which could cause changes in morphology and physiology. These changes could then open up further ecological opportunities. Initially, at least, none of these changes need involve genetic modifications (e.g. the changes in habitat and diet could be perpetuated by cultural means).

This idea of 'Ecological and Evolutionary Cascades' (EECs) is not entirely speculative. For example, the dietary shift by the Koshima troop of Japanese macaques to digging for peanuts buried on the beach, led to juveniles bathing, swimming and even diving for seaweed. One individual swam to a nearby island (Kawai, 1965; Tsumori, Kawai and Motoyoshi, 1965; Tsumori, 1967). By a small extension in dietary habits the troop had grated an additional way of life on to their previous mode. They were on the borderline of becoming partially marine organisms. Not only would this open up a whole new set of ecological opportunities it would also expose the troop to new series of physical pressures (i.e. the different mechanical requirements of swimming). The differential survival and reproduction of variants better able to cope with these pressures may reinforce the previous changes (see Bateson, this volume).

Lenton (1983, 1984) documents an interesting case where changes in a populations's ecology interacted with its members' life-history tactics to facilitate further evolutionary modifications. The recent expansion of oil plantations in Malaysia has produced a plague of rats. The barn owl, *Tyto alba*, which is normally a generalist feeder that hunts by flying over open areas, switched to being a 'perch and wait' predator specializing on the rats found in the forest-like plantations. As a consequence of the new glut of food its numbers have dramatically increased. The Malayan barn owls now lay more eggs than conspecifics found elsewhere in the world, and have two or three clutches a year instead of one. The rat population explosion has not just changed the owls' life history, it has also led to new social habits. The Malayan barn owls defend less rigid territories and the juveniles form roosts of up to 40 owls after foraging each night.

Jamieson and Craig (1987) offer a similar explanation for the evolution of 'helping' behaviour in populations of the Florida scrub jay. They note that in contrast to populations of the Western scrub jay, which do not show 'helping' behaviour, the Florida scrub jays live in an area with high population densities, reduced dispersal possibilities, and few vacant territories. Jamieson and Craig argue that these habitat differences mean that non-breeding birds in the Florida population are less able to disperse and hence much more likely to encounter and interact with begging nestlings. They claim that this stimulation induces these non-breeding birds to help the parents provide food for the chicks. Thus a change in habitat may have caused a change in life-history trajectory which has led to a change in social organization.

These cascading changes need not be initiated by environmental changes. Let us imagine that Slipjer's (1942) bipedal goat was the result of a mutation. As the goat developed it was able to accommodate this drastic morphological change by modifying its spine, bones and muscle insertions in a manner appropriate for bipedal locomotion. It is possible that in the wild the bipedal goat may have adopted new food types as a consequence of its different mode of locomotion. Shifts to new food types can be facilitated by physiological accommodations, such as the induction of detoxication enzymes (see Terrier, 1984), and their perpetuation across generations can be aided by both the relatively irreversible effects of early foraging experience and a variety of extra-genetic mechanisms (see Table 2).

Behavioural innovations such as the selection of new habitats, the adoption of new food types, the production of new calls or songs, and changes in mating behaviour could all play a major role in driving EECs. Wyles, Kunkel and Wilson (1983) advance a similar argument to account for the apparently rapid rate of anatomical evolution in birds. Unfortunately, their reification of taxonomic rank and their arbitrary measures of morphological distance mean that their data do not convincingly support this argument. However, the suggestion of Wyles *et al.* that evolution is an autocatalytic process, rather than a purely selectional one, provides a good description of the manner in which EECs may proceed.

In conclusion, the most important feature to note about EECs is that the interdependence of physiology, morphology, behaviour, life-history trajectories and ecology is not an obstacle to evolutionary change. The systems of feedback that integrate organisms and their environments can facilitate further change rather than impede it. In other words negative feedback can become positive feedback.

There might seem to be an obvious problem with the suggestion that these kinds of cascading effects are common in evolution — there are not lots of examples of EECs. I think this might simply be due to the fact that we have not conducted our research in the right way to discover EECs. Certainly such effects will not show up in population genetic models or experiments. They will also be invisible if we content ourselves with attempting to demonstrate the fitness advantages associated with a particular character or testing the predictions of optimal design models. What is required is research that actually sets out to document EECs and attempts to understand the rules they follow.

Some experiments[3] in this vein might involve an investigation of the following:

1. The ecological consequences of genetic or experimental changes in morphology and locomotory method, including the subsequent changes in experience that might result from these modifications.
2. The morphological, physiological and behavioural consequences of changes in the nutritional or mechanical properties of the diet.
3. The morphological, physiological and behavioural consequences of changes in humidity, salinity, temperature and turbulence. (These variables are known to induce these types of change in laboratory populations which parallel natural variations. For reviews of this evidence see Robson and Richards, 1936; Rosen and Buth, 1980; Ho, 1984.)
4. The effects of new food gathering modes, e.g. digging or swimming.
5. Controlled reciprocal transfer experiments both within a species and between closely related species. Changes in physiology, morphology, life history, behaviour and ecology resulting from the shift in geographical locality would be monitored and compared.
6. The range of factors (e.g. genes, transgenerational chemical cues, cultural traditions) that can perpetuate the effects of EECs and the interplay between these factors (see Bateson, this volume).

Panbiogeography

Many evolutionary biologists are placed in a dilemma by the predominantly theoretical criticisms of current evolutionary theory. If they accept these criticisms they are forced into either a methodological vacuum or impractical compromise solutions. Examining cascading ecological and developmental changes and the way in which they are perpetuated is part of an alternative methodology but it is far from a complete method for investigating the richness of the evolutionary process. What is needed is a method that ties actual evolutionary changes to particular places, times and events — a method that reveals the geological, ecological and developmental contingency of evolutionary transformations. Croizat's (1958, 1964) panbiogeography could be read as attempting such a project.

Croizat's view of evolution is encapsulated in his slogan 'earth and life evolve together' (Croizat 1964:46). In place of the usual organism–environment dichotomy Croizat presented a three-fold parallelism — space, time, form (to be understood not as absolute categories but as relations — see Craw and Page, this volume). His fundamental methodological tool was the track — a line on a map connecting all the localities of a taxon (Croizat, 1984:7). By analysing the distribution patterns of diverse taxa using the track method Croizat discovered that their tracks were highly repetitious (generalized tracks) and seemed to bear little relation to either current geography or the vagility of the taxa being examined. Current versions of Croizat's track methodology, minimal spanning tree biogeography (see Page, 1987), or cladistic biogeography (see Nelson and Platnick, 1981) can be used to investigate in a more rigorous way the degree to which taxa share a common evolutionary history. When this analysis is combined with geological information, evolutionary changes can be related to eco-geologic events such as the opening of ocean basins or the rise of mountain ranges.

Croizat's panbiogeography was far more than an attempt to construct a taxonomy of regions. It was an investigation of evolution. Thus Croizat attempted to synthesize his biogeographic analysis with ideas on phylogeny, development and ecology. While Croizat tended towards preformationist and internalist, orthogenetic explanations of evolution that no longer seem valid, the spirit of his synthesis can be applied today. There could be a process of 'reciprocal illumination' between biogeographic analyses and developmental and ecological studies. Biogeography could be used to test particular process explanations. This would be more informative than just cataloguing tracks. In turn, biogeographic analyses could suggest new process questions. Some of the questions that could be asked are: Have taxa that share a common history all changed together? Do ecologically related taxa show more congruent patterns of change than ecologically unrelated taxa? How do these relationships relate to geological and ecological events? Do taxa always change in response to major geological changes? If not, are there any differences between the kinds of taxa that do and do not change?

There is an ambiguity in Croizat's texts that can be resolved by this interplay between pattern analysis and process questions. Croizat's slogan 'earth and life evolve together' can be read in at least two ways. It could be read as claiming that geological changes are the major factor in evolution or it could be interpreted as suggesting a more symmetrical relationship between earth and life. Croizat frequently seemed to imply the former reading. 'Ecology is underpinned by geological factors which are the real prime mover of form-making' (Croizat, 1965:619). However, on other occasions he seems to deny that such a unitary explanation can be given:

> . . . all the factors in the equation, biogeography, biology, ecology, and finally geography, interlock of necessity so closely as to rule the possibility that one can be thought to come into being without the others jointly and simply proceeding *pari passu*. (Croizat, 1958: 801)

> . . . the 'island' no more came first than did the 'bird' dwelling on the island. (Croizat, 1958: 919)

The first reading seems to have been adopted by some proponents of vicariance biogeography. For example, Nelson and Platnick (1984) define vicarience as the passive response of taxa to changes in the environment. Cracraft (1985) argues that tectonic change is the main cause of biological diversification. The second reading is more in line with the arguments I have advanced for the reciprocal construction of organism and environment.

Process questions based on the idea of EECs could free historical analyses from the crude geological determinism implicit in the first reading by emphasizing the ecological and developmental contingency of evolutionary transformations. Such an emphasis would not deny the importance of geological events. Instead it would stress the dynamic interplay between geological and biological processes. Organisms do not generally experience changes in geology directly, but rather indirectly through changes in climate, altitude, food availability, gene flow and population size. To be relevant to a particular population geological events must be ecologically mediated, and thus the effects of geological changes will depend on the ecology and behaviour of the population. The ability of organisms to actively select new habitats and modify their environments means that they are not passive victims of autonomous geological forces. Indeed the activities of plants and animals can have a major influence on such geological processes as erosion, sedimentation, the formation of coral reefs and oil and chalk deposits, and by affecting climate, glaciation (see Tiffeny, 1985; Meadows, 1986)

The coupled co-evolutionary analysis of geology, biogeography, phylogeny and ecology contrasts with recent attempts to partition the effects of phylogeny and ecology, or heritage and habitat (see Cheverud, Bow and Lewtenegger, 1985; Dobson, 1985 and refs therein). In a manner reminiscent of efforts to partition phenotypic variance into genetic and environmental components these researchers have attempted to partition out the effects of allegedly 'intrinsic', inherited, historical factors from 'extrinsic', non-historical, ecological variables. If earth and life really do evolve together then ecological factors such as climate, diet and habitat also have an evolutionary history (see Croizat, 1964; Brooks, 1985; Ricklefs, 1987).

CONCLUSION

In this chapter I have attempted to analyse some of the problems in current evolutionary theory by focusing on a few underlying metaphors and methods. I have argued that two sources of confusion in recent disputes are, first, an essentialist view of Darwinism and, second, false interpretations of methodological problems as purely theoretical questions. I have argued that the functional method, despite its successes, inevitably constructs an impoverished view of organisms, organism–environment relations and the evolutionary process. Pluralistic compromises on the problems of the functional method (e.g. some things are adaptive and others are not) are unworkable in practice and tend to reinforce the same set of basic assumptions. Perhaps alternative approaches to evolution based on organism–environment reciprocity, ecological and evolutionary cascades, and panbiogeographic methods will prove too difficult to construct. Perhaps formalistic bookkeeping accounts of evolution as changing gene frequencies driven by utility are the best we can hope for. However, we will never know unless we risk going beyond 'transformed Paleyism' and the one-dimensional view of evolution it constructs. Attempting this shift will require not just a change in our thinking and metaphors, but also a change in our methods. For whatever reasons the major *Darwinian* themes of development, behaviour, ecology and biogeography (Ghiselin, 1969) have not been fully incorporated into the modern synthesis. And yet they have so much to offer. From development comes the idea that integrated systems can frequently respond appropriately to perturbations, from behaviour the idea that behaviour is not just the product of evolution but also an agent for further change, from ecology the inter-ramification of organism and environment, and from panbiogeography a method of historical reconstruction and the spirit of synthesis. The problem that confronts us now is the struggle to eliminate the pre-Darwinian legacies from our thought, and to create a truly modern synthesis that is both Darwinian and yet beyond Neo-Darwinism.

ACKNOWLEDGEMENTS

I am very grateful to R. Craw and R. Page for many helpful discussions on biogeography, evolution and the dynamics of sterile disputes in evolutionary theory. P. Bateson, D. Gordon, M.-W. Ho, T. Johnston, S. Oyama and P. Taylor sent me letters, preprints and unpublished manuscripts which provoked many of the questions I have addressed in this chapter. D. Brooks drew my attention to Hull's idea that Darwinism is an individual. D. Brooks, J. Craig, G. Dumbell, N. Gavey, J. Grehan, A. Hemmings, D. Hull, I. Jamieson, E. Minot, D. Paul and R. Rowe commented on an earlier version of the manuscript. R. Page drew the illustrations. N. Gavey, R. Moorhouse and H. Speed helped prepare the final manuscript. My research is funded by a UGC postgraduate scholarship, a University of Auckland research grant and a UGC research grant to M. Davison. I would like to dedicate this chapter to the memory of my brother Tony Gray.

NOTES

1. Of course this does not imply that the method is theory neutral.
2. For some interesting research that does emphasize the spatio-temporal organization of behaviour — the 'daily round' of harvester ants — see Gordon, 1987.
3. Some of these suggested experiments were derived from Johnston and Gottlieb, ms.

REFERENCES

Abe, M., and Iriki, T. (1978). Effects of diet on the protozoa populations on continuous culture or rumen contents, *Br. J. Nutr.*, **39**, 225–61.

Ahmad, S. (1983). Mixed-function oxidase activity in a generalist herbivore in relation to its biology, food plants and feeding history, *Ecology*, **64**, 235–43.

Al-Jaborae, F.F. (1979). *The influence of diet on the gut morphology of the Starling (Sturnis vulgaris L.* 1758), Ph.D. Thesis, Oxford University.

Arnold, G.W., and Maller, R.A. (1977). Effects of nutritional experience in early and adult life on the performance and dietary habits of sheep. *Appl. Anim. Ethol.*, **3**, 5–26.

Barash, D.P. (1982). *Sociobiology and Behaviour*, Second edition, Hodder & Stoughton, London.

Barthes, R. (1964). *Essais critiques*, Editions du Seuil, Paris.

Bateson, P. (1982). Behavioural development and evolutionary processes. In *Current Problems in Sociobiology* (eds King's College Sociobiology Group), Cambridge University Press, Cambridge, pp. 131–51.

Bateson, P. (1983). Genes, environment and the development of behaviour. In T.R. Halliday and P.J.B. Slater (eds), *Animal Behaviour. Genetics and Development*, Blackwell, Oxford, pp. 52–81.

Bateson, P. (1984). Sudden changes in ontogeny and phylogeny. In G. Greenberg and E. Tobach (eds), *Behavioural Evolution and Integrative Levels*, Lawrence Erlbaum Associates, Hillsdale, NJ, pp. 155–66.

Bateson, P. (1986). Functional approaches to behavioural development. In J.G. Else and P.C. Lee (eds), *Primate Ontogeny, Cognition and Social Behaviour*, Cambridge University Press, Cambridge, pp. 183–192.

Beecher, R.M., Corruccini, R.S., and Freeman, M. (1983). Craniofacial correlates of dietary consistency in a nonhuman primate, *J. Craniofacial Genet. Dev. Biol.*, **3**, 193–292.

Beer, G. (1985). *Darwin's Plots: Evolutionary Narrative in Darwin, George Elliot and Nineteenth Century Fiction*, Ark Paperbacks, London.

Bouvier, M., and Hylander, W.L. (1981). Effect of bone strain on cortical bone structure in macaques (*Macaca mulatta), J. Morphol.*, **167**, 1–12.

Bowler, P.J. (1983). *The Eclipse of Darwinism. Anti-Darwinian Evolution Theories in the Decades around 1900*, Johns Hopkins University Press, Baltimore.

Brandt, C.A. (1984). Age and hunting success in the brown pelican: influences of skill and patch choice on foraging efficiency, *Oecologia (Berl).*, **62**, 132–7.

Brattsen, L.B., Wilkinson, C.F., and Eisner, T. (1977). Herbivore–plant interactions: mixed function oxidases and secondary plant substances. *Science*, **196**, 1349–52.

Broekman, J.M. (1974). *Structuralism*, D. Reidel, Dordrecht, Holland.

Brooks, D.R. (1985). Historical ecology: a new approach to studying the evolution of ecological associations, *Ann. Mo. Bot. Gard.*, **72**, 660–80.

Brooks, D.R., Le Blond, P.H., and Cummings, D.D. (1984). Information and entropy in a single evolution model, *J. Theor. Biol.*, **109**, 77–93.

Brooks, D.R., and Wiley, E.O. (1986). *Evolution an Entropy: Towards a Unified Theory of Biology*, Chicago University Press, Chicago.

Burghardt, G.M. (1970). Chemical perceptions in reptiles. In J.W. Johnson, D.G. Moulton, and A. Turk (eds), *Communication by Chemical Signals*, Appleton-Century-Crofts, New York, pp. 241–308.

Caro, T.M. (1980a). The effects of experience on the predatory patterns of cats, *Behav. Neural Biol.*, **29**, 1–28.

Caro, T.M. (1980b). Effects of the mother, object play, and adult experience on predation in cats, *Behav. Neural Biol.*, **29**, 29–51.

Carrick, R. (1963). Ecological significance of territory in the Australian magpie, *Gynorhina tibicen, Proc. Int. Ornithol. Congr.*, **2**, 740–53.

Cassidy, M.D. (1978). Development of an induced food plant preference in the Indian stick insect, *Carausius morosus, Entomol. Exp. Appl.*, **24**, 87–93.

Chao, L.N. and Musick, J.A. (1977). Life history, feeding habits, and functional morphology of juvenile sciaenid fishes in the York River estuary, Virginia, *Fish. Bull. (Dublin)*, **75**, 657–702.

Cheverud, J.M., Dow, M.M., and Leutenegger, W. (1985). The quantitative assessment of phylogenetic constraints in comparative analyses: sexual dimorphism in body weight among primates. *Evolution*, **39**, 1335–51.

Cohen, J. (1979). Maternal constraints on development. In D.R. Newth and M. Balls (eds), *Maternal Effects in Development*, Cambridge University Press, Cambridge, pp. 1–28.

Collias, N.E., and Collias, E.C. (1976). External construction by animals. In *Benchmark Papers in Animal Behaviour*, Vol. 4, Hutchinson & Ross, Dowden.

Corbet, S.A. (1985). Insect chemosensory responses: a chemical legacy hypothesis, *Ecol. Entomol.*, **10**, 147–53.

Cracraft, J. (1985). Biological diversification and its causes, *Ann. Missouri Bot. Gard.*, **72**, 794–822.

Craw, R.C. (1984). Charles Darwin on 'Laws of growth', *Tuatara*, **27**, 19–20.

Craw, R.C. (1987). Panbiogeography: method and synthesis in biogeography. In A. Myers and P.A. Giller (eds), *Biogeographic Analysis: Methods, Patterns and Processes*, Chapman and Hall, London (in press).

Craw, R.C., and Gibbs, G.W. (1984). *Croizat's Panbiogeography and Principia Botanica: search for a novel biological synthesis*, Victoria University Press, Wellington (as *Tuatara*, **27** (1)).

Croizat, L. (1958). *Panbiogeography*, Vols 1, 2a, 2b, published by the author, Caracas, Venezuela.

Croizat, L. (1964). *Space, Time, Form: The Biological Synthesis*, published by the author, Caracas, Venezuela.

Croizat, L. (1965). An introduction to the subgeneric classification of *Euphorbia L.* with stress on the South African and Malagasy species 1, *Webbia*, **20**, 573–706.

Darwin, C. (1842). *The Structure and Distribution of Coral Reefs. Being the first part of the geology of the voyage of the Beagle*, Elder Smith, London.

Darwin, C. (1881). *The Formation of Vegetable Mould, Through the Action of Worms, with Observations on Their Habits*, Murray, London.

Dawkins, R. (1982). *The Extended Phenotype*, Freeman, Oxford, U.K.

Dawkins, R. (1983). Universal Darwinism. In D.S. Bendall (ed), *Evolution from Molecules to Men*, Cambridge University Press, Cambridge, pp. 403–25.

Dobson, F.S. (1985). The use of phylogeny in behaviour and ecology, *Evolution*, **39**, 1384–8.

Dobzhansky, T. (1956). What is an adaptive trait? *Amer. Natur.*, **90**, 337–47.

Dupuis, C. (1984). Willi Hennig's impact on taxonomic thought, *Ann. Rev. Ecol. Syst.*, **15**, 1–24.

Edwards, P.J., and Wratten, S.D. (1985). Induced plant defences against insect grazing: fact or artifact. *Oikos*, **44**, 70–4.

Evans, L.T. (1984). Darwin's use of the analogy between artificial and natural selection, *J. Hist. Biol.*, **17**, 113–40.

Ewer, R.F. (1968). *Ethology of Mammals*, Logos, London.

Fisher, J., and Hinde, R.A. (1949). The opening of milk bottles by birds, *Brit. Birds*, **42**, 347–57.

Fowler, S.V., and Lawton, J.H. (1985). Rapidly induced defences and talking trees: the devil's advocate position, *Amer. Natur.*, **126**, 181–95.

Galef, B.G. (Jr) (1985). Social learning in wild Norway rats. In T.D. Johnston and A.T. Pietrewicz (eds), *Issues in the Ecological Study of Learning*, Lawrence Erlbaum Associates, Hillsdale, NJ, pp. 143–66.

Galef, B.G. (Jr) and Henderson, P.W. (1972). Mother's milk: a determinant of the feeding preferences of weaning rat pups, *J. Comp. Physiol. Psychol.*, **78**, 213–19.

Gendron, R.P., and Staddon, J.E.R. (1983). Searching for cryptic prey: the effect of search rate, *Amer. Natur.*, **121**, 172–86.

Ghiselin, M.T. (1969). *The Triumph of the Darwinian Method*, University of California Press, Berkeley.

Ghiselin, M. (1974). A radical solution to the species problem, *Syst. Zool.*, **23**, 536–44.

Ghiselin, M. (1981). Categories, life and thinking, *Behav. Brain Sci.*, **4**, 269–313.

Ghiselin, M.T. (1986). The assimilation of Darwinism in developmental psychology, *Hum. Dev.*, **29**, 1–35.

Gibson, J.J. (1979). *The Ecological Approach to Visual Perception*, Houghton Mifflin Company, Boston, Mass.

Gilbert, J.J. (1966). Rotifer ecology and embryological induction, *Science*, **151**, 1234–7.

Gluck, E. (1984). Habitat selection in birds and the role of early experience, *Z. Tierpsychol.*, **66**, 45–54.

Gordon, D.M. (1985). Do we need more ethograms? *Z. Tierpsychol.*, **68**, 340–2.

Gordon, D.M. (1987). The dynamics of group behaviour. In P.P.G. Bateson and P.H. Klopfer (eds), *Perspectives in Ethology*, Vol. 7, Plenum Press, New York, pp. 217–31.

Gottlieb, G. (1976). Conceptions of prenatal development: behavioural embryology, *Psychol. Rev.*, **83**, 215–34.

Gould, S.J. (1983). The hardening of the modern synthesis. In M. Greene (ed), *Dimensions of Darwinism*, Cambridge University Press, Cambridge, pp. 71–93.

Gould, S.J., and Lewontin, R.C. (1979). The spandrels of San Marco and the panglossian paradigm: a critique of the adaptationist programme, *Proc. Roy. Soc. Lond. (B) Biol. Sci.*, **205**, 581–98.

Grant, J.W.G., and Bayley, I.A.E. (1981). Predator induction of crests in morphs of the *Daphnia carinata* King complex. *Limnol. Oceanogr.*, **26**, 201–18.

Gray, L. (1981). Genetic and experiential differences affecting foraging behaviour. In A.C. Kamil and T.D. Sargent (eds), *Foraging Behavior: Ecological, Ethological and Psychological Approaches*, Garland STPM Press, New York, pp. 455–73.

Gray, R.D. (1978a). Beyond labels and binary oppositions: what can be learnt from the nature/nurture dispute? *Riv. Biol-B. Forum*, **80**, 192–6.

Gray, R. (1987b). Faith and foraging: a critique of the paradigm argument from design. In A.C. Kamil, J.R. Krebs, and H.R. Pulliam (eds), *Foraging Behaviour*, Plenum Press, New York, pp. 69–140.

Greenwood, P.H. (1965). Environmental effects on the pharyngeal mill of a cichlid fish, *Astatoreochromis alluadi* and the taxonomic implications, *Proc. Linn. Soc. Lond.*, **176**, 1–10.

Grehan, J.R., and Ainsworth, R. (1985). Orthogenesis and evolution, *Syst. Zool.*, **34**, 174–92.

Grossman, M.I., Greengard, H., and Ivy, A.C. (1943). The effect of dietary composition on pancreatic enzymes, *Amer. J. Physiol.*, **138**, 676–82.

Hammen, L. van der (1985). A structuralist approach in the study of evolution and classification, *Z. Hededel. (Leiden)*, **59**, 391–409.

Hansell, M.H. (1984). *Animal Architecture and Building Behaviour*, Longman, London.

Harris, M.A., and Murie, J.O. (1984). Inheritance of nest sites in female Columbian ground squirrels, *Behav. Ecol. Sociobiol.*, **15**, 97–102.

Haukioja, E., and Niemela, P. (1977). Retarded growth of a geometrid larva after mechanical damage to leaves of its host tree. *Annals Zool. Fennici*, **14**, 48–52.

Heather, B.D. (1977). Foot-trembling by the black-fronted dotteral, *Notornis*, **24**, 1–8.

Heinrich, B. (1983). Do bumblebees forage optimally, and does it matter? *Amer. Zool.*, **23**, 273–81.

Helfman, G.S., and Schultz, E.T. (1984). Social transmission of behavioural traditions in a coral reef fish, *Amin. Behav.*, **32**, 379–84.

Hennig, W. (1966). *Phylogenetic Systematics*, University of Illinois Press, Urbana, Illinois.

Hinde, R.A.(1975). The concept of function. In G. Baerends, C. Beer, and A. Manning (eds), *Function and Evolution in Behaviour*, Clarendon Press, Oxford, pp. 3–15.

Hinde, R.A., and Bateson, P. (1984). Discontinuities versus continuities in behavioural development and the neglect of process, *Int. J. Behav. Dev.*, **7**, 129–43.

Ho, M.-W. (1984a). Where does biological form come from? *Riv. Biol. (Perugia)*, **77**, 147–79.

Ho, M.-W. (1984b). Environment and heredity in development and evolution. In M-W. Ho and P.T. Saunders (eds), *Beyond Neo-Darwinism: Introduction to the New Evolutionary Paradigm*, Academic Press, London, pp. 267–89.

Ho, M.-W. (1987). Genetic fitness and natural selection: myth or metaphor, In G. Greenberg and E. Tobach (eds), *Evolution of Social Behaviour and Integrative Levels*, Lawrence Erlbaum Associates, Hillsdale, NJ (in press).

Ho, M.-W., and Saunders, P.T. (1984). *Beyond Neo-Darwinism: An Introduction to the New Evolutionary Paradigm*, Academic Press, London.

Hull, D.L. (1976). Are species really individuals? *Syst. Zool.*, **25**, 174–91.

Hull, D.L. (1978). A matter of individuality, *Philos. Sci.*, **45**, 335–60.

Hull, D.L. (1984). Hypotheses that blur and grow. In T. Duncan and T.F. Stuessy (eds), *Cladistics: Perspectives on the Reconstruction of Evolutionary History*, Columbia University Press, New York, pp. 5–23.

Hull, D. (1985). Darwinism as a historical entity: a historiographic proposal. In D. Kohn (ed), *The Darwinian Heritage*, Princeton University Press (in association with Nova Pacifica), Princeton, pp.773–812.

Jamieson, I.G. (1986). The functional approach to behaviour: it is useful? *Amer. Natur.*, **127**, 195–208.

Jamieson, I.G., and Craig, J.L. (1987). Critique of helping behaviour in birds: a departure from functional explanations. In P.P.G. Bateson and P. Klopfer (eds), *Perspective in Ethology*, vol. 7, Plenum, New York, pp.79–98.

Jermy, T., Hanson, F.E., and Dethier, V.G. (1968). Induction of specific food preference in lepidopterous larvae, *Entomol. Exp. Appl.*, **11**: 211–30.

Johnston, T.D., and Gottblieb, G. (unpublished manuscript). The origins of variations: a development problem for evolutionary theory.

Jolly, A. (1972). *The Evolution of Primate Behaviour*, Macmillan, New York.

Jones, C.G. (1983). Microorganisms as mediators of resource exploitation. In P.W. Price, C,N. Slobodchikoff and W.S. Gaud (eds), *A New Ecology: Novel Approaches to Interactive Systems*, John Wiley, New York, pp. 53–99.

Jones, G.R., Aldrich, J.R., and Blum, M.S. (1981). Baldcypress allelochemicals and the inhibition of silkworm enteric microorganisms. Some ecological considerations, *J. Chem. Ecol.*, **7**, 103–14.

Kawai, M. (1965). Newly acquired pre-cultural behaviour of the natural troop of Japanese monkeys on Koshina islet. *Primates*, **6**, 1–30.

Kenwood, R.E., and Sibly, R.N. (1977). A wood pigeon *(Columba palumus)* feeding preference explained by a digestive bottleneck, *J. Appl. Ecol.*, **14**, 815–26.

Krebs, J.R. (1980). Optimal foraging, predation risk and territory defense, *Ardea*, **68**, 83–90.

Krebs, J.R., Charnov, E.L., and Houston, A.I. (1981). Some recent developments in optimal foraging. In A.C. Kamil and T.D. Sargent (eds), *Foraging Behavior: Ecological, Ethological and Psychological Approaches*, Garland STPM Press, New York, pp. 3–18.

Kruckeberg, A.R. (1969). The implications of ecology for plant systematics, *Taxon*, **18**, 92–120.

Krueger, D.A., and Dodson, S.I. (1981). Embryological induction and predation ecology in *Daphina pulex, Limnol. Oceanogr.*, **26**, 219–23.

Kuo, Z.Y. (1967). *The Dynamics of Behaviour Development: An Epigenetic View*, Random House, New York.

Lauder, G.V. (1982). Historical biology and the problem of design, *J. Theor. Biol.*, **97**, 57–67.

Lawton, J.H., Beddington, J.R., and Bonser, R. (1974). Switching in invertebrate predators. In M.B. Usher and M.H. Williamson (eds), *Ecological Stability*, Chapman and Hall, London, pp. 141–58.

Lehrman, D.S. (1965). Interaction between internal and external environments in the regulation of the reproductive cycle of the ring dove. In F.A. Beach (ed) *Sex and Behaviour*, John Wiley, New York, pp. 335–69, 378–80.

Lenton, G.M. (1983). Wise owls flourish among the oil palms, *New Scientist*, **97**, 436–7.

Lenton, G.M. (1984). The feeding and breeding ecology of Barn Owls *Tyto alba* in Peninsular Malaysia, *Ibis*, **126**, 551–575.

Lewontin, R.C. (1974). *The Genetic Basis of Evolutionary Change*, Columbia University Press, New York.

Lewontin, R.C. (1978). Adaptation, *Sci. Amer.*, **239**, 156–69.

Lewontin, R.C. (1980). Adaptation. In E. Sober (ed), *Conceptual Issues in Evolutionary Biology: An Anthology* (Reprinted from the *Encyclopedia Einaudi*, Milan, Italy, 1984), MIT Press, Cambridge, Mass, pp. 235–51.

Lewontin, R.C. (1981). On constraint and adaptation, *Behav. Brain Sci.*, **4**, 244–5.

Lewontin, R.C. (1983a). Darwin's revolution, *The New York Review*, 16 June, pp. 21–7.

Lewontin, R.C. (1983b). The organism as the subject and object of evolution, *Scientia*, **118**, 65–82.

Lewontin, R.C. (1983c). Gene, organism and environment. In D.S. Bendall (ed), *Evolution from Molecules to Men*, Cambridge University Press, Cambridge, pp. 273–85.

Lively, C.M. (1986). Predator-induced shell dimorphism in the acorn barnacle *Chthamalus anisopoma, Evolution*, **40**, 232–42.

Lovtrup, S. (1976). On the falsifiability of neo-Darwinism, *Evol. Theory*, **1**, 267–83.

Luthard-Laimer, G. and Roth, G. (1983). Reduction of visual inhibition to stationary prey by early experience in *Salamandra salamandra* (L.), *Z. Tierpsychol.*, **63**, 294–302.

Mattson, W.J. (Jr) (1980). Herbivory in relation to plant nitrogen content, *Ann. Rev. Ecol. Syst.*, **11**, 119–61.

Maynard Smith, J. (1978). Optimization theory in evolution, *Ann. Rev. Ecol. Syst.*, **9**, 31–56.

Maynard Smith, J., Burian, R., Kauffman, S., Alberch, P., Campbell, J., Goodwin, B., Lande, R., Raup, D., and Wolpert L. (1985). Developmental constraints and evolution, *Q. Rev. Biol.*, **60**, 265–87.

Mayr, E. (1975). The unity of the genotype, *Biol. Zentrabl.*, **94**, 377–88.

Mayr, E. (1983). How to carry out the adaptationist program? *Amer. Natur.*, **121**, 324–34.

McCleery, R.H. (1978). Optimal behaviour sequences. In J.R. Krebs and N.B. Davies (eds), *Behavioural Ecology: An Evolutionary Approach*, Blackwell Scientific Publications, Oxford, pp. 377–410.

Meadows, P.S. (1986). Biological activity and seabed sediment structure, *Nature*, **323**, 207.

Meyerriecks, A.J. (1971). Further observations on the use of the feet by foraging herons, *Wilson Bull.*, **83**, 435–8.

Miller, M.R. (1975). Gut morphology in mallards in relation to diet quality, *J. Wild. Manag.*, **39**, 168–73.

Mitter, C., and Brooks, D.R. (1983). Phylogenetic aspects of coevolution. In D.J. Futuyma and M. Slatkin (eds), *Coevolution*, Sinauer Associates, Sunderland, Mass, pp. 65–98.

Moore, P.J. (1984). Foraging and social behaviour of the white-faced heron at Pauatahanui inlet, *Notornis*, **31**, 285–99.

Moore, W.J. (1965). Masticatory function and skull growth, *J. Zool.*, **146**, 123–31.

Neal, E. (1948). *The Badger*, Collins, London.

Nelson, G., and Platnick, N. (1981). *Systematics and Biogeography:Cladistics and Vicariance*, Columbia University Press, New York.

Nelson, G., and Platnick, N. (1984). *Biogeography*, Carolina Biology Readers, Burlington, North Carolina.

Norton-Griffiths, M. (1968). *The Feeding Behaviour of the Oystercatcher (Haemotopus ostralegus)*, Ph.D. Thesis, Oxford University.

Oyama, S. (1985). *The Ontogeny of Information: Developmental Systems and Evolution*, Cambridge University Press, Cambridge.

Padilla, S.G. (1935). Further studies on the delayed pecking of chicks, *J. Comp. Psychol.*, **20**, 413–43.

Page, R.D.M. (1987). Graphs and generalized tracks: quantifying Croizat's panbiogeography, *Syst. Zool.*, **36**, 1–17.

Patten, B.C. (1982). Environs:relativistic elementary particles for ecology, *Amer. Natur.*, **119**, 179–219.

Piaget, J. (1968). *Le structuralism*, Presses Universitaires de France, Paris.

Pienkowski, M.W. (1983). Surface activity of some intertidal invertebrates in relation to temperature and the foraging behaviour of their shorebird predators, *Mar. Ecol. Prog. Ser.*, **11**, 141–50.

Pollard, J.W. (1984). *Evolutionary Theory: Paths to the Future*, Wiley, New York.

Polsky, R.H. (1977). The ontogeny of predatory behaviour in the golden hamster *(Mesocriecetus a. auratus)*. 1. The influence of age and experience, *Behaviour*, **61**, 26–57.

Pulliam, H.R., and Caraco, T. (1984). Living in groups: is there an optimal group size? In J.R. Krebs and N.B. Davies (eds), *Behavioural Ecology: An Evolutionary Approach*, 2nd edition, Sinauer, Sunderland, Mass., pp. 122–47.

Pyke, G.H. (1984). Optimal foraging: a critical review, *Ann. Rev. Ecol. and Syst.*, **15**, 523–75.

Rabinowitch, V.E. (1965). *The role of early experience in the development and retention of food habits in some birds*, Ph.D. Thesis, University of Wisconsin.

Rand, A.L. (1956). Foot-stirring as a feeding habit of wood ibis and other birds, *Amer. Midl. Nat.*, **55**, 96–100.

Rapport, D.J. (1981). Foraging behavior of *Stentor coeruleus*: a microeconomic interpretation. In A.C. Kamil and T.D. Sargent (eds) *Foraging Behavior: Ecological,*

Ethological and Psychological Approaches, Garland STPM Press, New York, pp. 77–93.

Ricklefs, R.E. (1987). Community diversity: relative roles of local and regional processes, *Science*, **235**, 167–71.

Robson, G.C., and Richards, O.W. (1936). *The Variation of Animals in Nature*, Longmans, Green and Co., London.

Rosen, D.E., and Buth, D.G. (1980). Empirical research versus neo-Darwinian speculation, *Syst. Zool.*, **29**, 300–8.

Rosenberg, A. (1985). *The Structure of Biological Science*, Cambridge University Press, Cambridge.

Ruse, M. (1975). Charles Darwin and artificial selection, *J. Hist. Ideas*, **36**, 339–50.

Schoener, J.W. (1987). A brief history of optimal foraging ecology. In A.C. Kamil, J.R. Krebs, and H.R. Pulliam (eds), *Foraging Behavior*, Plenum Press, New York, pp. 5–67.

Schuh, R.T., and Polhemus, J.T. (1980). Analysis of taxonomic congruence among morphological, ecological, and biogeographic data sets for the leptopodomorpha (Hemiptera), *Syst. Zool.*, **29**, 1–26.

Shaw, R., and Turvey, M.T. (1981). Coalitions as models for ecosystems: a realist perspective on perceptual organization. In M. Kubouy and J.R. Pomerantz (eds), *Perceptual Organization*. Lawrence Erlbaum Associates, Hillsdale, NJ, pp. 343–415.

Slipjer, E.J. (1942). Biologic–anatomical investigations on the bipedal gait and upright posture in mammals, with special reference to a little goat, born without forelegs. I and II, *Proc. K. Ned. Akad. Wet.*, **45**, 288–95, 407–15.

Smith, J.N.M., and Dawkins, R. (1971). The hunting behaviour of individual great tits in relation to spatial variations in their food density, *Anim. Behav.*, **19**, 695–706.

Smith, W.H. (1965). Observations on the flora of the alimentary tract of animals and factors affecting its composition, *J. Path. Bacteriol.*, **89**, 95–122.

Sober, E. (1980). Evolution, population thinking, and essentialism, *Phil. Sci.*, **47**, 350–83.

Sober, E. (1984a). Force and disposition in evolutionary theory. In C. Hookway (ed), *Minds, Machines and Evolution*, Cambridge University Press, Cambridge, pp. 43–61.

Sober, E. (1984b). *The Nature of Selection*, MIT Press, Cambridge, Mass.

Sober, E. (1985). Darwin on natural selection: a philosophical perspective. In D. Kohn (ed), *The Darwinian Heritage*, Princeton University Press (in association with Nova Pacifica), Princeton, pp. 867–99.

Stent, G. (1981). Strength and weakness of the genetic approach to the development of the nervous system. In W.M. Cowan (ed), *Studies in Developmental Neurobiology*, Oxford University Press, New York, pp. 288–321.

Stoner, A.W., and Livingstone, R.J. (1984). Ontogenetic patterns in diet and feeding morphology in sympatric sparid fishes from seagrass meadows, *Copeia*, **1984**, 174–89.

Strong, D.R., Lawton, J.H., and Southwood, R. (1984). *Insects on Plants*, Blackwell Scientific Publications, Oxford.

Taylor, P.J. (1987). Historical versus selectionist explanations in evolutionary biology, *Cladistics*, **3**, 1–13.

Terrier, L.C. (1984). Induction of detoxication enzymes in insects, *Ann. Rev. Ecol. Syst.*, **29**, 71–88.

Tiffeny, B.H. (1985). Geological factors and the evolution of plants. In B.H. Tiffeny (ed), *Geological Factors and the Evolution of Plants*. Yale University Press, New Haven and London, pp. 1–10.

Tinbergen, N. (1962). Foot-paddling in gulls. *Br. Birds*, **55**, 117–19.

Tsumori, A. (1967). New acquired behaviour and social interactions of Japanese monkeys. In S.A. Altman (ed), *Social Communication Among Primates*, University of Chicago Press, Chicago, pp. 207–19.

Tsumori, A., Kawai, M., and Motoyoshi, R. (1965). Delayed response of wild Japanese monkeys by the sand-digging method: 1, case of the Koshima troop, *Primates*, **6**, 195–212.

Van Denburgh, J. (1914). The gigantic land tortoises of the Galapagos archipelago, *Proc. Calif. Acad. Sci. San Francisco*, 4th ser., **2**, 203–374.

Waddington, C.H. (1975). *The Evolution of an Evolutionist*. Cornell University Press, Ithaca, New York.

Wankowski, J.W.J. (1979). Morphological limitations, prey size selectivity, and growth response in Juvenile Atlantic salmon, *Salmo salar, J. Fish Biol.*, **14**, 89–100.

Webster, G. (1984). The relations of natural forms. In M.-W. Ho and P.T. Saunders (eds), *Beyond Neo-Darwinism: An Introduction to the New Evolutionary Paradigm*, Academic Press, London, pp.193–217.

Webster, G., and Goodwin, B.C. (1982). The origin of species: a structuralist approach, *J. Soc. Biol. Struct.*, **5**, 15–47.

Weinrich, J.D. (1977). Human sociobiology: pair-bonding and resource predictability, *Behav. Ecol. Sociobiol.*, **2**, 91–118.

Werner, E.E. (1974). The fish size, prey size, handling time relation in several sunfishes and some implications, *J. Fish. Res. Board, Can.*, **31**, 1531–6.

Williams, G.C. (1985). Comments by George C. Williams on Sober's *The Nature of Selection, Biol. Phil.*, **1**, 114–22.

Woolfenden, G.E., and Fitzpatrick, J.W. (1978). The inheritance of territory in group breeding birds, *Bioscience*, **28**, 104–108.

Wyles, J.S., Kunkel, J.G., and Wilson, A.C. (1983). Birds, behavior, and anatomical evolution, *Proc. Nat. Acad. Sci. USA*, **80**, 4394–7.

Young, R.M. (1971). Darwin's metaphor: Does nature select? *Monist*, **55**, 442–503.

Zach, R., and Falls, J.B. (1978). Prey selection by captive oven birds (Aves: Parulidae), *J. Anim. Ecol.*, **47**, 929–43.

Evolutionary Processes and Metaphors
Edited by M.-W. Ho and S.W. Fox

CHAPTER 12

Behaviour changes — finding the rules

DEBORAH M. GORDON

ABSTRACT

Behaviour changes, in response to changes in environmental conditions, and in the course of development. Moreover, the rules governing such changes may themselves be altered: in ontogeny, in varying ecological contexts and in evolutionary time. Empirical investigations of animal behaviour can be designed to discover the rules for behavioural change. For example, a study of the dynamics of group behaviour in harvester ants shows that: (1) changes in the behaviour of one worker group cause predictable changes in the activities of other groups in the colony; (2) the rules for changes in colony behaviour are modulated by alterations in environmental conditions; and (3) the rules for changes in colony behaviour are themselves transformed as a colony grows older. These results differ from those to be obtained from a research programme primarily concerned with static aspects of social insect behaviour. The investigation of rules for changes in behaviour, and of the ecological, developmental and evolutionary dynamics of such rules, is an intriguing new area of research in animal behaviour.

TRANSFORMATION AND CHANGE

Behaviour always entails change. Behaviour is what we see animals doing, and when an animal behaves, something happens; something changes. Behaviour and change are linked in an obvious way because behaviour takes up time and entails some activity. But behaviour and change are also linked in other ways.

First, patterns of behaviour change in the course of development. A juvenile baboon walking along with its group moves in ways that are different from those of an adult (Altmann and Altmann, 1970). Second, behaviour changes in response to changing environmental conditions. In sparse habitats, the feeding behaviour of hummingbirds is different from the feeding behaviour in rich ones (Feinsinger, 1976). Third, rules for change may themselves change in response to altered conditions. For example, suppose that dominance relationships in a certain group of animals are mediated by access to food. In a particular habitat, dominance relationships are established, destroyed or maintained — that is, such relationships change — according to rules that depend on the availability of food. If the habitat changes, altering food availability, new rules may come to govern initiation, destruction or perpetuation of particular roles in the dominance hierarchy. In other words: in altered conditions, there may be new rules for the transformation of dominance relations.

Once we take it as given that behaviour patterns change, a practical problem arises about how to conduct empirical studies of behaviour. Is it possible to recognize that nature is in flux, that behaviour is always changing, and still make any headway in understanding it? To show that such an endeavour is both possible and interesting, I will describe an example, from my own work on harvester ants, of an investigation of social behaviour that examines behavioural transformations. The investigation begins with behaviour that is capable of change and that invites further questions about change.

An animal's behaviour is a continuous, flowing stream. To describe behaviour, and to explain it, we view it as a sequence of discrete acts or movements. For example, an ethologist watching an ant might describe a segment of its stream of behaviour as a sequence such as 'walks along foraging trail, antennates passing ant, picks up seed, drops seed, walks along trail'. In a description of a sequence of behaviour the terms may be movements ('head-cock'), vocalizations ('chick-a-dee call'), or activities identified by function ('nest building' or 'territorial display').

The choice of which behaviour to investigate is a crucial step in any research programme, because it determines the kinds of questions that can be asked. If the object of study is invariant, or static, it is awkward to use it to investigate transformations. McBride's (1975) review of research questions about social organization is a good source of examples of questions concerned with static aspects of behaviour. Some are: What distinguishable categories of behaviour are there? what social distances are maintained between members of a group, classified by type of relationship? Which behaviours are used to accomplish *x*? What is the size and composition of subgroups? These are all questions about what behaviour *is*, rather than how it changes, because they consider certain aspects of behaviour to be fixed. The questions are posed as though categories of behaviour, social distances, types of relationship, sizes of subgroups, are being considered to be unchanging attributes.

Another familiar research question about a unit of behaviour *x* is, why is *x* adaptive? As Levins and Lewontin (1985) and others have argued, many optimality arguments proceed from a static world-view. Such arguments often

TABLE 1 Non-additive effects of combined perturbations. The table shows the direction of change, relative to control colonies, in numbers of ants engaged in nest maintenance of foraging. *Indicates a direct effect of perturbations. For the combined perturbation (toothpicks plus barriers), the table shows the direction of change relative to the hypothetical, additive effect obtained by summing the responses to toothpicks and barriers when each perturbation was deployed singly

	Activity	
Experimental perturbation	Nest maintenance	Foraging
Toothpicks	+*	−
Barriers	+	−*
Toothpicks and barriers (relative to hypothetical additive effect)	−	+

suppose that, given a set of conditions, it is possible to assess what an animal's best response is. (One could conceive of asking optimality questions about a dynamic situation. Given a network of changing conditions, what are the best rules ordering a range of responses?) But research questions presupposing the existence of a fixed optimum suppose a stable world.

Research questions that treat behaviour as invariant are sometimes a necessary first step. Watching animal behaviour means watching many complicated processes, from the biochemical to the social. To understand this complexity, one way to proceed is first to select invariant aspects of behaviour as objects of study. Later on, such a procedure assumes, it may be possible to examine how the behaviour changes, develops and evolves. Ethology began with such a research programme. Viewing patterns of behaviour as preprogrammed fragments, each an ineluctable response to a particular stimulus, the task at hand was to focus on what these patterns are, to explore their components and mechanisms, and to look for generalizations about the sequences in which they 'typically' occur. This procedure led to ethology's concern with ethograms, with releasers and fixed action patterns, and with techniques for deriving regularities in sequences of behaviour in which the components are apparently unrelated to each other.

Is there an alternative procedure? Once we recognize that behaviour is context dependent, historically contingent and flexible, we have to consider transformations. Behaviour is connected to a changing world. What are the connections between behaviour and its setting, or context, and how do these connections themselves change? This perspective calls for research questions that examine the rules for changes in patterns of behaviour. I will argue that such questions can lead to new insights about social organization.

THE ORGANIZATION OF HARVESTER ANT COLONIES

My work on social organization reflects two concerns. One, an investigation of the rules for changes in behaviour, is the subject of this chapter. The other is a concern with behaviour at the level of the group rather than of the individual; I have discussed this aspect elsewhere (Gordon, 1986a).

The colony's daily round is a temporal pattern of behaviour. Harvester ants, which I observe in the desert in the southwestern US, engage in various activities outside the nest: travelling to or from the nest on cleared trunk trails to collect seeds; clearing the mound of vegetation and carrying sand out of the nest; patrolling the nest area for intruders, disturbances and new food sources; and building and rearranging the colony's refuse pile, or midden. Each of these four activities, foraging, nest maintenance, patrolling or midden work, is done each day in a characteristic sequence. That is, given an activity, the number of workers engaged in it, as a function of time of day, reaches a clear, predictable peak at a certain time. At any given time, each task is pursued by a distinct group of workers. Thus the daily round of a colony is a pattern of change over time in the intensity of different activities.

Daily rounds are both colony specific (Gordon, 1983) and species specific (Gordon, 1984a). Although it is predictable and regular, under ordinary circumstances, the daily round is not rigid. On some days colonies do more nest maintenance work; on others they do foraging. The daily round embodies a colony's interaction with its environment. Changing environmental conditions lead to changes in the daily pattern of colony behaviour. I have begun a series of perturbation experiments to investigate how the daily round responds to environmental change. These experiments alter the daily round by altering the conditions under which it occurs; thus, they investigate the rules for how the pattern itself changes.

The first question to arise about the dynamics of the daily round is, does a change in one activity cause other activities to change? I did a series of perturbations that directly affect only one activity at a time, and that are analogous to naturally occurring changes in a colony's environment, brought about by wind, rain, and the activities of other ants. For example, wind and rain wash debris on to the nest mounds and foraging trails; ants of other species go on to the nest mounds of harvester ants. Colonies' responses to experimental perturbations are the result of the ways colony organization has evolved to meet naturally occurring changes in environmental conditions.

I devised three kinds of perturbations, each of which directly affects only one activity, foraging, nest maintenance, or patrolling. Barriers on the trunk trial decrease foraging intensity by causing foragers to go around or over the barrier, instead of going directly to the seed source. Toothpicks on the nest mound increase nest maintenance intensity by creating extra work for nest maintenance workers, who carry the toothpicks away before going on with their regular tasks. Strange objects (cardboard cylinders) and ants of other species placed on the nest mounds increase patrolling intensity by creating a disturbance that patrollers eventually ignore (the cylinders) or dispel (the intruders). Each of such peturbations, then, directly affects only one of the normal activities: foraging, nest maintenance, or patrolling.

The perturbations cause changes in the temporal patterns of other activities, besides the ones directly affected by experiments (Gordon, 1986b). Thus, the first rule of the dynamics of harvester ant organization: *different groups of workers, engaged in different tasks, interact with one another*. The intensity of effort and time devoted to different activities are not independent. Because perturbations cause relatively immediate changes, within hours, the time scale of interactions must be quite short. Worker groups must assess, almost on a moment-to-moment basis, alterations in the behaviour of other worker groups.

The behaviour of a particular worker group, such as the foragers, cannot be fully accounted for in terms of the factors affecting only foragers, such as seed availability or chemical trails of recruitment pheromones. Instead, the behaviour of foragers depends also on the ways that other worker groups, such as patrollers and nest maintenance workers, are responding to changes in their micro-environments.

When two or more perturbations are imposed simultaneously, there arise interesting non-additive effects (Table 1). For example, barriers and toothpicks alone have the same affect. Barriers first decrease foraging, and eventually cause nest maintenance to increase. Toothpicks first increase nest maintenance and eventually also cause foraging to decrease. Thus both these single perturbations have the same ultimate effect: decreased foraging and increased nest maintenance. The additive effect of deploying both toothpicks and barriers simultaneously would be an even stronger decrease in foraging, and an even stronger increase in nest maintenance. The actual result, when colonies receive toothpicks and barriers at the same time, is very different. Relative to the hypothetical additive effect, foraging increases and nest maintenance decreases.

This shows that the interactions between worker groups cannot be described in terms of single, all-or-nothing responses by worker groups. When the foragers encounter barriers, foraging decreases. However, if foragers encounter barriers on the same day that nest maintenance workers had to clear away toothpicks, foraging decreases much less. When the two perturbations are done simultaneously, instead of allowing foraging to decrease as it does when just barriers are present, the colony recruits *more* foragers to compensate for the obstruction caused by barriers. Thus, in one situation, single perturbations, a pertubation of a particular activity causes various worker groups to react a certain way. In another situation (combined perturbations), the same perturbation of that activity causes a very different response. This suggests

TABLE 2 How consistent is response to perturbations? The table compares the responses of older (about 5 years) and younger (about 2 years) colonies to perturbations. The plus and minus signs show the directions of change, relative to control colonies, in numbers of ants engaged in nest maintenance, foraging, or patrolling. Shown for each perturbation are the results for three separate experiments, each on different groups of colonies. Older colonies were more likely to respond the same way to a particular perturbation, in successive experiments, than were younger colonies

Perturbation	Experiment	Activity		
		Nest maintenance	Foraging	Patrolling
Older colonies				
Toothpicks	1	+	−	+
	2	+	−	+
	3	−	−	+
Barriers	1	+	−	+
	2	+	−	+
	3	−	−	+
Younger colonies				
Toothpicks	1	+	+	−
	2	+	−	−
	3	+	−	+
Barriers	1	+	+	−
	2	+	−	+
	3	−	−	+

another role of colony response to altered conditions: *Interactions between worker groups are modulated by the environment, not invariant.*

Colony response to perturbations depends on the age of the colony. A colony is founded by a single mated queen who digs a nest and begins to produce sterile workers. The colony lasts the lifetime of the queen, up to fifteen years and perhaps more. I did a series of perturbation experiments on young colonies about two years old, and on well-established larger colonies, at least five years old (Gordon, 1987a). The perturbations were those described above, affecting foraging, nest maintenance, and patrolling, singly and in various combinations.

There were two interesting differences in interactions between worker groups, depending on colony age. Both suggest that older colonies have homeostatic responses to environmental change that are stronger than those of younger colonies. First, in successive experiments older colonies respond more consistently to the same perturbation than younger ones do (Table 2). For example, perturbations increasing nest maintenance caused all of three different groups, each of seven colonies, in three different weeks, to respond in very similar ways. The same perturbation caused three groups of younger colonies, in different weeks, to respond in divergent ways. Older colonies seem to be more stable than younger ones to environmental change.

The second difference in response to perturbations, depending on colony age, was apparent in the non-additive effects of combined perturbations. As stress was increased, older colonies emphasized foraging more. The non-additive effects of combined perturbations led older colonies to decrease foraging less than younger colonies did. Older colonies responded to the increased stress of combined perturbations by doing almost as much foraging as control, undisturbed colonies.

In older colonies, both the emphasis on foraging in stressful conditions and the consistent response to successive experiments suggest another rule in the dynamics of the daily round: *As a colony gets older, its response to environmental change becomes more strongly homeostatic.*

Individual ants live only a year, while the experimental colonies were at least three years apart in age. This means that age differences in response to perturbations cannot be caused by the accumulated learning of particular, older individuals. Instead these results show that the organization of the colony itself changes as the colony gets older. The colony's single queen lasts the life of the colony, but it is very unlikely that the queen's behaviour affects a colony's moment-to-moment response to perturbations. The queen is deep inside the nest, surrounded by workers engaged in broad care. It is mechanically implausible that exterior workers could get back and forth to the queen through nest passageways quickly enough for the queen to mediate moment-to-moment fluctuations in the intensity of various activities. The word 'queen' often leads us to attribute too much authority to an ant that in most species does little more than lay eggs. Colonies get larger as they get older. Experiments are underway to investigate the functions of colony age and colony size (in terms of numbers of workers) in determining the ontogeny of the dynamics of colony behaviour.

Experiments with marked individuals (Gordon, 1987b) reveal some of the mechanisms of interactions between worker groups. For each of the four activities, ants engaged in it were marked with a unique colour and released back into the nest. Then perturbation experiments were carried out on colonies with marked ants. The question was, when perturbations directly affecting one activity cause changes in other activities, is that because ants are switching tasks? For example, toothpicks placed near the nest entrance cause an increase in nest maintenance and a decrease in foraging. Is this because foragers become aware of a current need for more nest maintenance workers, and switch tasks to do nest maintenance instead of foraging? The experiments show clearly that the question must be answered in the negative.

Response to perturbation, in terms of changes in temporal patterns of numbers of ants engaged in the various activities, is not mediated by task switching. Instead, it is mediated by recruitment of reserve workers from inside the nest. This implies a continuous exchange of information between different worker groups about the current state of the colony's environment. How does the recruitment of extra workers to help with nest maintenance, when the colony is faced with a pile of toothpicks, cause a decrease in the activity of regular foragers? How are these responses modulated to lead to the non-additive effects described above? Perturbation experiments raise intriguing new questions about the dynamics of interactions among different worker groups in the harvester ant colony.

To recapitulate this section in general terms: I began with a pattern of change in numbers of workers engaged in various activities, the daily round of colony activities. Perturbation experiments reveal some of the rules governing transformations of this pattern in response to changing environmental conditions. Responses to perturbations in colonies of different ages show that daily round flexibilities, that is, rules of colony response to environmental change, themselves change as colonies mature. In my current work, I am investigating how these processes affect interactions between neighbouring colonies.

It is clear that this investigation of harvester ant behaviour is concerned with change, from the initial pattern of change, or daily round, to interactions within a changing environment. A further step is to take into account ontogenetic change (in this case, a colony ontogeny). The results are novel not merely because harvester ant colonies are organized in interesting ways. Asking new kinds of questions, about transformations of behaviour, reveals new rules for behavioural processes. A research programme based on questions about static phenomena might not uncover these rules.

For example, much research in social insect behaviour is based on the concept of 'caste'. In some ant species, the workers in each colony fall into several distinct size classes. Wilson (e.g. 1978) showed that in some such polymorphic species, ants of particular sizes tend to do certain tasks. Subsequent investigations of polymorphic species, made by many different people, have led to similar results. In other, monomorphic species, an ant's task changes as it gets older. Ants of a particular age/size class are said to belong to a certain 'caste', which is assigned to a particular task.

In this view, what a colony *does*, is determined by which kinds or castes of ants are present, and the numbers of ants of each type. The distribution of workers into castes determines the behaviour of the colony. This leads to a research programme based on the questions: What is a colony's distribution into various castes? Why is this distribution adaptive? (e.g. Wilson, 1985).

These questions are concerned with static aspects of behaviour. 'Caste' itself is a static concept; Webster's dictionary defines 'caste' as 'a system of rigid social stratification characterized by hereditary social status' (Webster, 1977), and in the literature on social insects, the term carries a connotation of permanence through hereditary origin. The research programme assumes that there *is* a fixed caste distribution, and that this distribution can be used to predict a colony's behaviour. For example, Oster and Wilson (1978) characterize the foraging abilities of a colony in terms of the numbers of ants present in the foraging caste.

The research programme mentioned two paragraphs back has inspired many investigations of which individuals do which tasks in social insect colonies. Ant species may be ranked along a continuum. At one end of the continuum are species with extreme morphological specializations and strong task specificity. For example, workers of the leaf-cutter ant *Atta sexdens* may be sorted into eight distinct size classes. Each size or caste of worker tends to do a distinct task; the largest ants cut and retrieve leaves, while the smallest ants work inside the nest, inserting new bits of leaves into the colony's fungus garden (Wilson, 1980). Moving along the continuum, there are species with little morphological diversity, in which the age of a worker strongly determines its task (e.g. Otto, 1958). At the other end of the continuum are species in which all workers are morphologically similar, and a worker's age only partially determines its task (Gordon, 1984b).

The existence of this continuum of species, showing strong to weak task specificity in individual ants, has been well established. But the dynamics of task specificity are not yet understood. We know that in some species and in some situations, individuals tend to keep doing the same tasks. Now we can ask, in what species, how and when do workers change tasks?

It has been shown that the task performed by an individual may change if environmental conditions are altered (Meudec and Lenoir, 1982). That is, individuals switch tasks in response to moment-to-moment fluctuations of colony environment and colony needs. Thus the number of individuals of each physical type, such as age and size, is not the same as the number of individuals actually available at any instant to do various tasks. The actual distribution of workers into different task, that is, the numbers actually available to do various tasks, change in response to changed environmental conditions (Gordon, 1986a, 1987). My work, discussed above, also shows that the ways in which this actual distribution changes, are themselves subject to ontogenetic variation as a colony gets older (Gordon, 1987a).

It is becoming clear that at least in some species, an age/size class does not fully determine what its members will do. Instead, the tasks of age/size classes change, depending on environmental conditions. Thus caste, viewed as an inherent attribute that permanently adheres to individuals, may not be appropriate to describe the actual behaviour of individuals in an ant colony, because caste under-determines behaviour. The distribution of workers into different age/size classes does not necessarily generate the social organization of an ant colony.

Investigating the causes of transformations in the numbers of ants doing various tasks reveals that changes are predictable and rule governed. An investigation of the social organization of a colony merely in terms of caste would fail to register the dynamics of such changes which are clearly crucial to the way a colony functions. The study of caste distributions is important, but provides an incomplete explanation of colony behaviour. To understand colony organization, we need to invesitgate the dynamics, as well as the statics, of role behaviour. What are the rules that link the tasks performed by colony members with the colony's developmental status and environmental context? How do these rules change as the environment changes and as the colony gets older? These questions approach the understanding of colony organization very differently from questions about caste distributions.

CONCLUSIONS

The most direct way to formulate research questions about changes of behaviour, is to start out with patterns of behaviour that change in obvious ways. That is, the behaviour to be investigated can itself be a pattern of change. The literature on alternative tactics contains many examples of behaviour patterns that change in the course of development (Caro and Bateson, 1986) or in different social contexts (Lott, 1984). Two simple examples of behaviour patterns whose rules of change invite empirical investigation are movement patterns and cycles of behaviour.

First, movement patterns exist in many phyla and at various levels of organization. At the group level, schools of fish, flocks of birds and troops of primates are all known to have movement patterns. Schools of coral reef fish travel along regular, predictable routes to and from their feeding grounds. Such patterns change when environmental conditions, such as predators' behaviour, are altered (Wolf, 1985). The ontogeny of movement patterns in the history of a particular school, the extent and conditions of their flexibility, and their effects upon the movement patterns of neighbouring schools — all remain to be investigated. Bird flocks show annual migration patterns and daily patterns of space use, especially in the movements of small single- and mixed-species passerine flocks (e.g. Grzybowski, 1983). Primate troops in many species go on a 'daily march' which takes them along some predictable routes. Again, the development of a group's patterns, how the patterns change in response to environmental change, and the reciprocal dynamics of interactions between groups, are not yet understood for either birds or primates.

At the individual level, patterns of movement have been considered in studies of foraging behaviour (reviewed in Gray, 1986). Often these investigations do not go beyond the question: Is a particular pattern of individual movement adaptive? For example, does the movement pattern allow the animal to find the most food in the shortest possible time? Such questions assume the existence of a fixed optimum. Some further questions about the transformation of a forager's movement patterns would be: How does the pattern of a forager's movements change as the individual develops? How does the pattern change in response to changing conditions? How do changes in the foraging movements of one individual affect the movements of other individuals? When the dynamics of the behaviour are understood, one could go on to investigate the evolutionary consequences of variation in behaviour.

Second, daily cycles in social activities have been observed in many social groups (Harcourt, 1978). There are annual cycles of many kinds of behaviour, such as lek formation, migration and hibernation. The periodicity of some other cycles is not understood; for example, there are cycles of helping and reproducing in cooperatively breeding birds (see Woolfenden and Fitzpatrick, 1984), and cycles of population size and group behaviour in voles (Krebs, 1970).

In general, then, the first step is to choose some pattern of behavioural change. The next step is to discover how this pattern changes, developmentally and in response to changes in environmental conditions. Then we can examine how the dynamics of neighbouring individuals or groups interact. When these dynamics are understood, it will be possible to ask evolutionary questions.

Evolutionary questions about static entities are unrealistic. For example, it doesn't make sense to ask the evolutionary question, 'Why is it adaptive to do *x*?' while ignoring the ways that *x* changes. What evolves are the dynamic relations that an act such as *x* is part of, and from which a particular frozen instance of *x* cannot usefully be extracted. Other chapters in this book, and many other authors elsewhere, have emphasized that historical change, environmental change, context dependence — all are crucial in the formulation of evolutionary questions. To ask evolutionary questions about behaviour, we will first need to understand the rules for the transformation of behaviour; how behaviour develops, responds to environmental change, and interacts with changes in the behaviour of neighbours. Then we can look at how these rules for transformation themselves evolve.

ACKNOWLEDGEMENTS

I thank J.R. Gregg and R.C. Lewontin for their comments on the manuscript. The work was supported by a Junior Fellowship from the Harvard Society of Fellows.

REFERENCES

Altmann, S.A., and Altmann, J. (1970) *Baboon Ecology: African Field Research*, University Chicago Press, Chicago.

Caro, T.M. and Bateson, P. (1986). Organization and ontogeny of alternative tactics, *Anim. Behav.*, **34**, 1483–99.

Feinsinger, P. (1976). Organization of a tropical guild of insectivorous birds, *Ecol. Monog.*, **46**, 257–91.
Gordon, D.M. (1983). Daily activity rhythms in social activities of the harvester ant, *Pogonomyrmex badius*, *Psyche*, **90**, 413–23.
Gordon, D.M. (1984a). Species-specific patterns in the social activities of harvester ant colonies. *Ins. Soc.*, **31**, 74–86.
Gordon, D.M. (1984b). The persistence of role in exterior workers of the harvester ant, *Pogonomyrmex badius*, *Psyche*, **91**, 251–66.
Gordon, D.M. (1986a). The dynamics of group behaviour. In P.P.G. Bateson and P.H. Klopfer (eds), Plenum Press, New York.
Gordon, D.M. (1986b). The dynamics of the daily round of the harvester ant colony, *Anim. Behav.*, **34**, 1402–19.
Gordon, D.M. (1987a). Group-level dynamics in harvester ants: young colonies and the role of patrolling, *Anim. Behav.*, **35**, 833–43.
Gordon, D.M. (1987b). The dynamics of worker allocation in harvester ants. (in preparation).
Gray, R. (1986). Faith and foraging: a critique of the paradigm argument from design. In A.C. Kamil, J.R. Krebs, and H.P. Pulliam (eds), *Foraging Behavior*, Plenum Press, New York.
Grzybowski, J.A. (1983). Patterns of space use in grassland bird communities during winter. *Wilson Bull.*, **95**(4), 591–602.
Harcourt, A.H. (1978). Activity periods and patterns of social interaction: a neglected problem, *Behaviour*, **6**, 121–35.
Krebs, C.J. (1970). *Microtus* population biology: behavioral changes associated with the population cycle in *M. ochrogaster* and *M. pennsylvanicus, Ecology*, **51**, 34–52.
Levins, R., and Lewontin, R. (1985). *The Dialectical Biologist*, Harvard University Press, Cambridge, Mass.
Lott, D.F. (1984). Interspecific variation in the social systems of wild vertebrates, *Behaviour*, **88**, 266–325.
McBride, G. (1975). The study of social organizations, *Behaviour*, **59**(3), 96–115.
Meudec, M., and Lenoir, A. (1982). Social responses to variation in food supply and nest suitability in ants (*Tapinoma erraticum*), *Anim. Behav.*, **30**, 284–92.
Oster, G., and Wilson, E.O. (1978). *Caste and Ecology in the Social Insects*, Princeton University Press, Princeton, NJ.
Otto, D. (1958). Uber die Arbeitsteilung im Staate von *Formica rufa rufo-pratensis minor* Gossw. und ihre verhaltensphysiologischen Grundlagen: Ein Beitrag zur Biologie der Roten Waldamiese. *Wissenschaftliche Abhandlungen der Deutschen Akademie der Landwirtshaftswissenschaften zu Berlin*, **30**, 1–69.
Webster's New Collegiate Dictionary (1977). G. and C. Merriam Co, Springfield, Mass.
Wilson, E.O. (1978). Division of labor in fire ants based on physical castes (Hymenoptera: Formicidae: *Solenopsis*), *J. Kans. Ent. Soc.*, **51**, 615–36.
Wilson, E.O. (1980). Caste and division of labor in leaf-cutter ants (Hymenoptera:Formicidae:*Atta*). 1. The overall pattern in *A. sexdens, Behav. Ecol. Sociolbiol.*, **7**, 143–56.
Wilson, E.O. (1985). The sociogenesis of insect colonies, *Science*, **228**, 1489–95.
Woolfenden, G.E. and Fitzpatrick, J.W. (1984). *The Florida Scrub Jay*, Princeton University Press, Princeton, N.J.
Wolf, N. (1985). *Behav.Ecol. Sociobiol.*, **17**, 47–52.

Evolutionary Processes and Metaphors
Edited by M.-W. Ho and S.W. Fox

CHAPTER 13

Stasis, development and heredity

SUSAN OYAMA

ABSTRACT

Development is the result of integrated organism–environment systems. It is the heart of evolution, conceived as change in the distribution and constitution of developmental systems. Ontogenetic and phylogenetic changes are a function of active organisms in their changing surroundings. Despite this, evolutionary theory continues to be associated with a peculiarly static view of organisms. To a certain extent this is due to our models of changes and to the ways they inform some fundamental concepts in biology. Two models of change are examined: *variational* (differential perpetuation of fixed variants in a collection) and *transformational* (predetermined change in an entity or a collection). Though they are treated as alternative explanations, their synthesis allows a clearer representation of both evolutionary and developmental change than does either of them separately.

Three interrelated concepts are also discussed: natural selection, innateness and hereditary transmission. Each has a metaphorical aspect that is intimately tied to traditional models of change. To link evolution and development in the traditional way, by genetic transmission of genes or traits, is to embrace a set of untenable nature–nurture oppositions, among them the one between biology and culture. A heredity based on processes and resources rather than products eliminates these oppositions while restoring activity and change to the studies of development and evolution.

INTRODUCTION

Though contemporary evolutionary theory is said to rest on a synthesis, it is also based on a number of antitheses. Internal and external forces are opposed in explanations of ontogeny and of phylogeny, biology is contrasted with culture, organisms are separated from their environments. In so far as development is attributed largely to internal forces and evolution to external ones, these too are contrasted. But development is the very heart of evolution. To see why this is so, we must examine our assumptions about the nature of ontogenetic and phylogenetic change and adopt a more dynamic, holistic approach to biological processes.

Evolution is change in the distribution and constitution of developmental (organism–environment) systems. This often involves change in gene frequencies, but focusing exclusively on the gene level excludes from life processes the very richness and activity that commanded attention in the first place. Defining evolution, heredity and development in terms of genes also commits us to a set of nature–nurture oppositions that proliferate confusion where we can least afford it. This is especially the case when the organisms in question are human beings, and when biology is identified with 'propensities' that seem to leave little room for deliberation, choice and action.

I will look at two models of change and at three other concepts that are closely tied to them. The models have been called 'variational' (differential perpetuation of fixed variants in a collection) and 'transformational' (predetermined change in an entity or a collection of entities) (Lewontin, 1982). The concepts they inform are: (1) natural selection, (2) innateness and (3) hereditary transmission of traits. Contemporary thinking about each of these three is dominated by a ruling metaphor. Natural selection is treated as action by an external agent. Innate traits are attributed to internal 'programmes'. And, finally, heredity is modelled on the transmission of wealth and objects in human societies—thus, traits are 'passed' from one organism to another.

These metaphors distort the phenomena they describe and create a spurious set of connections among them. Because heredity is seen as the transmission of traits, or more precisely, genes for traits, evolution is reduced to changes in gene frequencies, while ontogeny must be largely explicable in terms of those genes. The whole story requires, then, that the genetic products of evolution pass through the bottleneck of a narrowly defined heredity; they can then re-create the living world by ontogenetic expansion. As in algebraic expansion, however, the solution is implicit in the first equation and must simply be revealed by routine operations. The idea of genetically created phenotypes then reinforces the idea of natural selection as 'operating' on static traits. We thus have a kind of reverberating circuit of ways of thinking about life processes. It is this circuit that must be broken if we are to understand the relations among selection, development and heredity. My aim is to explore the aspects of evolutionary theory that imply passivity and stasis in organisms, and to show that these implications can be denied without misrepresenting biological processes. We can make development central to evolution without being committed to notions of predestined change in either ontogeny or phylogeny.

TWO MODELS OF CHANGE

Richard Lewontin has described two models of change which have been treated as alternative explanations for both ontogeny and phylogeny (1982, 1983a). Each combines two ideas, however, which are not necessarily connected. In each case, one can accept the first idea and reject the second. In the 'transformational' model of evolution: (1) change in a collection is explained by change in its constituent entities, and (2) change in the entities is predetermined and uniform. In the 'variational' model: (1) only some of the variants in a collection are perpetuated, and (2) the variants themselves are static. Change is assumed to be internally driven in the former model, externally directed in the latter. Notice, though, that these assumptions stem from the second component in each — predetermined change and static variants. When these are dropped, the transformational and variational explanations are not only compatible, but form the basis for a more dynamic, holistic approach to biological processes than the one that is now prevalent. A population changes as its constituent developmental systems change and perpetuate themselves with different frequencies.

Elliot Sober has drawn on Lewontin's distinction for his discussion of Darwinian natural selection. Natural selection is a variational process. Some property of the variant acts as a 'positive causal factor' in increasing its probability of appearing in the next generation (1984, Chapter 8).[1] Focus is not on change in individuals but on differences among them. The individuals themselves may even be static. Lewontin says, 'It is only the collection that changes, not the individuals of which it is composed' (1982: 155).

Modern evolutionists, then, have rejected the transformational model of population change and embraced the variational one. This is often seen as the triumph of Darwinian over Lamarckian thought. Developmentalists have not been so absolute; they have tended to incorporate both models into their schemes. Behaviourism is an example of a selectional approach to ontogeny. Other examples are found in immunology and neurobiology (Edelman, 1978; Jerne, 1967). Because of the gene-centred logic described above, however, the transformational model of ontogenetic unfolding predominates whenever a feature is thought to have an evolutionary history.[2] In fact, Lewontin (1983b) asserts that contemporary evolutionary thought has combined the Mendelian idea, that internal factors make the organism, with the Darwinian one that external factors shape the population. The opposition of internal to external causes is, as he points out, a deeply problematic one. Insisting on the primacy of one or the other for either ontogeny or phylogeny is not as useful as constructing an alternative to the framework that seems to demand a choice. An alternative is available, but it entails rethinking some of the metaphors that reflect (and support) this attempt to allocate causal efficacy to internal or to external sources.

Though it has been true that an emphasis on natural selection has tended to promote a static view of organisms (Gray, this volume; Lewontin, 1982; Sober, 1984), it need not be so. A variational explanation need not preclude individual activity and development, and regular change in an entity need not be seen as the fulfilment of a plan. Moving beyond the alternatives, then, involves moving beyond the opposition of internal to external forces, towards a view of biological constancy and change as a function of interactive systems. I would like to examine the metaphors of selection by an agent, internal 'programmes' for innate traits, and heredity as trait transmission, and the ways each implies a kind of stasis. Interrelated as they are, they must change together if they are to change at all. Natural selection requires reliable life cycles, not static genetic programmes or organisms. Ontogeny is the contingent functioning of entire developmental systems. It is the systems, mobile networks of organism–niche relations, that are 'transmitted' (reconstructed) in heredity. These enlarged conceptions allow us to capture the relation between individual variability and population change without being committed to a static view of life.

NATURAL SELECTION: ACTION OR INTERACTION?

Selection as action by an agent

In the variational model of evolution, as we have seen, focus is on the changes in the collection, not the individual. Sober asserts that selection, a variational process, permits and even requires a kind of stasis in individuals (1984: 149). Natural selection is seen as a force external to the organism, 'acting' on arrays of traits; change in the traits seems at best irrelevant to the process and at worst a threat to it. Why is this so?

The notion of choices made from a static array, first of all, encourages a momentary denial of development. In order to make decisions about biological 'raw materials', the breeder on whom nature-as-selector is modelled typically surveys organisms in the same developmental state, so that they are as closely comparable as possible. If the adult character is of interest, earlier ones will be 'invisible' at least until the adult state is predictable.

Another kind of stasis is implied in conventional conceptions of heredity and development. Traits under selection must be 'inherited', and the use of the word to describe both population relationships and the ontogeny of certain phenotypic features encourages us to associate a sort of developmental stasis with a selectional history. The common belief is that there are two kinds of development, one driven from within by immortal genes and giving rise to fixed 'innate' traits, and the other caused from without and resulting in more flexible (though evolutionarily irrelevant) products. We will return to the problem of 'programmed' development later.

Not only does the organism at selection tend to be treated as a fixed bundle of (fixed) traits; it tends to be seen as passive as well. But organisms choose and transform their environments even as they are affected by them. Selection is the outcome of certain interactions between organism and milieu, not the act of an agent on a passive object. (See Bateson and Gray, both in this volume; Ho, in press; Lewontin, 1982; Taylor, 1987.)

Selection as interaction

These assumptions of stasis and of passivity make the image of selection by the environment misleading in important ways.[3] Nature is not a deciding agent, standing outside organisms and waving them to the right or to the left. However much we may speak of selection 'operating' on populations, 'moulding' bodies and minds, when the metaphorical dust has settled, what we are referring to is still the cumulative result of particular life courses negotiated in particular circumstances. 'Selection' is shorthand for certain kinds of changes in the distribution of interacting developmental systems. It refers to those population changes that occur when there is heritable variation (traits are differentially associated with lineages) and the variants interact with their environments in ways that confer on them different probabilities of being perpetuated. The mistake is to speak of a selecting *agent* at all.

As it is currently defined, evolution by natural selection occurs only when gene frequencies change, but one could be less restrictive. My preference is for a broader definition of evolution as change in developmental (organism–niche) systems (Oyama, 1985); the degree to which genetic change is involved would then be an open question. Lewontin (personal communication) uses a distinction made by Sober (1984): there is selection 'for' properties that enhance the likelihood of survival and reproduction, and selection 'of' the objects that recur in the next generation as a result. Lewontin suggests that selection is thus 'for' an instantaneous state but 'of' the developmental system.

As we come to understand the processes that produce transgenerational stability and change we may continue to speak of natural selection because the concept has expanded with our understanding (to accommodate reciprocal influences between organism and environment, for instance), or we may decide the term is too closely tied to undesirable implications (stasis, action by an agent on a passive object) and choose another. In so far as the former is true the term will have lost much of its metaphorical power. Some would argue that metaphors do not so easily relinquish their hold on our thought and action, but what is finally important is how we construe the concept and how we demonstrate its appropriateness. The developmental system includes not only the organism but also the features of the extra-organismic environment that influence development. By positioning the organism in the ecological context in which it exists and grows, this view overcomes the alienation of organism and environment of which Lewontin has complained (1982 and elsewhere; see also Gray, this volume) while restoring change and relation to the world we study.

INNATENESS: PERSISTENT ESSENCES OR RECONSTRUCTED SYSTEMS?

Programmes for natural states

The concept of the innate is a second part of the reverberating circuit mentioned earlier. One of the many metaphors attached to it is that of the interior plan for development. Sober has traced the descent of the related idea of the 'natural state' from Aristotelian essentialism. It appears in biology in several guises. Among them is the notion of a species essence, which will tend to be expressed as long as there is no interference; it is a 'zero-force state' (Sober, 1980). Innateness is used to refer to both species-typical traits and to certain variable ones (one of the many sources of confusion in the nature–nurture problem), so it is not surprising that individual variations can also be attributed to internal essences or plans. Thus a person's 'genetic propensity' or 'biological potential' is treated as that phenotype towards which he or she tends as long as extraneous influences do not intervene.[4]

In so far as natural selection is said to act only on inherited traits, and in so far as these are treated as natural states represented in and created by the genes, selection implies a kind of developmental stasis. 'Innate' characters are thought to be internally generated and trebly static: immutable in individuals, uniform across generations and/or universal in individuals. They are attributed to genetic programmes 'for' those features, and are assumed to be inevitable or changeable only at heavy cost. Form and causal agency are placed in the nucleus, protected from commerce with the shifting world outside by Weismann's barrier (the segregation of germ cells from body cells in some organisms).[5] Thus the innate is opposed to the acquired as the necessary and timeless is opposed to the contingent and fluid. Though departures from the species or individual natural state are acknowledged, they are seen as *deviations* from the natural (even desirable) condition.

The importance of natural selection in evolution is a matter of dispute. (It may be less common than assumed; Sober's 1984 criteria for a positive causal factor seem very stringent; see also Hailman, 1982.) But in so far as a selectional history *is* indicated, it does not require developmental fixity (Lehrman, 1970) or any other traditional marker of innateness.

Persistence and reconstruction

It is true that if one takes the snap-shot attitude mentioned above, of stopping time to create arrays of features for comparison, transgenerational stability can give the impression of immobility; many photographs of similar objects may seem to be images of the same object. We speak of 'the shark', for instance, remaining 'the same' for millions of years, as though it were a single shark, not a succession of them. Interestingly enough, George Williams, seeking to rebut Sober's criticism of his gene-level account of evolution, makes this very point: that phenotypes do not actually *persist* across generations, but rather *recur*. Genes, he says, persist. They are thus the real objects of selection. He counters the obvious objection, that genes themselves recur by replication, by detaching the notion of the gene from its physical embodiment, characterizing it instead as a 'weightless package' of information coded in the secondary structure of DNA (Williams, 1985: 121). But even leaving aside the complexities of overlapping and discontinuous genes and the problems they pose for the concept of coding, the information-as-sequence must be reconstituted as well. Disembodiment seems not to solve the problem of discontinuity. Williams seems, in fact, to be saying that physical persistence is not important after all. Reliable availability in each life cycle, however, is. Though I agree with him, I think this admission undermines his insistence on the primacy of the genetic level.

Williams writes of 'natural selection *of information for its effects*' (1985: 121) and says the logic of selection would permit other mechanisms of information transmission besides DNA replication.[6] If 'information' is defined by its contribution to development, as Williams' phrase implies, one need not look far for alternative means of transmission; organisms have a variety of media at their disposal. Cell structure (both membrane and cytoplasm) is part of the developmental system that connects the generations, as are chemicals derived from maternal DNA and a myriad of extracellular and extra-organismic influences (Cohen, 1979). Many of these are provided by parental physiology and behaviour, while others are reliable aspects of the larger environment, including the social environment. The 'informational' function of any influence is determined by the role it plays in the developmental system as a whole. Regularity of gene function is thus a result of developmental regularity as well as a cause of it, and the impact of stimulation depends on the state of the organism. Any biologically interesting notion of information must be interactively defined in this way, and what is crucial is not permanence but availability at the appropriate time. Activity, not stasis, and relation, not autonomy, are central to this conception. Persistence is beside the point in accounting for the reliability of both genetic and non-genetic effects.

It is possible for features to evolve without appearing in each generation or in all natural environments, and even the most reliable features have an ontogeny. If developmental courses are regular, they can for some purposes be taken for granted. (It does not follow that taking them for granted makes them regular, and taking them for granted is not the same as understanding them.) One can explain a bed of red flowers by pointing out that only certain kinds of bulbs were planted, even though bulbs have no flowers, the flowers are fleeting, and under other conditions the bulbs might have produced blooms of a different shade, or none at all.

Dynamic stability, then, is needed if natural selection is to affect a population. It is a contingent matter whether and how the conditions for construction of a feature are correlated with the conditions for its use. The variants must recur, and so must the causal background against which a given phenotypic difference makes for a difference in fitness. But, as we have already observed, people often reverse this reasoning and assume developmental stability from some evolutionary argument or other. Having concluded that a feature is in some sense adaptive, they may assume that it evolved by natural selection (as traditionally defined, by genes), and thus tacitly assume this kind of long-term stability even when the assumption may not be justified. Or they may project the stability into the future, predicting fixity without knowing whether the conditions it requires will be present (Kitcher, 1985, this volume).

HEREDITY: TRAITS, DIFFERENCES OR SYSTEMS?

Basic to much of this confusion about the relationship between evolution and development is the concept of heredity itself. The frequent conflation of population and individual levels is very much related to the dogma that evolution is fundamentally a matter of selection for genes that make organisms. Thus emphasis is placed sometimes on population properties (variation or lack of it) and sometimes on phenotypic properties thought to reflect degree of genetic causation (absence of learning, presence at birth, etc.). Genes that go to fixation in a population are somehow assumed to fix traits in individuals or genetically determined variance is thought to imply 'genetically determined' traits.

We have seen that a variational mechanism permits stasis; Sober claims that Darwinian selection requires stasis both in individuals and across generations (1984: 149–52). But I have argued that selected entities need not be static. What stability across generations requires is heritable variation. Heritability is often thought to imply stasis of some sort (our third metaphor, the transmission of traits across generations, on the model of property inheritance), but what it really involves is recurrence in lineages. Offspring must resemble their relatives more than they resemble non-relatives (Hailman, 1982; Lewontin, 1978; Sober, 1984: 151). This in turn involves dynamic regularity, which depends on order both outside and inside the organism. As reliable as these processes may be under some circumstances, their repetition is always a contingent matter, and though it is often assumed that variations in conditions must be large in order to deflect an evolved developmental course, small changes, especially outside the normal range, can have significant effects.

Individuals: inheritance of traits

What then is inherited? In seeking to come to terms with transgenerational regularity of living forms and functions, one is tempted to employ the metaphor of hereditary transmission of traits (Oyama, in press b). One focuses on similarity between generations or on some aspect of the phenotype that justifies saying the feature was 'passed on' from parent to offspring. An apparent benefit of this tack would seem to be that everyone 'knows' what these words mean. Some things are in the genes in some unspecified (unspecifiable, really) way, while others are imposed from the outside. Unfortunately, the multitude of meanings that travel together under this rubric are neither logically required by each other nor reliably linked empirically. A trait that is present at birth, for instance, is not necessarily immune either to prenatal or postnatal variation in conditions (birthweight is related to a variety of prenatal influences, and does not always predict later size). One that involves no apparent learning is not always invariant in or between generations (insect caste membership can be induced by a variety of factors). One that shows parent–offspring similarity need not be longitudinally stable (a brunette and her dark-haired daughter may both have been blond as infants). Reliance on this ready-made set of assumptions, then, is risky.

Since all aspects of the phenotype must develop, they are all 'acquired' in ontogeny, whatever their distribution in the population. They are all 'environmental', since particular conditions are required for development, and since these conditions enter into the formation of the organism from the beginning. Phenotypes are all 'inherited' and 'biological' as well, if by this one means some sort of causal role of the genes in their development. If one seriously accepts the origin of phenotypes in causal interaction, as everyone seems to, judging by their writings, no distinction between inherited and acquired components of the organism is defensible.

Populations: inheritance of differences

When the statistical meaning of heritability is used, workers sometimes resort to locutions like 'inherited differences' or 'innate variation' to refer to the relationship between genetic and phenotypic variance in the population. This has the appeal of being technically sophisticated. By referring to variation, not to traits, it theoretically frees the user from unwanted nature–nurture implications. But it is quite hard to think of inheriting *differences* or *changes* (G. Bateson, 1979: 167–8). G. Bateson observes that a difference has no location; it refers to a relationship (p. 109). It is even harder to know what it would mean to attribute selfishness to one (see P. Bateson, 1986). The word, 'inherit', after all, customarily refers to the passing of objects, land, money — roughly, *resources*, from one individual to another.

Heritability statistics are dependent on the constitution and distribution of developmental systems in a specific population at a specific time, and correlations among relatives may change (Hailman, in press). Human behavioural geneticists seek to document age changes in heritability (Plomin, 1986). This relativity contradicts many of the implications of static, internal essence that inheritance has traditionally had (which is a good thing), but the developmental implications nevertheless tend to be attached to the variants in population discussions (which is not a good thing).

Inheritance of developmental systems

I have suggested we should see heredity not as the transmission of traits between organisms, or even as the transmission of differences, but rather as the ways in which developmental resources become available to the next generation. Focus on process allows us to be true to the regularity these metaphors are meant to convey while recognizing the organized and organizing activity that creates both regularity and variety.

Traits do not pass from one organism to another, but must be constructed in ontogeny. This is true whether or not they are invariant in the population or have a traceable phylogenetic history. Most people's response to this obvious truth is to assert that what is literally transmitted is the genes (or information) that makes the traits, so that the metaphor is innocuous. If, however, one believes that what is inherited is whatever it takes to make a trait, why be satisfied with a partial account? If one wishes, that is, to identify heredity with developmental *means* (the prerequisites for ontogenetic construction) one should include all of them; the organism then inherits (has available to it) a highly complex, widely ramified developmental system.

Some seek to solve the problem of the developmental insufficiency of the genes by including in the hereditary package some cytoplasm, transcripts of maternal genes and other cellular features. This is marginally better than focusing on naked DNA. It tends to pass muster because it remains comfortably within the boundary of the egg, the traditional vehicle of heredity. But no egg can develop in a vacuum, and other ontogenetic requirements must be provided in some way. Many are reliably present in the wider environment, but many others may have to be supplied by the organism itself, by parents and other conspecifics, or even organisms of other species. (Consider mutualisms, in which members of one species feed, protect, transport or pollinate members of another. Systems can also be nested, as with endosymbionts.) The developmental system includes, then, not just genes, but whatever else in the living or non-living environment contributes to or supports development. Its constituents and configuration shift with its activity. In this view many parts of the system are themselves generated by it. Hormones produced by the developing organism, for instance, serve as chemical environment for further chemical interactions. Activity may provide necessary self-stimulation or social contacts. Whether the interaction is within the organism or between it and its niche, both external and internal conditions are important, and the dynamic exchange characteristic of biological processes finally makes the division artificial.

If, on the other hand, one believes that what is inherited is that which makes one trait different from some other one (the 'inherited differences' approach), one must inherit not only the influence that makes the difference but the system in which it makes that difference. There are also many non-genetic influences that can play this role of difference-maker, generation after generation (Cohen, 1979; Gray, this volume). Consider the ability of customs and social structure to maintain differences among humans. Whether one speaks of traits or differences, then, what one must ultimately reckon with is a full developmental system.

Development is thus basic to all notions of inheritance. What I have in mind is most emphatically not the sort of unfolding of predestined fate involved in the transformational model. This would be revelation, not what I call development. Forms emerge. They are neither imposed from without nor 'translated' from within. Nor do the genes give the basic outlines, while the environment provides the details. Just as unsatisfactory is a dual model (some things unfold, some do not) or a continuum (there are degrees of programming). All development involves contingency and ordered (but real) change. It is a process generated by interaction of a heterogeneous mix of influences, and its mobile unity and order derive from its embeddedness in this matrix of causes, not from insulation against it.

CONSTRAINTS AND POSSIBILITIES

Sometimes the focus in heredity is not on what is generated, but on what may *not* be. Nature–nurture disputes are increasingly being recast in the language of constraints. But to see biological nature in terms of internal limits on extraneous developmental variation is simply to resurrect (albeit in attenuated form) the old concepts of natural states and developmental necessity that fuelled the nature–nurture debate. At another level, scientists argue about developmental constraints on natural selection. Again, the concern is with what may not happen, and once more the opposition is between an internal resistance to change and an arbitrary external force.

In both cases the dispute is ultimately about the possibilities of developmental systems: about the ontogenetic possibilities of a genotype or the possibilities for variation in successive developmental systems. No rules, however, specify the range for either. There is no reason to suppose that all conceivable alterations are possible, or that all possible ones are equally probable. Non-random possibilities for change would seem to be part of our concept of a structured system. But structure has a history; it comes into existence through activity and may change with further activity. It simultaneously generates change and directs or constrains it — *these are not opposing forces but two perspectives on dynamic order*. Like potentials, constraints are most usefully conceptualized as relational, not 'endogenous', and as emerging in processes, not as prior to them. Possibilities for change evolve; they are generated in interaction (Oyama, 1981). To oppose necessity (physical, biological or developmental) to history, then, is to misrepresent both.

Developmental systems can vary in many ways. Sometimes a variation enhances the likelihood that the new system will recur. If the perpetuation of variations is a joint function of active organisms and surrounding conditions, and if development is structured but not predestined, then the variational and transformational accounts of population change can be synthesized (Gray, this volume).

BIOLOGY AND CULTURE

I have touched on some ways the environment has been personified in descriptions of natural selection. I have also maintained that the treatment of the 'innate' as produced by static internal plans is more a product of conceptual confusion than an accurate empirical description. The role of the metaphor of trait transmission in joining these views of evolution and development has also been examined.

These mutually reinforcing metaphors often lead us to associate evolution with stasis — a paradox, since evolution is defined by change, depends on change, and makes other kinds of change possible. Stasis (or, better, some stable set of conditions and relations) is projected backward or forward in time. Thus, as mentioned earlier, developmental fixity is inferred from a phylogenetic history, and a history of natural selection is assumed when a trait is pronounced to be adaptive or 'biological'. Though heredity does link development to evolution, the relation is misconstrued when the processes are misconstrued. I have proposed a wider definition of inheritance than is generally used — one based on process rather than product, on the 'transmission' of developmental resources rather than of traits.

None of the assumptions of stasis associated with evolution by natural selection is justified. Stasis at selection is permitted by a selection mechanism but not required by it, even though our observational practices tend to freeze time in cross-sections. Our assumption that organisms under selection are passive reflects our tendency to think about these interactions as choices made by an outside agent. The stability requirement, that a selected entity actually manifest itself in successive generations with some regularity, entails reliable ontogenetic reconstruction, and this is a matter of process, not immobility (Ho, in press and this volume).

Natural states, natural conclusions

The usual way to distinguish biology from culture is to attribute 'innate' characters to genetic evolution, while 'acquired' or learned ones are chalked up to 'Lamarckian' cultural evolution. This is to rely on an old and deeply entrenched version of the nature–nurture opposition. If inheritance is defined in terms of developmental systems, no such distinction between innate and acquired characters is necessary or even possible. Yet, even those who insist that developmental and evolutionary questions are not to be confused, who denounce the entire nature-nurture controversy as senseless, tend to associate the 'biological' with deep, immutable truths, usually emotional or motivational ones, and the 'cultural' with contingent detail, perhaps learned by imitation or conditioning. (However casually 'imitation' is invoked in such discussions, the term embraces considerable diversity; see Galef, in press.)

Though the essentialist idea of a privileged developmental pathway and phenotype is not defensible, it is very much alive in biology, psychology and anthropology. Generally expressed in terms of biological 'bases' or 'propensities', it is a version of the natural state model described earlier. When I plead for increased attention to development in evolutionary studies, this model is just what I do *not* intend.

If one believes in natural states and thinks they are properly termed 'biological', perhaps it is natural to conclude that anything that has been dubbed 'biological' (because it has been associated with a brain locus, occurs in some phylogenetic relative, appears to be difficult to influence in individuals, seems adaptive, is universal or correlated with genotype in some population, etc.) represents one of these states. It may be natural for the species or for an individual. If humans are 'naturally' polygynous, common tendencies are at issue, while if a person is a 'natural' athlete, he or she is distinguished from the rest of us. One reads of 'genotypic' aggression (Buss, 1984; Konner, 1982: 203–7). Aggression, then, can be in the nucleus, regardless of the behaviour of the *organism* (Klama, 1988; Oyama, in press a). Ultimate biological reality is somehow represented in the DNA, no matter what the outcome of any particular developmental course.

In addition to this vision of a hidden reality lurking in the genes, the natural state involves the idea of a tendency to approach the state. Sober observes that Aristotelian essences were 'causal mechanisms of a certain sort' (1984: 162). They are what we will be if nothing interferes. So we are told we must carefully '*teach* our children altruism, for we cannot expect it to be part of their biological nature', a nature defined by their selfish genes (Dawkins, 1976: 150, emphasis in original).

Allied to this notion of natural-state-as-vector is the assumption that such tendencies will be hard to deflect. Attempts to do so are likely to fail or to incur unwanted consequences. (Kitcher, 1985, Chapter 4, speaks of precluded states and states that preclude important desiderata. See Oyama, 1985, Chapter 6, for other examples.) Natural tendencies will press for expression, and we may suffer if they are denied. Instincts and psychological drives fit well here. We are even told what kinds of political systems are incompatible with our biological nature (Wilson, 1978: 190). Many social scientists have been quick to accept biologists' offer of a definition of a universal, fixed human nature and to draw conclusions about possible social arrangements (see, for instance, Peterson and Somit, 1978). One question never addressed in these treatments is why it is only *deviation* from the programme that is thought to entail stress — a peculiar omission unless one assumes that the natural is also good and easy. This seems a small point until one realizes that inordinate focus on the putative costs of changing may divert attention from the costs of *not* changing.

The natural state model encourages a sequence of inferential leaps among the present, the past and the future. It assumes just the conclusions we should be questioning and takes for granted just the complicated developmental and social systems we should be investigating. (These systems will not necessarily be easy to alter, but invoking or demonstrating an evolutionary history does not tell us one way or another.) Finally, it diverts attention from the many kinds of regularity that would have been required as preconditions for such evolution.

Many questions about evolution, adaptiveness, or the role of genes in development are not only technical questions. They are also about the degree to which we feel we ought to view aspects of our lives as necessary, even good. When human customs and activities are at issue, the developmental stability we must assume extends far beyond what is required to produce recognizable morphology.[7] It even extends beyond 'behaviour', if by this we mean the relatively well-defined patterns of movements observed in many other animals. Social life is made up not of *movements*, but of *acts*, which are largely defined by intent and meaning.[8] Intentions are construed, and meanings constructed, by agents interacting in a social and physical context that itself influences the significance of the interaction. We need a view of development that takes the organism and its intimate interchanges with its surroundings (including its social surroundings) seriously. How do the nested systems work? A given child-rearing practice may have different meanings in different places or at different times, and thus, different implications for the child's development and its relations with others (to say nothing of its reproductive success). The human mind, in all its subtlety and variety, is social from its inception (a developmental psychology that is not social is no psychology at all).

Unified systems, multiple questions

In a discussion of behavioural development, Gilbert Gottlieb (1976) suggested some questions to guide research. How does experience facilitate, induce or maintain development? Patrick Bateson (1983) added a predisposing role, and pointed out that the same question could be asked of internal influences. Such questions undermine the attribution of internal necessity to some processes (and of 'mere' external contingency to others) by some *a priori* criterion. At the same time, they direct attention to the causal factors that, in combination, could actually help us understand the phenomena that give rise to our concerns about developmental chance and necessity in the first place — degrees of reliability over time and across individuals. How do these constructions occur, and how do previous processes influence the likelihood of subsequent ones? How, in other words, do developmental possibilities and constraints arise?

If we move one level up, from individual ontogeny to what one might dub Iterated and Interacting Developmental Systems (companions to Gray's Ecological and Evolutionary Cascades, this volume — IIDSs and EECs, pronounced, naturally, 'ids' and 'eeks') that constitute evolution, perhaps we can ask similar questions, with similar benefits. To account for phenotypic regularity across generations we need to account for change and constancy in developmental processes themselves, including those involving social interaction. For a feature to increase in frequency in a population, the conditions required for its development must be associated with increasing frequency. How are variations in developmental systems generated and perpetuated? Some perturbations will have no noticeable impact on an organism–niche system, while others will induce a change in it. In some cases the change will damp out in one or a few generations — in others it will be maintained and in still others, amplified. Other factors may predispose the system towards certain kinds of change or facilitate it when it occurs.[9] Genetic and non-genetic changes may be entrained. Gray observes that current research practice is not geared to discovering EECs (this volume). Laboratory researchers and breeders typically control conditions for successive generations. Such control may well minimize the probability of observing the kinds of novelties that might lead to a cascade. Whether one finds damping, maintenance or amplification will depend not just on the source of perturbation or the nature of the change. It will also depend on the available developmental and ecological resources — the possible organism–niche complexes. The relationship between an influence and the rest of the system is crucial.

Focus on the ontogenetic construction of phenotypes eliminates the opposition between biology and culture. What we need is *not* ever more sophisticated ways to prize them apart, but rather a view of life and history that is rich enough to integrate the genetic, morphological, psychological and social levels (each biological, each with a history) in such a way that we are not tempted to indulge in phenotype partitioning at all.

In any developmental system stability depends on integration of lower levels into very broad ecological ones. For cultural stability, this means certain kinds of relationships among institutions, beliefs, values and acts. It is an error to assume that cultural facts can change quixotically, while 'biological' ones, however defined, cannot. To know whether a system is apt to continue to perpetuate itself, does it make more sense to extrapolate from past stability (or infer it from inappropriate evidence) or to look at the structure of the system itself?

Activity and agency

At a recent conference I watched a behaviour geneticist diagram human development. Seeking to counter charges of oversimplification, she had filled a slide with arrows and loops. There were causes, influences and moderators galore — but amid the maze of lines and labels, there was no *person* to be found. Just as evolutionists frequently regard organisms as evanescent epiphenomena to the real action in the gene pool (Dawkins, 1976), so developmentalists, in their analytic zeal, have often ignored the unifying role of the individual, rendering it an abstract nexus of inner and outer causation. To see the organism–niche complex as both source and product of its development is to acknowledge the role of activity in life processes — not genes organizing inert raw material into beings, and not environments shaping or selecting passive bodies and minds, but organisms assimilating, seeking, manipulating their worlds, even as they accommodate and respond to them.

To insist upon the organism as a locus of agency is not to deny causality or to render causal investigation useless; it is not, that is, to embrace some concept of free will as uncaused cause. Nor is it to attribute human mentality to organisms that lack it. Still less is it to deny the social integration of individuals. It is, simply, to reinstate living beings in our analytic schemes. One benefit of this is that it makes clear where we must look if we wish to know what the possibilities for change are, for an individual or for the species: not at some set of disembodied constraints or rules or programmes, but at relations within the organism and between it and its surround.

Biology is often contrasted with history, especially cultural history. Human history, however, is fully biological, not because it is predestined, but because it is the chronicle of the activities of living beings. Moreover, there is no 'biological' character that does not have developmental history. Missing from many views of cultural change and stability is the recognition that values, beliefs and ways of behaving are ontogenetic constructions of a subtlety that is not captured by the metaphor of transmission, even 'transmission by learning'. The same is true of 'genetically transmitted' characters. Both transmission metaphors deny real development in so far as they attribute form and directive causality to some shaping agent — genes or culture — rather than allowing them to emerge through the activity of the organism itself.

Much of cultural change occurs as individuals come to interpret old messages (including messages about biology and history) and practices in a new way; this in turn affects their impact on others. What comes of the chemical, mechanical and social–psychological resources an organism inherits depends on it and its relations with the rest of the world. It is in this way that it makes its own present and prepares its future, never out of whole cloth, always with the means at hand, but often with the possibility of putting them together in novel ways.

NOTES

1. The variational model is thus broader than the selectional one, including not only selection but phenomena like genetic drift as well (Sober, 1985). In exploring the structure of these models Lewontin and Sober have voiced many of the criticisms presented here, and the former's concept of the interpenetration of organism and environment (Lewontin, 1982) fits well with my concept of the developmental system.
2. Sober substitutes 'developmental' for 'transformational'. He considers the theories of Piaget and Chomsky to be 'developmental' because of their emphasis on 'preordained' sequences and 'endogenous constraints' (Sober, 1984: 153–5). Not all developmental theories involve the unfolding of predestined form, of course, and though both Piaget and Chomsky are concerned with universals, their views of development are quite different.

   Though it is the variational approach to evolution that is associated with stasis in these accounts, the transformational model involves a kind of stasis as well. In psychology these preformationist connotations are present in instinct theories. They are also evident in traditional conceptions of maturation — the ontogenetic unfurling of pre-existing forms (Oyama, 1982).
3. Sober argues that it is mistaken to speak of a metaphorical relationship between natural and artificial selection. Rather, he says, the latter is natural selection occurring in a particular niche (1984: 19). Even if one takes this position, however, it is reasonable to point out the special qualities of this niche — something Darwin himself was careful to do. If the image of a deliberating, manipulating breeder with fixed goals influences our thinking about evolution, it is perhaps worth while to reflect on this influence. Robert M. Young (1985, Chapter 4) argues that the anthropomorphism of Darwin's descriptions of natural selection had the paradoxical effect of increasing the acceptance of his ideas for just the wrong reasons: by making evolution seem to be the work of a creating agent.
4. Just as a ball seeks its natural position by rolling down an incline, so, by this logic, does a phenotype approach its natural condition by maturation. Waddington's (1957) image of an epigenetic landscape, through which a ball-like organism rolls, has unfortunately lent itself all too easily to this model.
5. Research on the permeability of Weismann's barrier is both interesting and important (see other chapters in this volume). While the reformulation I am proposing is entirely consistent with such permeability, it does not require it. In fact, exclusive concentration on feedback to DNA structure risks legitimizing the tendency to define evolution in terms of genes alone.
6. If the particulars of gene replication are not necessary to the logic of natural selection, one could argue that evolution would proceed even if the genes only materialized a nanosecond before they were used. Though DNA does not behave this way, the chemical complexes for transcription and translation *are* assembled on the spot, just as many other aspects of the developmental system are.

7. This fact is easy to ignore when 'evolved', 'biological' and 'physical' are treated as synonyms. Lest I imply a false dichotomy between body and mind, let me point out that both morphology and physiology often reflect variations in the activities and experiences of the organisms.
8. This statement probably doesn't do justice to much of animal behaviour, but the point, I think, stands.
9. These questions are usually asked about genetic changes, but they are appropriate for non-genetic ones as well. *Damping out* is common in human affairs (traditions can disappear), and the change that fails to recur in offspring is a prototype of the 'acquired' character. *Maintenance* is seen in Douglas-Hamilton's account of radically altered behavioural patterns in elephants following intensive hunting. The formerly tame animals became highly aggressive and nocturnal, remaining so four generations later (cited by Bonner, 1980: 177, whose examples of transgenerational regularity are more useful than his opposition between genetic and behavioural trait transmission). Denenberg and Rosenberg (1967) showed that early handling of female rats affected behaviour and weaning weight of their grandoffspring. In a discussion of the perpetuation of range selection, feeding preferences and modes of predator avoidance in a wide range of vertebrates, Galef (1976) observed that the opposition between inheritance and individual acquisition of behaviour had hindered recognition of such social processes. *Amplification* is seen in Ho's (1984) description of cumulative cytoplasmic effects in fruitflies. Many social changes in humans show positive feedback characteristics; the rich may get richer *and* beget children who are richer still.

ACKNOWLEDGEMENTS

I am grateful to P. Bateson, R. Gray, J. Hailman, M.-W. Ho, P. Kitcher, R. Lewontin, J. Ricker, E. Sober, D. Symons and P. Taylor for their valuable comments on earlier drafts of this essay.

REFERENCES

Bateson, G. (1979). *Mind and Nature: A Necessary Unity*, Bantam Books, Toronto.
Bateson, P.P.G. (1983). Genes, environment and the development of behaviour. In T.R. Halliday and P.J.B. Slater (eds), *Animal Behaviour*, Vol. 3, *Genes, Development, and Learning*, Blackwell, Oxford, pp. 52–81.
Bateson, P.P.G. (1986). Sociobiology and human politics. In S. Rose and L. Appignanesi (eds), *Science and Beyond*, Blackwell, Oxford, pp. 79–99.
Bonner, J.T. (1980). *The Evolution of Culture in Animals*, Princeton University Press, Princeton, NJ.
Buss, D.M. (1984). Evolutionary biology and personality psychology: towards a conception of human value and individual differences, *Amer. Psychol.*, **39**, 1135–47.
Cohen, J. (1979). Maternal constraints on development. In D.R. Newth and M. Ball (eds), *Maternal Effects in Development*, Cambridge University Press, Cambridge, pp. 1–28.
Dawkins, R. (1976). *The Selfish Gene*, Oxford University Press, Oxford.
Denenberg, V.H., and Rosenberg, K.M. (1967). Nongenetic transmission of information, *Nature*, **216**, 549–50.
Edelman, G.M., and Mountcastle, V.B. (1978). *The Mindful Brain*, MIT Press, Cambridge, Mass.

Galef, B.G. (Jr) (1976). Social transmission of acquired behavior: a discussion of tradition and social learning in vertebrates. In J.S. Rosenblatt, R.A. Hinde, E. Shaw and C. Beer (eds), *Advances in the Study of Behavior*, Academic Press, New York, pp. 77–100.

Galef, B.G. (Jr) (in press). Imitation in animals: history, definition, and interpretation of data from the psychological laboratory. In T. Zentall and B.G. Galef (Jr) (eds), *Comparative Social Learning*, Erlbaum, Hillsdale, NJ.

Gottlieb, G. (1976). Conceptions of prenatal development: behavioral embryology, *Psychol. Rev.*, **83**, 215–34.

Hailman, J.P. (1982). Evolution and behavior: an iconoclastic view. In H.C. Plotkin (ed.), *Learning, Development, and Culture*, Wiley, Chichester, pp. 205–54.

Hailman, J. (in press). The heritability concept applied to wild birds. In R.F. Johnston (ed.), *Current Ornithology*, Vol. 4, Plenum, New York, pp. 71–95.

Ho, M.-W. (1984). Environment and heredity in development and evolution. In M.-W. Ho and P.T. Saunders (eds), *Beyond Neo-Darwinism: Introduction to the New Evolutionary Paradigm*, Academic Press, London, pp. 267–89.

Ho, M.-W. (in press). Genetic fitness and natural selection: myth or metaphor. In E. Tobach and G. Greenberg (eds), *Evolution of Social Behavior and Integrative Levels*, Erlbaum, Hillsdale, NJ.

Jerne, N.K. (1967). Antibodies and learning: selection versus instruction. In G.C. Quarton, T. Melnechuk, and F.D. Schmitt (eds), *The Neurosciences: A Study Program*, Rockefeller University Press, New York, pp. 200–5.

Kitcher, P. (1985). *Vaulting Ambition: Sociobiology and the Quest for Human Nature*, MIT Press, Cambridge, Mass.

Klama, J. (in press). (Pseudonym for J. Durant, P. Klopfer and S. Oyama, eds.; authors are J. Durant, E. Honore, N. Nur and S. Oyama). *The Myth of the Beast Within: Aggression Revisited*, Longman, London.

Konner, M. (1982). *The Tangled Wing: Biological Constraints on the Human Spirit*, Holt, Rinehart and Winston, New York.

Lehrman, D.S. (1970). Semantic and conceptual issues in the nature–nurture problem. In L.R. Aronson, E. Tobach, D.S. Lehrman, and J.S. Rosenblatt (eds), *Development and Evolution of Behavior*, Freeman, San Francisco, pp. 17–52.

Lewontin, R. (1978). Adaptation, *Sci. American*, **239**, 3212–30.

Lewontin, R. (1982). Organism and environment. In H.C. Plotkin (ed), *Learning, Development, and Culture*, Wiley, New York and London, pp. 151–70.

Lewontin, R. (1983a). Darwin's revolution, *New York Review of Books*, **30**, 1021–7.

Lewontin, R. (1983b). Gene, organism and environment. In D.S. Bendall (ed.), *Evolution from Molecules to Men*, Cambridge University Press, Cambridge, pp. 273–85.

Oyama, S. (1981). What does the phenocopy copy? *Psychol. Rep.*, **48**, 571–81.

Oyama, S. (1982). A reformulation of the idea of maturation. In P.P.G. Bateson and P.H. Klopfer (eds), *Perspectives in Ethology, Vol 5*, Plenum, New York, pp. 101–31.

Oyama, S. (1985). *The Ontogeny of Information*, Cambridge University Press, Cambridge.

Oyama, S. (in press a). Essentialism, women and war: protesting too much, protesting too little. In C. Flamenbaum, A. Hunter, and S. Sunday (eds), *Genes and Gender VI, The Societal Origins of Peace and War: A Challenge to Genetic Determinism*, Gordian Press, New York.

Oyama, S. (in press b). Transmission and construction: levels and the problem of heredity. In G. Greenberg and E. Tobach (eds), *Evolution of Social Behavior and Integrative Levels*, Gordian Press, New York.

Peterson, S.A., and Somit, A. (1978). Sociobiology and politics. In A.L. Caplan (ed.), *The Sociobiology Debate*, Harper and Row, New York, pp. 449–61.

Plomin, R. (1986). *Development, Genetics, and Psychology*, Lawrence Erlbaum, Hillsdale, NJ.
Sober, E. (1980). Evolution, population thinking, and essentialism, *Phil. Sci.*, **47**, 350–83.
Sober, E. (1984). *The Nature of Selection*, Bradford/MIT Press, Cambridge, Mass.
Sober, E. (1985). Darwin on natural selection: a philosophical perspective. In D. Kohn (ed.), *The Darwinian Heritage*, Princeton University Press, Princeton, NJ, pp. 867–99.
Taylor, P.J. (1987). Historical versus selectionist explanations in evolutionary biology, *Cladistics*, **3**, 1–13.
Waddington, C.H. (1957). *The Strategy of the Genes*, Allen & Unwin, London.
Williams, G.C. (1985). Comments by George C. Williams on Sober's *The Nature of Selection, Biol. Phil.*, **1**, 114–22.
Wilson, E.O. (1978). *On Human Nature*, Harvard University Press, Cambridge, Mass.
Young, R.M. (1985). *Darwin's Metaphor*, Cambridge University Press, Cambridge.

Evolutionary Processes and Metaphors
Edited by M.-W. Ho and S.W. Fox

CHAPTER 14

Sociobiology: a house built on sand

Peter T. Saunders

ABSTRACT

Sociobiologists claim that their theory must be taken seriously because it is based on the synthetic theory of evolution. In fact this is precisely the source both of its failings as a scientific discipline and, more importantly, of its dangers when applied to issues with political implications.

Neo-Darwinist explanation consists largely of suggesting selective advantages for various traits of organisms. Because the theory is unfalsifiable, with a little ingenuity such an explanation can always be found. This causes severe methodological problems in the study of physiological evolution, but the situation is even worse when we are concerned with behaviour. For a claim to have explained how a trait has evolved is by implication also a claim to have confirmed that the trait is innate, indeed that it is a trait at all. Thus Neo-Darwinist sociobiology can be used to give pseudo-scientific support to what are actually mere prejudices.

Sociobiology, like the rest of Neo-Darwinism, involves the *a priori* assumption that all characters of organisms have arisen out of the competition for survival and reproduction. Consequently, while the theory can supply an explanation for any behavioural trait, it lends itself most naturally to those which are essentially selfishness in one form or another. We are, therefore, led to the dismal view of ourselves as inherently selfish and aggressive not because of the nature of humans but because of the nature of Neo-Darwinism.

INTRODUCTION

Sociobiology is the study of the evolution of social behaviour by natural selection. To be sure, in the book which is responsible for much of the current interest in the subject, Wilson (1975: 6) first defines it as 'the systematic study of the biological basis of all social behaviour', but he adds, 'This book makes an attempt to codify sociobiology into a branch of evolutionary biology, and particularly of modern population biology'. In other words, by 'sociobiology', we are to understand 'Neo-Darwinist sociobiology'. This is what the term has come to mean, and it is how I shall use it here. I shall argue that the whole enterprise is misguided and, when applied to the study of human social behaviour, potentially dangerous.

Many workers, both in the natural and in the social sciences, are uneasy about sociobiology. They consider the approach too simplistic and facile, and they find many of the conclusions unacceptable. Some of them have attacked sociobiology in very strong terms indeed. Yet while the critics have pointed out flaws in individual sociobiological accounts, and have identified such characteristic errors as reductionism, anthropomorphism and adaptationism (e.g. Kitcher, 1985), comparatively little has been written about the theory itself, and why these errors should be so common.

This is largely because most of the critics of sociobiology accept the Neo-Darwinist account of biological evolution. They then find it hard to explain why the same theory should not apply to human behaviour, or why the synthetic theory can be so successful in biology and yet have nothing to contribute to the social sciences. Not only has this made it difficult to resolve the problems, it has also raised the temperature of the debate, since there is an implication that what is wrong with sociobiology is the very poor standard of its practitioners.

Now there is certainly some very shoddy work being done in sociobiology. All the same, I find it wholly implausible that the theory could be basically sound, but that somehow no one has ever managed to apply it properly. Besides, not all those who call themselves sociobiologists display the same cavalier attitude as is found in the worst examples; some of them are clearly aware that there are pitfalls to be avoided, even if they have not succeeded in bringing about the improvements they seek. If sociobiology is failing, as I and many others believe it is, there must be something fundamentally wrong with the theory itself.

In fact, the shortcomings of Neo-Darwinist sociobiology arise from its foundation in the Neo-Darwinist theory of evolution. They are present in other applications of that theory as well. Neo-Darwinism is not the hugely successful and powerful theory that its supporters would have us believe it to be. It can explain far less than is frequently claimed for it, and what it does explain are only some of the minutiae of evolution rather than the major features: changes in the coloration of moths, but not how there came to be moths in the first place. If, however, we accept the synthetic theory at face value, we accept the basic premises of sociobiology, and we are no longer in a position to demonstrate its overall weaknesses.

Because so many of the opponents of sociobiology are unable to formulate a general critique, they are frequently accused of being more concerned with politics than with science, so anxious to believe that humans are not selfish, or aggressive, or easily indoctrinated, or whatever, that they cannot look at the evidence dispassionately. And in attempting to refute this charge, some weaken their case still further by appearing to adopt the extreme environmentalist position that biology has nothing to do with human behaviour, which is patently not true.

There are, however, good scientific reasons for rejecting the claims that sociobiology makes about human nature and for seeing the whole enterprise as politically dangerous. Within Neo-Darwinism, explanation consists largely of the search for plausible accounts in terms of selective advantage. Since the theory is unfalsifiable, with ingenuity this search will always be successful, and for those who are content with such explanations (sometimes called 'Just So Stories') a satisfactory account can always be found.

The danger in applying this research programme to psychological and sociological phenomena is that in purporting to explain how a trait arose it gives credence to the idea that the trait exists, and so provides pseudo-scientific backing for what are often no more than prejudices. Nor is it even-handed, contributing to any concept of human nature with equal force. Its reductionist approach leads to a view of behaviour, as indeed of physiology, as a mere collection of independent characteristics, rather than as a complex whole. What is more, the picture of a world of selfishness, aggression and struggle, which Darwinism inherited from its origins in nineteenth-century England, has a very great influence on the sorts of traits to which contemporary sociobiologists choose to apply their theory and hence to the view of human nature which they perpetrate.

NEO-DARWINISM

Since the fundamental weakness of sociobiology is in its foundation, the 'synthetic' or 'Neo-Darwinist' theory of evolution, we must first be clear what that theory is. There does not appear to be a canonical definition, but it has been described quite well by a staunch Neo-Darwinist, Futuyma (1978). In the introduction to his textbook *Evolutionary Biology* he writes of ' . . . [the] neo-Darwinian view, in which the joint action of mutation, whereby variation arises, and selection, whereby it is shaped into coherent adaptive form, is considered sufficient explanation of the evolutionary process'.

The claim to be a sufficient explanation is a very strong one, implying as it does that for Neo-Darwinists no other form of explanation is needed in evolution, that any trait of any organism is to be accounted for as the consequence of the natural selection of random variations. It does, however, reflect what Neo-Darwinists actually do and, which is more important, it is an essential part of the structure of the theory. All the same, Neo-Darwinists do sometimes try to back away from it, at least in their more reflective moments. For example, the same author quoted above, when defending the synthetic theory against criticism, has written (Futuyma, 1984): 'A simplistic, false, but remarkably widespread view of this synthetic theory is that it was a simple union of Darwin's theory of evolution with the mechanistic theory of Mendelian genetics. In this conception, neo-Darwinism is equated with "natural selection of random variations".' But the view which he attacks as false and simplistic is precisely the one he gives in his own book. It is widespread because he and his fellow Neo-Darwinists teach it to their students. It forms the basis for research into evolution within the Neo-Darwinist paradigm[1] and it is, therefore, the foundation on which sociobiology is built.[2] And because the theory it defines is not falsifiable, which is to say not a scientific theory at all in the sense of Popper's (1934) criterion of demarcation, the foundation is of sand.

The reason that the claim to sufficiency is so important for Neo-Darwinism is that without it the theory is nothing but the mere statement that the natural selection of random variations can bring about adaptive changes. This, as Maynard Smith (1969) points out, is true enough, but it is not a theory of evolution. It is, in fact, a tautology.

We can see this clearly in the common expression of the principle of natural selection as 'the survival of the fittest'. The problem is that there is no way of specifying the 'fittest' individuals other than by the property that they are the ones that are best at surviving (and leaving offspring, or at least ensuring that their near relatives do) and so the argument is circular.

Some years ago there were several attempts to destroy the tautology by defining fitness independently. The suggestions that were made included efficiency in the use of scarce resources (MacArthur, 1962) and homeostasis (Slobodkin, 1964). None of these gained general acceptance, and it now seems generally agreed that no single independent definition exists. Many workers, however, argue that it is still possible to separate the concepts of 'adapted' and 'fit'. The idea is that we are to think of the organism as having to solve a number of problems posed by the environment. We can then judge its solutions on external criteria such as comparison with what an engineer might have done. (See, e.g. Lewontin, 1978; Dunbar, 1982; Riddiford and Penny, 1984; Sober, 1984: 84.)

Whatever value this approach may have as a research strategy, it cannot get us out of the tautology. It is quite wrong to think of an organism as facing a collection of individual problems which it can deal with separately. For example, when Orians (cited by Lewontin, 1978) found that certain birds do not forage in what would appear to be the most efficient way, he did not argue that they are, therefore, less than optimally adapted and predict that they will eventually be replaced by better adapted variants. Instead, he suggested that the birds are obliged to compromise between optimal foraging and defence of their nests. This is plausible enough, but it also makes it clear what the real definition of adaptation is, viz., contribution to surviving and leaving offspring. So we are right back in the circle.[3] (One could, I suppose, argue that the birds are less well adapted than an optimally foraging variant but nevertheless fitter on account of their tendency to defend their nests. This alters the status of the principle of natural selection from tautologically true to false, and would therefore seem unlikely to commend itself to Neo-Darwinists as a solution to the problem. Burian (1983), however, seems surprisingly close to advocating just such a position.)

The principle of natural selection is not, however, trivial, like the statement that all white horses are white. On the contrary, it is tautological in the same sense as Euclidean geometry or any other branch of mathematics. The theorems are logically entailed by the axioms and, therefore, contain nothing essentially new. Since, however, many of them are far from obvious, their discovery can be a significant accomplishment and an important contribution to our understanding, though this still does not make them scientific theories.

IS NEO-DARWINISM A SCIENTIFIC THEORY?

While the *principle* of natural selection is a tautology, albeit a non-trivial one, the *theory* of natural selection is not. This is because it includes something more, viz., the claim to be a sufficient explanation of evolution. The synthetic theory is not, however, falsifiable and, therefore, according to the criterion of demarcation, it is not a scientific theory. Now just as being a tautology does not make the principle of natural selection vacuous, so being unfalsifiable does not make the theory of natural selection worthless. It does, however, profoundly affect the ways in which it can be used, and makes its extrapolation to sociobiology not only unsound but actually dangerous.

Because the part of the synthetic theory that rescues it from being a tautology is the claim to sufficiency, it is on this that any potential falsifier should be based. It is, however, inherently very difficult to suggest an observation which could falsify the claim, and so defenders of Neo-Darwinism have concentrated on other aspects of the theory. I shall deal briefly with some of them here; a more detailed account is given in Saunders and Ho (1982).

Some claim that the postulates (that organisms multiply, that progeny generally resemble their parents, that occasionally new variants appear, that Mendelian genetics is an adequate account of heredity) are falsifiable and that this makes the theory falsifiable too. Thus Ruse (1981) argues that 'if we all just budded off one and only one offspring asexually, evolutionary theory [i.e. neo-Darwinism] would be false'. But this would be a test of Neo-Darwinism only if the postulates were peculiar to the theory, not just part of what Popper calls the 'background', i.e. previously well-established theories and observations which are neither new nor controversial. Otherwise the criterion of demarcation would also make astrology a science, since the arrangement of the heavenly bodies could in principle be other than the one that we observe and on which astrological predictions are based. Besides, a theory can be correct even if the postulates are not universally valid, providing that the exceptions are not relevant to the class of phenomena the theory is supposed to explain. Thus while some Neo-Darwinists (e.g. Maynard Smith, 1969) have argued that their theory is falsifiable because a demonstration of the inheritance of acquired characters would be fatal to it, the reaction to the claim that this has been done (Gorczynski and Steele, 1980) was that the theory could survive because the phenomenon described was ‘rather special’ (Ruse, 1982; Maynard Smith, 1980).[4] It should, however, be pointed out that the work of Gorczynski and Steele, despite the furore it aroused, is by no means the only known case of the inheritance of an acquired character. Even leaving aside cytoplasmic inheritance and maternal effects,[5] there is a clear example in flax (see Cullis, this volume) which involves a direct and heritable change in the genome. Yet the synthetic theory survives.

Other writers argue that such and such is a prediction of Neo-Darwinism, that it is true, and that therefore Neo-Darwinism is both falsifiable and verified. But the criterion requires not just a prediction but one which, were it to be false, would compel us to reject the theory. So Williams's (1973) claim that there should be — and are — populations in every stage of transition from one species to two is not a test, because if this were not observed Neo-Darwinists would simply conclude that natural selection had reached its logical conclusion with every organism ideally adapted to its environment. As Ruse (1982: 314) writes, acknowledging the existence of ‘living fossils’ like the coelacanth and the crocodile: ‘If conditions are stable and there are no virtues in change, then so be it’.

Predictions in science depend not just on a theory but on other information as well. A potential falsifier of the theory must not just be a test of the additional information. For example, many evolutionists currently believe that *Stegosaurus's* fins were used for cooling, and Ruse (1981, 1982) claims that Neo-Darwinism would be falsified if this turned out to violate the laws of thermodynamics. Of course, such an outcome would only be taken — correctly — as showing that the fins were not used for cooling, and Neo-Darwinists would search for some other plausible function. Until quite recently it was generally thought that the fins were used for defence, or possibly courtship, and the consensus appears to have altered without anyone claiming that the synthetic theory has been falsified. What was shown to be wrong was a particular hypothesis about adaptation, not the synthetic theory of evolution (cf. Maynard Smith, 1982).

It has also been argued (e.g. Maynard Smith, 1969; Ghiselin, 1969) that Neo-Darwinism would be falsified if it could be shown that there existed so much as one unequivocally non-adaptive trait. This seems to get to the very heart of the matter, but it doesn't work. In the first place, it is very hard to prove that a trait has no adaptive significance whatsoever. It would be difficult enough to establish such a negative result conclusively if Neo-Darwinists required that selective advantages must actually be measured; since all that they demand is that some plausible function can be suggested, the task becomes well-nigh impossible. As Dobzhansky (1956) wrote, 'All that can be said is that one's failure to see the adaptive significance of a trait is no proof that it has none'.

Sometimes it does appear to be especially difficult to imagine what conceivable function some trait might serve, but there are other arguments that can be brought into play to save the situation. One may claim that the character conferred a selective advantage in the past (as in the case of the precise arrangement of body hair on humans), that it arose through pleiotropy, or that it survives on account of heterozygote advantage.

If all else fails one can even avoid falsification by denying that the character in question is really a trait at all. As Waddington (1969) pointed out, no one would seriously claim that Neo-Darwinism is falsified if we cannot suggest any conceivable selective advantage the two deep sea fish pictured on the frontispiece of *Animal Biology* (Haldane and Huxley, 1927) gain from their remarkable resemblance to the two authors. And if that example seems a bit far fetched, is the chin a trait which requires explanation (DuBrul and Sicher, 1956), or an epiphenomenon which does not (Gould and Lewontin, 1979)?

According to orthodox Neo-Darwinism, evolution proceeds by the slow accumulation of minute, advantageous random variations. So we might expect that a potential falsifier would be the absence of intermediate forms in the fossil record. Unfortunately, since by no means all organisms leave fossils and not all the fossils that were formed will be found, there will always be gaps in the record, and so it will always be possible to avoid falsification. It is also frequently argued that the gaps are not significant because gradual changes can accumulate into major ones over a period of time which is long in evolutionary terms but short from the point of view of a geologist. And Turner (1983) writes, 'The "gradualism" of the new synthesis certainly does not exclude from evolution mutations of a rather large phenotypic effect'. Such a gradualism is clearly unfalsifiable.

Other tests have also been suggested (see Saunders and Ho, 1982, for examples) but they all fail. Indeed, that there have been so many attempts to demonstrate that Neo-Darwinism is falsifiable is in itself significant; one is left with the distinct impression that each writer is no more convinced by the other's suggestions than I am. Thus while the synthetic theory of evolution is not a tautology it is also not falsifiable. In Popper's terms, it is not a scientific theory. This does not, however, mean it can have no role in the study of evolution, because not all explanation in science must necessarily be based on falsifiable theories.

This is because falsification depends not on explanation but on prediction. Now, as Scriven (1959) has pointed out, the two are not equivalent; we can have either without the other. The sudden fall of a barometer is a very reliable predicator of a coming storm, but does not explain it. Conversely, while we may be satisfied that we understand why the dodo became extinct, that is not to say that we could have predicted it, nor that we can foretell which of the currently endangered species will survive and which will not.

Without predictions, however, falsification is not possible. Hence we can have explanation in science in terms of theories which are, according to the criterion of demarcation, non-scientific. This apparent paradox is essentially semantic; it would be better to think of 'scientific theory' as all one word, not to be confused with the quite different concept of 'a theory which may legitimately be used in scientific explanation'. It does, however, mean that Popper's criterion of demarcation cannot be used as an absolute test of what is or is not acceptable in science, especially in the biological and social sciences, where the complexity of the phenomena and the large number of factors which can significantly influence them are most likely to decouple explanation from prediction.

The criterion of demarcation is still important, however. Explanation in science consists largely of the fitting of phenomena into an accepted framework. For example, Popper (1963: 63) writes that 'in pure science . . . explanation is always the logical reduction of hypotheses to others which are of a higher degree of universality'. If a theory is testable, and if it survives some crucial tests, then it may be taken as a secure framework for explanation. An untestable, and therefore untested, theory may also be used, but as with any untested device, what it produces must be viewed with caution and subjected to careful scrutiny.

This point of view is by no means novel, even if it is sometimes lost sight of. Popper (1963: 37) himself acknowledges that much of what Freud and Adler wrote is 'of considerable importance, and may well play its part one day in a psychological science which is testable'. The psychoanalytic theories of Freud and Adler fail to satisfy the criterion of demarcation not because they are vacuous nonsense but because they are incomplete. They may — and probably do — explain certain aspects of human behaviour in the same way that some future testable and tested theory will. The explanations will not then suddenly change from being 'superstition' to being 'scientific truth', nor will clinicians suddenly become entitled to draw on ideas which were previously taboo — but the confidence with which they can be employed will be greatly increased.

The same is true of the synthetic theory. It may provide a framework for understanding some aspects of evolution. It leads to predictions which, while they are not tests of Neo-Darwinism itself, do allow us to test certain particular hypotheses about evolution (cf. Maynard Smith, 1982). But it is not itself testable, and it is quite wrong to think that it forms as secure a foundation for sociobiology as Newton's laws do for classical physics or as the atomic theory does for chemistry.[6] (See Maynard Smith, 1984. For more on the comparison between Newtonian mechanics and Darwinism, see Barzun, 1941, Chapter I.)

The Neo-Darwinist mode of explanation does not give an adequate account of evolution, but it is a part of the story, and of course population genetics has been a fruitful subject quite apart from its relation to the study of evolution. It is hardly surprising that there should be a reluctance to reject all this, and suspicion of any new idea that seems to require that this be done. But this is not what I am advocating. Perhaps Neo-Darwinists would not strive so hard to refute the charges of tautology and unfalsifiability if they realized that both tautologies and unfalsifiable theories can still be useful in science.

SOCIOBIOLOGY

While the major flaws in sociobiology arise from, and are therefore already present in, the Neo-Darwinist theory of biological evolution, they are even more serious in the extrapolation to the social sciences. So even though I do not agree with workers like Maynard Smith (1982) and Kitcher (1985, and this volume) who believe that the whole programme is sound from pre-biotic evolution right up to ethology and suspect only when applied to human behaviour, I think it is important to discuss some of the ways in which the problems become worse.

In the first place, the connection between the genes and the characters they are supposed to determine is even longer and more complicated for behavioural traits than for physical features. It is not surprising that classical population genetics can account for the changes in the relative frequency of the melanic form of the peppered moth, because it is quite plausible that everything really does depend on the choice between two alleles at a single locus, and the selective advantage involved is large and has actually been measured (Kettlewell, 1973; but see Sargent, 1974, and Lambert, Millar and Hughes, 1986, for critical appraisals of even this classical case of Neo-Darwinism in action). But as the traits to be explained become more and more complex, and where the selective advantages are very much smaller and indeed more often surmised than actually measured, the accounts become correspondingly less convincing. And when it comes to details of behaviour like the incidence of various human sexual practices (Weinrich, 1977) or the tendency of grandparents to annoy their children by spoiling the grandchildren (Fagen, 1976), one wonders how anyone can take this seriously.

Behaviour is also very greatly affected by the environment, which is another way of saying that learning is very important. It is all too easy to see as instinctive behaviour which is in fact learned. Anyone who is familiar with more than one society will know that some aspects of behaviour which are so firmly fixed and universal in one as to appear innate are absent in others. And if a trait is not innate, there is little point in postulating a gene for it and speculating on how the appropriate allele has been selected.

To be sure, sociobiologists do not generally claim that our social habits are determined in precisely the same way as the colour of our eyes, at least not explicitly. Most of them would deny that they assume 'one gene/one character', and would insist that expressions like 'a gene for altruism' are to be understood as a shorthand. Indeed, Bateson (1982) urges that sociobiologists should avoid using such expressions altogether, on the grounds that they lead both supporters and critics to see the subject as essentially preformationist, with the developmental process having been dismissed as irrelevant.

Bateson is right to stress the dangers of what he and other sociobiologists sometimes refer to as the shorthand version, but preformation cannot so easily be eliminated from the subject. It is not simply a matter of an unfortunate choice of words leading to confusion about what sociobiologists are really trying to say. For example, Dawkins (1976: 66) writes: 'The details of the embryonic developmental process, interesting as they may be, are irrelevant to evolutionary considerations. Konrad Lorenz has put this point well'. Dawkins is wrong (see, e.g. Saunders, 1984) but what he says is clear enough. Nor is he alone in this view; Maynard Smith and Holliday (1979) tell us that the gift of Weismannism to evolutionary theory is that the mode of development is irrelevant.

What do sociobiologists actually assume? What is 'a gene for altruism' a shorthand for? According to both Dawkins (1976: 66) and Bateson (1982) (the latter apparently speaking for all the members of the King's College Sociobiology Group) it is still supposed that a behavioural trait is determined by a single gene. This means not that the gene is the sole cause of the behaviour, but that, other things being equal, it makes the difference between an animal behaving in the given way or not. It is also supposed, though neither Dawkins nor Bateson states this explicitly, that other things *are* equal, at least equal enough and for long enough for natural selection to work on the crucial locus.

Thus while it is accepted that there are other factors involved in the causation of behaviour, it is assumed that they are all spear carriers in the evolutionary play, necessary for the action to proceed but having no influence on its direction. Like other modern Neo-Darwinists, sociobiologists are happy to allow that there are lots of things beside genes in evolution, so long as they aren't actually expected to take any of them into account. Not 'one gene/one character' perhaps, but 'one gene that makes any difference/one character'.

In any case, it is still a matter of one gene. So we are left with the question of whether it is possible, even in principle, to identify a single gene as the crucial determinant of a given behavioural trait. For if it is not, the whole programme is in imminent danger of collapse.

Dawkins (1982: 23) acknowledges the problem, but argues that it is nevertheless perfectly legitimate to talk about such genes. 'Reading', he says, 'is a learned skill of prodigious complexity, but this provides no reason in itself for scepticism about the possible existence of a gene for reading. All we would need in order to establish the existence of a gene for reading is to discover a gene for not reading, say a gene which induced a brain lesion causing specific dyslexia.' He points out that such a gene is not implausible, and that it is more accurately called a gene for not reading than would be, say, a gene for total blindness because at least in present circumstances inability to read would be its major effect.

He then triumphantly concludes, ' . . . it follows from the ordinary conventions of genetic terminology that the wild-type gene at the same locus, the gene that the rest of the population has in double dose, would properly be called a gene "for reading". If you object to that, you must also object to our speaking of a gene for tallness in Mendel's peas, because the logic of the terminology is identical in the two cases.' The question, however, is not whether the internal logic of population geneticists' conventions ever allows them to ascribe a meaning to the expression 'a gene for reading' but whether the concept has anything whatsoever to do with evolution, and, in particular, in explaining the origin of the human capacity to learn to read.

In fact it does not, as Dawkins recognizes. For in discussing the similar question of eggshell removal behaviour in black-headed gulls, he admits, 'It most definitely does not follow that this particular locus "for" eggshell removal was one of the ones on which natural selection worked during the evolution of the adaptation. On the contrary, it seems much more probable that a complex behavioural pattern like eggshell removal must have been built up by selection on a large number of loci, each having a small effect in interaction with the others.'[7] And a bit later on he adds, with commendable candour, 'It is too bad if geneticists usually are forced to concentrate on loci that are convenient rather than evolutionarily important'.

Thus the best justification that Dawkins can find for the concept of a single crucial gene for altruism, or reading, or eggshell removal, is on his own admission irrelevant to evolution. Yet this in no way inhibits him or other sociobiologists from putting forward explanations of the origins of various aspects of human nature which depend crucially on the existence of such genes.

The implicit reliance on the one gene/one character assumption can be seen in, for example, the large amount of work being done in applying classical population genetics to sociobiology (e.g. Cavalli-Sforza and Feldman, 1981; Lumsden and Wilson, 1981). Among the issues that are discussed are what difference it would make if the gene for altruism were dominant (O'Donald, 1982) or whether the gene for being altruistic towards siblings is linked with the gene for recognizing siblings (see Rubenstein, 1982). Such analyses are meaningful only if one gene/one character is being taken literally. The case that it is merely a shorthand does not stand up.

Even where the arguments are not strictly in terms of population genetics, we still have to ask how applicable they would be if the idea that many genes and other factors can be involved were taken seriously. If a trait really is under the control of a single gene it is easy to see how natural selection could operate in the way that the theory demands. But what if the character is controlled by genes at a number of different loci and the effects are not (as in general they will not be) simply additive? (See Lewontin, 1974; Bateson and D'Udine, 1986.) How will selection be affected if some of those genes also significantly influence other characters? What if there are strongly non-linear interactions?

I would not go so far as to claim no human behavioural traits have evolved in the way that sociobiologists claim. It may be that some are determined by a single gene and we have not so far discovered this, possibly because there is no variation left in the population. It may be that in some cases the arguments used in sociobiology apply even though there is no single determining gene. But the burden of proof is firmly on the sociobiologists. Before they start thinking up their 'Just So' stories, they ought to provide evidence that their assumptions are valid. It is surely this, rather than the question of altruism, that is — or should be — the central theoretical problem of sociobiology.

Sociobiologists also fall foul of the fallacy of misplaced concreteness. Not only is it hard to establish that a given behavioural trait is genetically determined, it is hard to be certain that it is a 'trait' at all. The same problem arises in biological evolution — one may question whether the chin is something whose origin is to be explained in terms of its own selective advantage or whether it is just an epiphenomenon arising out of the interaction between two growth fields — but at least we all agree that humans have chins. In the social sciences, however, while there are certainly forms of behaviour which we may describe as 'aggressive' or 'altruistic', that is not to say that there are traits called 'aggression' and 'altruism' which have evolved by the action of mutation and natural selection and are, therefore, immutable parts of our nature.

No one can doubt that there are heritable biological factors which influence behaviour. The problem, and it is a fascinating and challenging one, is to discover what these are. The Neo-Darwinist approach is too superficial, because what is innate is unlikely to be what we might call the surface phenomena, like spoiling grandchildren. We should be looking for characteristics of the brain at much deeper and more fundamental levels, which in different cultures and in different circumstances manifest themselves as the traits we can observe directly.

To borrow terms from computer science, sociobiologists fail to distinguish between hardware and software. A computer has innate, determined features such as its internal wiring. It also has features which are variable: the programs that are written for it by the user. Now these are not independent; the properties of the computer obviously have an influence on the sorts of programs that will be written. Consequently, if you analyse enough programs you may be able to make some inferences about the nature of the hardware. But it would be silly just to glance at a small amount of output from one user and then claim to have proved that the computer was designed specifically to be able to solve (say) quadratic equations, with supporting evidence in the form of plausible accounts of how that would help the manufacturer to beat the competition.

Why is sociobiology dangerous?

Scientists who study mind and behaviour have a particular duty to be cautious, just like those who work with pathogens or radioactive materials. A poor theory in the social sciences can provide backing for unjustifiable views about our fellow human beings and how we should act towards them. It can even lead governments to follow what are later recognized to be mistaken policies — consider the effect on English secondary education of Cyril Burt's now discredited studies of the heritability of intelligence. But in fact the danger in Neo-Darwinist sociobiology is more fundamental than that, for it arises out of the very structure of the subject.

The pattern of Neo-Darwinist explanation is to fasten our attention on some trait of an organism and try to fit it into the theory (cf. Ruse, 1982: 123). Where nothing is known or likely to be known about the genes involved — which is generally the case in sociobiology — this amounts to finding an adaptive significance. And where — as is also the case in sociobiology — it is not possible actually to measure selective advantages, the whole enterprise reduces to postulating a plausible one.

Unfortunately, with a little ingenuity it is possible to imagine a selective advantage for just about anything. There are many examples in biological evolution (for a particularly amusing one see Eden, 1967) and sociobiologists are no less adept: they have even been able to produce Neo-Darwinist accounts for such apparently intractable cases as homosexuality (Trivers, 1974; see also Wilson, 1975: 555).[8]

Only the problem is more acute in sociobiology, and it has political ramifications as well. Whatever our differences may be about explanation in biological evolution, it is generally possible to agree that an organism does possess such and such a character and to determine whether or not it is heritable.

There is nothing like the same sort of agreement about which human behavioural traits are innate — anthropology would be a far less interesting subject if there were. This does not, however, discourage the sociobiologists. They simply fasten on some trait which 'everyone knows' is innate: male dominance, aggression, the tendency of males to be promiscuous and females to be monogamous. They then try to think of a plausible selective advantage for the trait, and, because the theory is unfalsifiable, with a little ingenuity they soon find one.

When they have thus succeeded in explaining, to their own satisfaction, how the trait has evolved, they thereby confirm, again to their own satisfaction, that it is a trait and that it is innate. Thus sociobiology can provide support for the existence of any trait whatsoever and so be used to put a scientific gloss on whatever prejudices the investigator began with.[9] Since our views on 'human nature' are bound to have political repercussions, such a theory is potentially dangerous.

In fact the situation is even worse than this, because sociobiology does not just confirm its practitioners' view of humanity, it also contributes towards forming it. Darwinism had its roots in the harsh economic climate of Victorian England, with its emphasis on individual success or failure, on competition, on struggle. The same ideas survive in contemporary Neo-Darwinism. Time and time again we see the same metaphors: cost-benefit analysis, investment, conflict of interest and so on. (See Ho, 1986 and also this volume. Compare also Ghiselin, 1974, who describes his own book on evolution as 'a cross between the *Kama Sutra* and the *Wealth of Nations*'.)

These are the ideas that permeate the synthetic theory, and when the theory is extended into the social sciences they come with it. Thus a particular set of notions about society have found their way into biology and then back into the social sciences. Unfortunately, in the process the provenance of the ideas has been largely forgotten, and because they are now seen as originating in the natural sciences, and therefore to be politically neutral, they have gained in credibility. It is rather like the laundering of ill-gotten gains by depositing the money in a foreign bank and transferring it back home later on.

For sociobiologists, the essential characteristic of humans, as indeed of all organisms (and, of course, genes as well) is selfishness. 'Scratch an "altruist" ', writes Ghiselin (1974: 247), 'and watch a "hypocrite" bleed'. Not a very edifying picture of humanity, but you can see how it arises. The structure of the theory, and also those ideas which are closely associated with it even if not part of the formal structure (i.e. the 'model'; see Hesse, 1966) all point towards selfishness as fundamental. So the sociobiological literature is disproportionately concerned with such topics as aggression, male dominance, war, rape, conflict and so on. And because the theory is unfalsifiable, once its adherents have been led to this dismal view of human nature, they find ample confirmation of it.

There is a certain irony in all this. In the period before Darwin, there was an influential school of 'natural theologians', the most famous of whom was William Paley, who argued that what we would now call the adaptedness of organisms cannot have been brought about by mere accident, and must imply the existence of a Creator. Like many Christians before them, the natural theologians were faced with the problem of explaining why, if the God who created the universe is good, there is so much evil. Paley's solution (Paley, 1819) was to ascribe what he called 'cases of apparent evil' to the occasional 'thwartings and crossings' of general laws whose effects were for the most part benevolent.

The 'argument from design' was a major obstacle to the acceptance of the idea of evolution. In the end, of course, it was refuted by Darwin, since the theory of natural selection provides a mechanism by which organisms can become adapted without divine intervention or some sort of vital force. But just as Paley had to deal with the problem of evil, when applied to human nature, Darwin's theory leaves us with the opposite problem: how to account for good. Wilson (1975: 3) calls altruism the 'central theoretical problem' of the subject and he and other sociobiologists ascribe it to the occasionally subtle workings of a general law whose effect is for the most part selfish.

What is wrong with sociobiology is that an unfalsifiable theory is being used as though it had all the authority of one that satisfies the criterion of demarcation. Sociobiologists may believe, with Maynard Smith (1984), that their work is based on a theory with the same status as Newtonian mechanics, but they are wrong. Perhaps Neo-Darwinist evolutionary theory does have something useful to say about human behaviour, but anything it suggests must be carefully substantiated and checked against the evidence from the social sciences themselves. Ethology and comparative psychology have long lives ahead of them yet as disciplines, and if and when they are superseded it will not be by Neo-Darwinist sociobiology, despite Wilson's (1975: 6) confident claim.[10]

I should stress that in criticizing sociobiology as a politically dangerous theory, I do not mean to say that all or even the majority of sociobiologists hold undesirable political views. Some of them may, others do not; that is not the primary issue. It is the theory itself that is politically unacceptable, and for good scientific reasons. All the same, we might ask whether it is perhaps not entirely surprising that some of us who have done considerably better than the average inhabitant of this planet would be attracted to a theory which ascribes so much to our genes, thereby reassuring us that our favoured position is largely a consequence of our innate superiority rather than merely a reflection of the much better than average environment in which we have had the good fortune to be raised.

CONCLUSION

From the claims that are made for it, one could easily form the impression that Neo-Darwinism has largely succeeded in explaining the evolution of life on earth, leaving only some details to be cleared up. In fact, quite the reverse is true. The synthetic theory can account for some of the details, but it has little to say about the major questions. It appears to be successful only because it is unfalsifiable: if we are content with 'Just So' stories, the theory can provide them.

Now even if Neo-Darwinism were the immensely successful theory that its supporters believe it to be, its extension into the social sciences would still have to be treated with caution. Given the inadequacy of the synthetic theory even as an account of biological evolution, however, the whole enterprise must be considered suspect.

The attraction of Neo-Darwinism in the social sciences, as in biology, is that it appears to offer easy and simple solutions to what are actually very difficult and complex problems. It is becoming more and more clear that this is an illusion even in biology, and that the study of evolution must be pursued largely by the same methods as are used in the rest of science (see Ho and Saunders, 1984, for examples). Since the role of natural selection in the social sciences is even more problematic, the most important lesson to be learned from evolutionary biology is that to the understanding of human behaviour, as to geometry, there is no royal road. And if this sounds like a rather negative conclusion, surely there is more than ample compensation in that we are liberated from the dismal Neo-Darwinist world-view in which selfishness is paramount.

ACKNOWLEDGEMENTS

Many of the ideas in this chapter grew out of discussions with Mae-Wan Ho. I am also grateful to the other participants in the Hawk's Cay meeting for helpful conversations.

NOTES

1. Four leading Neo-Darwinists, Dobzhansky, Ayala, Stebbins and Valentine, have published what they call 'a reasonably comprehensive account of the theory of evolution' (Dobzhansky *et al.*, 1977). In the preface, they write of genetics and developmental biology on apparently equal terms as 'two fields of fundamental importance for evolutionary studies'. Yet development is hardly mentioned in the book: in almost twenty pages of subject index there are no entries under either development or epigenetics, which we may contrast with 60 under 'gene' and 73 under 'chromosome'. One can argue about definitions, but this makes it abundantly clear what the synthetic theory really is. As Bernard Shaw put it, what a man believes may be ascertained not from his creed, but from the assumptions on which he habitually acts.
2. A well-known apologist for Neo-Darwinism (Ruse, 1980) writes: 'I see sociobiology, the study of animal social behaviour from an evolutionary perspective, as a natural and unforced growth and development from orthodox and established neo-Darwinian evolutionary biology. This being so, I suggest that because neo-Darwinian biology is a genuine and fruitful branch of science, the respect that it deserves should automatically be transferred to sociobiology.' These two sent-

ences, and the fact that I agree with the first and disagree with the second, more or less sum up this chapter.

3. For more on optimal foraging theory, see Gray (this volume).
4. Dawkins (1982: 164ff.) has another way of coping with these potentially embarrassing results; if Neo-Darwinism is in danger of being falsified then he moves the goal posts. He notes that Steele (1979) has suggested an explanation for his observations based on the hypotheses of selection within somatic cells and conveyance of genes into the germ cells by proviruses (Temin, 1974). This would appear to constitute a violation of Weismann's doctrine of the independence of the germline, but Dawkins avoids this by redefining the germline to include any gene in a somatic cell which is a candidate for proviral conveyance into a germ cell.
5. But why should we leave cytoplasmic inheritance and maternal effects aside? For an adaptation to become fixed in a population constitutes the inheritance of an acquired character whether there is a change in the genome or not. There is more to heredity than DNA (cf. Ho, 1986). Some Neo-Darwinists (e.g. Dobzhansky *et al.*, 1977) *define* evolution to be change in gene frequencies, which avoids this difficulty but leaves open the question of how much their account of what they call evolution has to say about the problem of explaining how the organisms we see about us came to be and why they are the way they are, which is surely what we are really trying to do. What is more, in trying to avoid Scylla in this way they set course firmly for Charybdis; to accept the gene-frequency definition of evolution may well force them to accept that molecular drive (Dover, 1986) is more important than natural selection as a mechanism of evolution.
6. It is, of course, true that any theory can be protected by a large enough collection of *ad hoc* hypotheses (cf. Lakatos, 1978). In the case of Newtonian physics, however, this is a logical possibility rather than an everyday occurrence. Compare Lakatos's amusing example of how astronomers might react to a planetary orbit that could not be explained by Newtonian mechanics with what actually happened in the case of the advance of the perihelion of Mercury. It was not seen as a falsification, although it later figured as one of the crucial tests in favour of General Relativity, but it was recognized an anomaly, i.e. something which could be tolerated for the time being in the hope that it would eventually be explained by the theory. As Lakatos (and, for that matter, Popper too) argue, if scientists were not willing to accept occasional anomalies in otherwise successful theories, every theory would be rejected quickly.

   In the synthetic theory, *ad hoc* hypotheses are not the exception but the rule. There is a world of difference between tolerating a few anomalies because of a large number of predictions that have been confirmed, and accepting a collection of *ad hoc* accounts with little but each other for support. The style of Neo-Darwinism is that of a degenerate research programme (Lakatos, 1978): rationalizations of known observations rather than predictions of unexpected phenomena.
7. As Fox (this volume) points out, it is tempting but hazardous to work backwards from the present. Evolution is an irreversible process, and while analysis may provide important clues to how synthesis occurred, it can also lead us wildly astray.
8. The problem is that homosexuality presumably reduces the probability of reproduction, and this makes it difficult to explain within the framework of synthetic theory. Wilson offers two possible explanations. The first is the all-purpose Neo-Darwinist escape from falsification, heterozygote advantage. Alternatively, homosexuals might have used their freedom from parental obligations to act as helpers to close relatives and 'genes favoring homosexuality could then be sustained at a high equilibrium level by kin selection alone'. (Compare Note 6.)

9. In the last century, some Darwinians used their theory to account for the low intelligence of the Irish. Today, of course, no one seriously believes that the Irish are less intelligent than anyone else, although some have passed on the intellectual wooden spoon to other groups. There is a moral here. It is also instructive to see how odd these accounts appear when what the theory so confidently explains is something which does not tally with our own particular prejudices or those of our age and society.
10. After writing this section, I noticed the following quotation in Henry Drummond's Lowell Lectures (Drummond, 1910): 'The first step in the reconstruction of Sociology will be to escape from the shadow of Darwinism — or rather to complement the Darwinian formula of the Struggle for Life by a second factor which will turn its darkness into light.'

REFERENCES

Barzun, J. (1941). *Darwin, Marx, Wagner*, Little, Brown & Co, Boston.

Bateson, P.P.G. (1982). Behavioural development and evolutionary processes. In *Current Problems in Sociobiology* (eds King's College Sociobiology Group), Cambridge University Press, Cambridge, pp. 133–51.

Bateson, P.P.G., and D'Udine, B. (1986). Exploration in two inbred strains of mice and their hybrids: additive and interactive models of gene expression, *Anim. Behav.*, **34**, 1026–32.

Burian, R.M. (1983). Adaptation. In M. Grene (ed.), *Dimensions of Darwinism*, Cambridge University Press, Cambridge, pp. 287–314.

Cavalli-Sforza, L., and Feldman, M. (1981). *Cultural Transmission and Evolution*. Princeton University Press, Princeton.

Cullis, C.A. (1984). Environmentally induced DNA changes. In J.W. Pollard (ed.), *Evolutionary Theory: Paths into the Future*, Wiley, London, pp. 203–16.

Dawkins, R. (1976). *The Selfish Gene*, Oxford University Press, Oxford.

Dawkins, R. (1982). *The Extended Phenotype*, W.H. Freeman, Oxford.

Dobzhansky, Th. (1956). What is an adaptive trait? *Amer. Natur.*, **90**, 337–47.

Dobzhansky, Th., Ayala, F.J., Stebbins, G.L., and Valentine, J.W. (1977). *Evolution*, Freeman, San Francisco.

Dover, G. (1986). Molecular drive in multigene families: how biological novelties arise, spread and are assimilated, *TIG*, **2**, 159–64.

Drummond, H. (1910). *The Ascent of Man*, Hodder & Stoughton, London.

DuBrul, E.L., and Sicher, H. (1956). *The Adaptive Chin*, Charles C. Thomas, Springfield.

Dunbar, R.I.M. (1982). Adaptation, fitness and the evolutionary tautology. In *Current Problems in Sociobiology* (eds King's College Sociobiology Group), Cambridge University Press, Cambridge, pp. 9–28.

Eden, M. (1967). Inadequacies of neo-Darwinian evolution as a scientific theory. In P.S. Moorhead and M.M. Kaplan (eds), *Mathematical Challenges to the Neo-Darwinian Interpretation of Evolution*, Wistar Institute Press, Philadelphia, pp. 5–12.

Fagen, R.M. (1976). Three generation family conflict, *Anim. Behav.* **24**, 874–9.

Futuyma, D.J. (1978). *Evolutionary Biology*, Sinauer, Sunderland.

Futuyma, D.J. (1984). Neo-Darwinism in disfavor, *Science*, **226**, 532–3.

Ghiselin, M. (1969). *The Triumph of the Darwinian Method*, University of California Press, Berkeley.

Ghiselin, M. (1974). *The Economy of Nature and the Evolution of Sex*, University of California Press, Berkeley.

Gorczynski, R.M., and Steele, E.J. (1980). Inheritance of acquired immunological tolerance to foreign histocompatibility antigens in mice, *Proc. Nat. Acad. Sci. USA*, **77**, 2871–5.

Gould, S.J., and Lewontin, R.C. (1979). The spandrels of San Marco and the Panglossian paradigm: a critique of the adaptationist programme, *Proc. Roy. Soc. Lond.*, **B205**, 581–98.

Haldane, J.B.S., and Huxley, J.S. (1927). *Animal Biology*, Clarendon, Oxford.

Hesse, M.B. (1966). *Models and Analogies in Science*. University of Notre Dame Press, Notre Dame, Indiana.

Ho, M.-W. (1986). Genetic fitness and natural selection — myth or metaphor? In E. Tobach (ed.), *Proceedings of the Third T.C. Schneirla Conference on Behaviour and Evolution* (in press).

Ho, M.-W., and Saunders, P.T. (1984). *Beyond Neo-Darwinism: An Introduction to the New Evolutionary Paradigm*, Academic Press, London.

Kettlewell, H.B.D. (1973). *The Evolution of Melanism*, Clarendon Press, Oxford.

Kitcher, P. (1985). *Vaulting Ambition*, MIT Press, Cambridge, Mass.

Lakatos, I. (1978). *The Methodology of Scientific Research Programmes* (P. Worrall and G. Currie, eds), Cambridge University Press, Cambridge.

Lambert, D.M., Miller, C.D., and Hughes, T.J. (1986). *Riv. Biol. — B. Forum*, **79**, 11–49.

Lewontin, R.C. (1974). *The Genetic Basis of Evolutionary Change*, Columbia University Press, New York.

Lewontin, R.C. (1978). Adaptation, *Sci. American*, **239**, 157–69.

Lumsden, C.J., and Wilson, E.O. (1981). *Genes, Mind and Culture*, Harvard University Press, Cambridge, Mass.

MacArthur, R.H. (1962). Some generalized theorems of natural selection, *Proc. Nat. Acad. Sci.*, **48**, 1893–7.

Maynard Smith, J. (1969). The status of neo-Darwinism. In C.H. Waddington (ed.), *Towards a Theoretical Biology 2: Sketches*, Edinburgh University Press, Edinburgh, pp. 82–9.

Maynard Smith, J. (1980). Regenerating Lamarck, *Times Literary Supplement*, 24 October.

Maynard Smith, J. (1982). *Evolution and the Theory of Games*, Cambridge University Press, Cambridge.

Maynard Smith, J. (1984). Game theory and the evolution of behaviour, *Behav. Brain Sci.*, **7**, 95–125.

Maynard Smith, J., and Holliday, R. (1979). Preface. In J. Maynard Smith and R. Holliday (eds), *The Evolution of Adaptation by Natural Selection*, The Royal Society, London, pp. v-vii.

O'Donald, P. (1982). The concept of fitness in population genetics and sociobiology. In *Current Problems in Sociobiology* (eds King's College Sociobiology Group). Cambridge University Press, Cambridge, pp. 65–85.

Paley, W. (1819). *Natural Theology* (16th edn), F.C. and J. Rivington, London.

Popper, K.R. (1934). *Logik der Forschung*, Vienna. (English translation: *The Logic of Scientific Discovery*, Hutchinson, London, 1959.)

Popper, K.R. (1963). *Conjectures and Refutations*, Routledge and Kegan Paul, London.

Riddiford, A., and Penny, D. (1984). The scientific status of modern evolutionary theory. In J.W. Pollard (ed.), *Evolutionary Theory: Paths into the Future*, Wiley, London, pp. 1–38.

Rubenstein, D.I. (1982). Risk, uncertainty and evolutionary strategies. In *Current Problems in Sociobiology* (eds King's College Sociobiology Group). Cambridge University Press, Cambridge, pp. 91–111.

Ruse, M. (1980). Neo-Darwinism, *Theoria to Theory*, **14**, 17–25.

Ruse, M. (1981). Darwin's theory: an exercise in science, *New Scientist*, **90**, 828–30.

Ruse, M. (1982). *Darwinism Defended*, Addison-Wesley, Reading.

Sargent, T.D. (1974). The evolution of melanism by Bernard Kettlewell, *J. Lepid. Soc.*, **28**, 176–8.

Saunders, P.T. (1984). Development and evolution. In M.-W. Ho and P.T. Saunders (eds), *Beyond Neo-Darwinism*, Academic Press, London, pp. 241–63.

Saunders, P.T., and Ho, M.-W. (1982). Is neo-Darwinism falsifiable? — and does it matter? *Nature and System*, **4**, 179–96.

Scriven, M. (1959). Explanation and prediction in evolutionary theory, *Science*, **130**, 477–82.

Slobodkin, L.B. (1964). The strategy of evolution, *Amer. Scientist*, **52**, 342–57.

Sober, E. (1984). *The Nature of Selection*, MIT Press, Cambridge, Mass.

Steele, E.J. (1979). *Somatic Selection and Adaptive Evolution*, Williams & Wallace, Toronto.

Temin, H.M. (1974). On the origin of RNA tumour viruses. *Ann. Rev. Ecol. and Systematics*, **8**, 155–77.

Trivers, R.L. (1974). Parent–offspring conflict. *Amer. Zool.*, **14**, 249–64.

Turner, J.R.G. (1983). The gradualist–saltationist schism. In M. Grene (ed.), *Dimensions of Darwinism*, Cambridge University Press, Cambridge, pp. 129–69.

Waddington, C.H. (1969). Paradigm for an evolutionary process. In C.H. Waddington (ed.), *Towards a Theoretical Biology 2: Sketches*, Edinburgh University Press, Edinburgh.

Weinrich, J.D. (1977). Human sociobiology: pair bonding and resource predictability (effects of social class and race). *Behav. Ecol. Sociobiol.*, **2**, 91–118.

Williams, M.B. (1973). Falsifiable predictions of evolutionary theory. *Phil. Sci.*, **40**, 518–37.

Wilson, E.O. (1975). *Sociobiology*, Harvard University Press, Cambridge, Mass.

Evolutionary Processes and Metaphors
Edited by M.-W. Ho and S.W. Fox

CHAPTER 15

Imitating selection

PHILIP KITCHER

ABSTRACT

I consider the most persuasive argument for human sociobiology, the reasoning that begins from the claim that people are products of the evolutionary process and concludes that our social behaviour and social institutions can be understood by tracing a history of modifications that have been guided primarily by natural selection. In response, I argue that the explanation of human social behaviour and human social institutions requires the development of a multi-levelled theory, in which autonomous social and psychological variables play important roles. I attempt to indicate some of the complexity that must be introduced, and to forestall dangerous oversimplifications that will lead to new versions of naive adaptationism, and I conclude by trying to show how parts of human history may imitate selection.

Consider the following argument. Human beings, like other animals, are products of the evolutionary process. Human behaviour is simply one aspect of the human phenotype. Hence patterns of human behaviour have evolutionary explanations. Human societies are constituted by individual human beings and by their individual forms of behaviour. Hence there are evolutionary explanations of human social institutions. Since natural selection is the dominant force in the evolutionary process, both the individual patterns of behaviour of individual human beings and human social institutions are shaped by natural selection. We can thus explain the ways in which people behave and the kinds of social institutions that we find in human societies by identifying the selective advantages that those forms of behaviour and those social institutions have conferred on the people who engage or participate in them.

Some contemporary scholars accept this argument, and even view it as the key that will open up genuine social *science*. For present purposes, I want to detach the basic line of reasoning from particular suggestions that often accompany it. Sociobiologists frequently continue by assuming that the construction of an evolutionary explanation for some facet of human social behaviour shows that our genes constrain our culture (see, for example, Wilson, 1975, 1978; Barash, 1979; Lumsden and Wilson, 1981, 1983). Sometimes they add on the supposition that the identification of selective advantage is to be achieved by particular kinds of optimality analysis (Trivers, 1971, 1974, 1984; Daly and Wilson, 1978). Finally, of course, there is a multitude of particular claims about the evolutionary explanations of specific items of human behaviour — appeals to inclusive fitness calculations that are alleged to account for xenophobia and the persistence of the avunculate under certain conditions, models of mate desertion that are touted as explaining male and female stereotypes, suggestions that human resistance to incest has been shaped by the costs of inbreeding, and so forth (Barash, 1979; Wilson, 1975, 1978; Alexander, 1979; Kurland, 1979; Dawkins, 1976; Daly and Wilson, 1978; van den Berghe, 1983; Shepher, 1983; Lumsden and Wilson, 1981). I have criticized these additions to the basic argument elsewhere (Kitcher, 1985, 1986a). I shall not repeat those criticisms here.

There are three obvious lines of objection to the initial argument. The first criticizes the identification of historical explanations of human behaviour with evolutionary explanations (Bock, 1980). The second protests against the atomization of human behaviour, or, for that matter, the atomization of phenotypes generally (Gould and Lewontin, 1979). The third questions the identification of human social institutions with patterns of individual behaviour (Kitcher, 1985 Chapter 10, 1986b; this is a simple transference to the present context of well-known ideas in the debate about methodological individualism in the social sciences). I shall consider each in turn.

It should be uncontroversial that there are historical explanations for patterns of individual human behaviour and for human social institutions. But historians and anthropologists may legitimately doubt whether the actions and attitudes of human beings towards one another in situations of conflict — whether they concern an axe fight among Yanomamo Indians (Chagnon and Bugos, 1979) or a dispute over zoning in an American suburb — are to be understood by recognizing an adaptive disposition to behaviour that is triggered by the prevailing ecological conditions. *One* way to articulate the point is to propose that the force of selection can be countered (or reinforced) by forces involving systems of cultural transmission, so that the initial argument errs through failing to note all the historical forces that have combined to produce the behaviour which is to be explained (Durham, 1979; Lumsden and Wilson, 1981; Boyd and Richerson, 1985; there are important differences among these authors, and, in particular, Lumsden and Wilson seem only to introduce cultural transmission in order to minimize its importance; see Kitcher 1985 Chapter 10, 1986b).

Such modifications of sociobiology contrast with another proposal. According to this idea, the forms of human behaviour are not to be understood by appreciating their coincidence with evolutionary expectations, whether such evolutionary explanations involve one or many systems of transmission, cultural 'selection' as well as natural selection. Instead, the evolutionary analysis of the behaviour should be preceded by a proximate analysis that is designed to lay bare the details of ontogeny. Biological forces shape *something*, and that something gives rise, through development within the context of the prevailing environment, cultural as well as purely biological, to mechanisms that underlie patterns of adult behaviour.

This approach can be motivated by considering the second objection to our original argument. We cannot simply pick out a form of behaviour that happens to strike our fancy and propose that that piece of behaviour is an adaptation to some environment (past or present). Two reasons underlie the contention. First, the mechanism that generates the behaviour may well have further behavioural manifestations, and, if anything is to be counted as adaptive, then it ought to be the entire spectrum of behavioural effects, not any of the individual items that compose that spectrum. The second point is analogous. The process of development that gives rise to the behavioural mechanism in question may only be modifiable at the cost of generating alternatives to other behavioural mechanisms, so that what has been shaped by selection is not a single mechanism but an entire pattern of development involving a complex of behavioural mechanisms, each with its attendant spectrum of manifest behaviour. Unless these obvious possibilities can be excluded, we shall simply mislocate our explanatory project. This, I suggest, is what has gone wrong in one of the best sociobiological studies in anthropology, Dickemann's analysis of patterns of infanticide, where a fascinating descriptive account is distorted by the curious claim that selection has favoured a very specific form of behaviour — the murder of daughters by wealthy males in pre-Colonial Northern India (Dickemann, 1979; for criticism, see Kitcher, 1985 Chapter 9).

The third criticism of our original argument protests against the idea that social institutions can be treated as frequency distributions of individual human behaviour. In the version I favour, there is no invocation of mysterious social entities — the state, the proletariat, or whatever. It is conceded that a given society at a given time is constituted by its members at that time and that the facts about that society are supervenient[1] on facts about those individuals, their psychological properties, the relations among them, their relations to other individuals, past and present, and so forth. The crucial point is to appreciate that, among the structural relations that constitute the society, are relations that determine the transmission of items of information (the systems of cultural transmission studied by Boyd and Richerson, 1985, and, in more detail, by Cavalli-Sforza and Feldman, 1981) and also relations that affect the assignment of individuals within the society to different developmental environments, relations that constrain the development of adult phenotypes by lowering the probabilities that individuals will acquire certain combinations of psychological states, and perhaps others. So, for example, a society may be characterized by a particular system of educating the young in some aspects of behaviour (how to fish or whom to associate with), it may be characterized by a system in which members with a certain phenotypic property (say skin colour) have a very high probability of developing in environments in which some stimuli do not occur or in which the order of stimuli is at variance with the order in the rest of the population, or it may have a system of distributing benefits that makes certain options appear unobtainable for some subgroup of the population. *Perhaps* it will be possible to provide an analysis of such institutions in terms that are acceptable to methodological individualists, but until we have some idea of how the reductions might proceed, it would seem more responsible to treat social institutions as autonomous variables rather than ignoring them.

Even if all these objections to the opening argument are accepted, as I believe that they should be, it will remain possible for champions of that argument to insist on its fundamental correctness. After all, the criticisms work by assembling lists of potentially disruptive factors, and if these can be dismissed as rare perturbations of a general trend, then the identification of selective pressures may be viewed as a useful, if not infallible, guide to understanding patterns of human behaviour and human social institutions.[2] There are obviously important empirical issues here. The accuracy of the method of seeking the selective value of human behaviour and social institutions cannot be determined simply through conceptual analysis. But I think that considerable conceptual clarification needs to be done if we are to pose the empirical questions precisely. Moreover, though the separate objections I have offered tell against the argument with which we began, until we have a general picture of how various kinds of biological, cultural and developmental forces interact in the production of behaviour and institutions, the proponent of the 'selective method' may legitimately charge that there is no alternative strategy for providing historical explanations of behaviour and society.[3]

Thus we arrive at my main project. How can we use the separate insights I have claimed to be scattered among the three objections to construct an integrated account of the historical modifications that may occur in the society and behaviour of an organism as complicated as *Homo sapiens*? One source of inspiration for any such account is population genetics, and I shall take the crucial task to be that of providing an analogue for the ideal life cycle that Lewontin presents in his lucid introduction to the main problems of contemporary population genetics (Lewontin, 1974 Chapter 1).[4] An adequate sketch of the process through which human social behaviour is modified should proceed in three stages: (1) a descriptive account of the kinds of variables that can be expected to operate in a single generation; (2) a life-cycle schema, of the form provided by Lewontin; (3) a mathematical specification of the course of modification of abstractly characterized populations under the influence of specified combinations of the potential forces. I shall pursue (1) and (2), attempting to show how various unresolved empirical and conceptual issues arise at various stages of the project and how extant approaches to 'cultural evolution' or 'dual inheritance' abstract from particular aspects of the basic model. I claim that these simplifications cannot be justified without begging some of the outstanding questions.

BEHAVIOUR AND ITS MECHANICS

When we consider the adult behavioural phenotype, we may focus either on the patterns of overt behaviour or on the underlying mechanisms. It is natural to think of the activities of an organism as resulting from dispositions that have been set by development and which are triggered by a stimulus. There are many activities of non-human organisms and some behavioural reactions in humans that we can quite naturally treat in this way (think of the knee's jerk under the doctor's hammer). But in other instances, the idea of a fixed disposition that is both the result of pre-adult development and the proximate condition of the piece of behaviour seems quite inappropriate. Instead we attempt to think in terms of a more serious psychology — perhaps by postulating a rich system of inner states that stand in causal relations to one another, to stimuli and to items of behaviour. I propose to think of the mechanics of behaviour in terms of a system of inner states. The first step in any explanation of an individual action is to trace it to the combination of environmental trigger and psychological condition of the subject. If our goal is to explain general regularities in the behaviour of many agents, then we shall expect an explanation in terms of common facets of the environment and a common structure in the system of inner states.

Ultimately, psychology should provide us with the conceptual framework for describing the system of inner states, and the ultimate taxonomy may be very different from what we now envisage. However, for the sake of concreteness, I shall operate with a very simple model of the springs of action, assuming that there are two kinds of inner states, beliefs and desires, and that the first level of mechanistic explanation of an action consists in identifying the belief and the desire that jointly caused the action (typically a belief that the action would bring about the goal identified in the desire).[5] Let us now borrow some useful terminology from contemporary psychology and say that a state is cognitively penetrable if input from the environment through our sensory apparatus can lead to the modification of the content of that state (or, if you like, to the replacement of that state by a state of the same genus but with different content). Otherwise, the state will be said to be cognitively impenetrable. Within the class of cognitively penetrable states it will be valuable to distinguish channels of penetrability, corresponding to the kinds of stimuli that can bring about modifications (or replacements) of the states in question. In particular, I shall want to distinguish those states that can be affected by the utterances of others and, further, to consider those that can be modified through the utterances of people who are perceived as having a particular phenotypic property.[6]

Let us now idealize the life cycle by dividing the life of an individual into two distinct phases. The *developmental phase* covers the passage from zygote to adult, and I assume that this phase culminates in capacities for forming and modifying beliefs and desires. The capacities thereafter remain fixed to the end of the subject's life (throughout a period I'll call the *adult learning phase*). The second stage in the explanation of the action under study is the tracing of the causal history of the belief and desire proximate to the action back to the state of the subject at the end of the developmental phase. If the states are cognitively impenetrable, then I shall assume that the appropriate form of explanation is that those states were induced in the subject by the end of the developmental phase and that they persist simply because the subject has survived without being seriously debilitated. If the states are cognitively penetrable, then I shall suppose that we explain their presence by delineating a tree-like structure (probably extremely complex) whose nodes include states that were present at the end of the developmental phase and stimuli that impinge on the subject between the end of the developmental phase and the time of the action (a period that is an initial segment of the 'adult-learning phase').[7]

Three main categories of capacities affect state modification during the adult-learning phase. The first comprises those capacities that govern the modification of states in response to the actions of others which, if successful, transmit information. Among these capacities would be the ability to respond to verbal instruction, the ability to learn from the demonstrations given by others, perhaps the ability to learn from reading. The second consists of the capacities to modify states in the presence of other external stimuli. Perceptual competence obviously belongs here. Finally, there are capacities for modifying states in response to other internal states. One example would be the ability to adjust the total system so as to produce whatever coherence is necessary for its functioning. Others might be various forms of creative ability which generate novel beliefs or desires.

DEVELOPMENT

We can think of the process of development as one in which the zygote interacts with a sequence of environments. The sequence, together with the genotype, determines the form of the adult phenotype.[8] Because interactions with other organisms in the population, including organisms that are developing at the same time, may have important effects on the development of any individual, it should not be hard to see that any attempt to treat the ontogeny of any individual as determined by its genotype and a set of initial environmental conditions will be extremely difficult — for there may be bifurcations in the developmental pathway which are very sensitive to the behaviour of surrounding organisms. Hence I suggest that we take relatively gross features of the subject's initial environment — the background social institutions, the parental phenotypes, the position in birth order, the distribution of phenotypes in the surrounding population, and possibly others — as specifying a genotype-dependent probability distribution over a class of developmental sequences that are viewed as available to the subject.

The process of development culminates in the acquisition of those beliefs and desires with which the subject enters the adult-learning phase, together with the capacities that will guide the modification of the system of inner states during that phase. For each state or capacity in which we are interested, we can identify the invariant classes among the developmental sequences, that is, the classes of sequences that produce the same outcome. If there is a single invariant class for a state or capacity, then that state or capacity is invariant under development, and I shall say that it is *genetically determined*. Note that it is not necessary that a genetically determined state (in this sense) be cognitively impenetrable (or, more generally, immune to modification during the adult-learning phase), nor need we conclude that cognitively impenetrable states are genetically determined. Finally, even if a state is genetically determined, there may be differences in ontogeny in different individuals (see Bateson 1982 on equifinality).

There may be interesting regularities in the composition of the invariant classes. For example, a state or capacity may come to be fixed in only those individuals who have encountered environments with some specific combination of the variables that are specified in the initial probability distribution. I shall assume that such regularities can indeed be found — that the process of development is governed either by probabilistic or by universal laws that relate the outcome of ontogeny to genotypes and to such gross variables as social institutions, parental phenotype, character of the asocial environment, and so forth. Then the second stage in the explanation of the presence of a pattern of behaviour will consist in the subsumption of the capacities and beliefs identified at the end of the first stage under such laws of development.

GENETICS

At last we have reached a point at which we can straightforwardly consider the action of evolutionary forces. Waiving any qualms we might have about the hegemony of selection, let us suppose that we are to provide an explanation in terms of natural selection for the presence of the genotypes identified at the end of the second stage. Unfortunately, we are not at the end of our complications.

How do we trace the evolution of the genetic basis that underlies a particular item of behaviour? First, we must identify the evolutionary genetic unit that is the basis for the behaviour. Suppose that we begin with a single locus such that allelic substitutions at that locus modify the probability distribution governing the development of states and capacities that are causally relevant to the behaviour. Given the prevalence of pleiotropy, we must consider the ways in which the substitution of an alternative allele affects the probability distributions that are causally relevant to other phenotypic traits. Now we must also consider the genetic bases of these other traits. It is obvious that the process iterates — although the iteration may terminate in some combination of loci and phenotypic traits which is stable in the sense that the set of loci comprises all and only the loci that are relevant to exactly those phenotypic characteristics. From the perspective of selection the set of traits acts as a unit which governs the changes in gene frequency that occur at the corresponding loci.[9] It is easy to see that considerations of linkage and epistasis introduce further complications into the picture.

THE FORCES OF CULTURAL CHANGE

From the perspective of the argument with which we began, natural selection is the only (or, at least, the dominant) force in the shaping of human behaviour. More enlightened scholars (Durham, Richerson and Boyd) add forces that derive from cultural transmission. But, both in the initial picture and in the refinement of it, phenotypes are taken to be transparent both to genotypes and to the action of the cultural forces. I want to suggest that the picture I have outlined in previous sections enables us to discern the possibility that forms of human behaviour may be the product either of a more inclusive set of forces or of indirect actions of forces that have been previously recognized.

Start with the biological forces. Let us continue to assume that the dominant biological force is natural selection. Then, I claim, *even in the absence of any action from cultural forces*, the kind of analysis of behaviour that I have sketched is crucial to the proper application of evolutionary ideas. Suppose (implausibly) that northern Indian high-ranking males actually did maximize their inclusive fitness by murdering their infant daughters. Does this result have any bearing on the explanation of their behaviour? I claim that it may well be quite irrelevant. For an alternative story is that evolution equips all human beings with a capacity for forming desires that leads, in the context of a class of environments that includes that encountered by the relevant males, to the formation of a desire for amassing as much wealth as possible. The *capacity* not its behavioural manifestation is what should be given an evolutionary explanation. If that capacity has been shaped by selection then we may expect that, on average, it will lead to behaviour that, on average, has higher fitness than the behaviour that would have resulted from any rival capacity. But this is compatible with the existence of instances in which the behaviour reduces the fitness of individuals who engage in it, as well as with the existence of cases in which the behaviour is fitness-maximizing. Moreover, as my rudimentary exploration of parts of the aetiology of behaviour should make quite clear, the alternative story I have told is only the first step in adding realistic complexity to the description of the case. Even if we think only in terms of selection, genetics and development, there is plenty of scope for rival models.[10]

Let us now turn our attention to some cultural forces. One obvious causal factor in the production of beliefs and desires is the action of the environment. (We have here a kind of direct 'induction by the environment' that Darwinism repudiates). On the account I have given above, the subject enters the adult-learning phase equipped with a set of beliefs and desires and with capacities for modifying those beliefs and desires. If the members of a pastoral community regularly find an optimal (or near-optimal) way to make use of their land, then the correct explanation of their doing this may be that each member of the population has initially (arbitrarily) chosen a method for sowing seeds and adjusted it in successive seasons. Even though this is unlikely (because it neglects the obvious role of tradition and communication) it is surely more probable than the idea of some peculiar tropism that selection has honed to achieve the optimal method. Moreover, it is easy to imagine that the process of belief acquisition is constrained to preclude the adoption of the best method; suppose, for example, that background beliefs about the operations of the natural world — or about the supernatural — are incompatible with the belief that maximal return will be obtained by sowing in what is, in fact, the optimal way.

A second type of cultural force is the direct transmission of items of information from one member of the population to another. The common way of treating forces of this kind is to consider transmission in terms of the passing on of atoms of culture (Cavalli-Sforza and Feldman, 1981; Boyd and Richerson, 1985). I suggest that the interdependence of psychological states is better captured by supposing that a system of cultural transmission is given by specifying a function that takes as arguments an initial set of psychological states of the subject who is undergoing enculturation and a description of the roles and psychological characteristics of the cultural 'parents' or 'models'; the value of the function is the modified set of psychological states of the subject. The limiting case is that in which we simply consider a single state and relate its content to the contents of the states of the 'models'.

Boyd and Richerson (1985) have explored the ways in which the dynamics of the distribution of phenotypes is affected by the form of the function. They consider the difference between 'blending' and 'non-blending' inheritance, a certain kind of linkage among traits, correlations between reproductive success and cultural influence, and so forth. While I applaud their care, I want to emphasize that their scheme only allows for three different types of causes of adult behavioural phenotypes: (1) determination of traits by genotype in combination with the asocial environment (tacitly taken to be equivalent for all members of the population); (2) insertion of traits into the adult phenotype through cultural transmission; (3) modification of traits through a very simple learning rule, according to which the value assigned to a parameter is taken to be a compromise between the value assigned by tradition and the value that the subject acquires through observation of the relevant parts of nature. I shall now explore some ways in which the explanatory schema that I have sketched complicates the picture offered by Boyd and Richerson.

My own descriptions of the processes of individual learning and of enculturation are designed to allow for interdependencies within the system of psychological states. They yield a more general class of learning rules and a more general class of systems of cultural transmission. However, even if we set aside the concessions to holism that I have recommended, there are three other points that deserve to be noted.

First, any realistic application of the Boyd–Richerson models has to consider iterated interactions within a (biological) generation. One of the obvious features of human behavioural development is that iterated interactions among peers will enable a new psychological trait to sweep through an entire population in a single generation. Consider the changes in attitudes towards homosexuality in the past fifteen years in the populations of various Western nations. Now, quite obviously, such changes in attitudes are not independent of modifications of other psychological states, but, even if they were, the dynamics of attitudes towards homosexuals (and the associated behaviour) can hardly be understood by supposing that there is a single system of cultural transmission that is involved in a once-in-a-generation modification of adult pheontype. Because some 'cultural forces' work on a much faster time scale than natural selection, we can have a phenomenon that appears from the biological perspective as 'instant drift': a psychological trait appears in a population in the generation in which it becomes fixed! (Pulliam and Dunford, 1980 and Boyd and Richerson, 1985 make similar observations.)

Second, psychological innovation is plainly a far more powerful force than its biological analogue, mutation. Moreover, innovation is not a *classic* Darwinian force in that the rates and directions of innovation are influenced both by external contingencies (the new demands of the environment) and by the prior psychological state of the subject. (However, some recent developments in molecular genetics have indicated that mutation may not be a classic Darwinian force either; see Hunkapiller *et al.*, 1982 and Campbell, 1982.) I suggest that there may be probabilistic laws relating innovations to features of the antecedent system of psychological states, or to features of development and/or adult environments. If this is so, then the full explanation of the psychological traits of the adults should make reference to such laws, as well as to the forces of gene–bioenvironment determination, cultural transmission and adult learning.

So far I have been suggesting that the best available model of the evolution of culture is defective because it restricts the classes of systems of transmission and modes of learning, because the realistic optima result from iterations of the forces it recognizes, and because it leaves out an important cause of psychological change. Now I want to turn to what I take to be a deeper point. While the system of biological inheritance is likely to be shared by all members of the population (barring occasional occurrences of segregation distortion or meiotic drive), I believe that it is a genuine possibility that the kinds of systems of cultural transmission that have been studied in the recent literature are highly variable across natural populations. Perhaps all that is shared is a common developmental pattern that leads the members of different groups to be susceptible to very different forms of cultural transmission. (As John Dupré suggests in a forthcoming paper, it may be illuminating to consider this situation by thinking in terms of 'cultural species'.)

A system of transmission can only be effective if the individuals to be enculturated have dispositions to modify their psychological states in response to the actions and advice of others. Let us call such dispositions *sensitivities*. The idea that there is a common system of cultural transmission in force in a population (with respect to the items within a particular cultural field) thrives on the notion that all members of the population will develop common sensitivities. But is seems all too probable that this will not be the case. Assume, as seems necessary for any of the more interesting models for cultural transmission to be applicable, that some sensitivities are cognitively penetrable. Then *one* obvious way for different sensitivities to develop in different groups, is to suppose that sensitivities are tuned by individual learning. The tuning may be seen as a process in which the influential value of a phenotype for a subject is assessed by noting the past success rate of the modifications induced by advice from people with that phenotype and generalizing. Individual variation may occur here as a result of: (1) what constitutes success in the subject's environment (something which cannot, I think, be understood without considering the entire system of the subject's desires at the time at which the assessment is made); (2) what phenotypes are perceived as offering advice that promotes the relevant kinds of success; and (3) perhaps, what capacities for generalization the subject has at this developmental stage.

An obvious example of this phenomenon is the reluctance of black adolescents to accept information transmitted to them by mainstream white culture (e.g. the public school system). But I want to suggest that this is only the tip of the iceberg. *Perhaps* all human beings start with a general sensitivity to social stimuli from the people around them, but it seems to me to be likely that this sensitivity is made more particular in different ways in different social groups, and, with respect to some phenomena, it may even be critically variable from individual to individual. Thus, far from being a fixed element in the explanation of patterns of human behaviour, the polymorphic system of cultural transmission itself may stand in need of developmental explanation.

SOCIAL INSTITUTIONS

So far I have done little more than mention social institutions and offer a plea for taking them seriously. The time has now come to understand what they are, how they might affect patterns of human behaviour, and how they might result from patterns of human behaviour.

Consider two examples of social institutions — a system of racial segregation and a system for recruiting young people within a particular age range to form a standing army. Both systems correspond to group properties of particular populations at particular times. But there is nothing more to either population than the collection of individuals with all their properties, including, of course, their relational properties. So, in what sense can we talk of a 'group property' here?

My answer is that the existence of such systems can correspond to any of a heterogeneous collection of individual properties. In each particular case, we can talk of the *realization* of the group property in the population. Here the army recruitment system is realized in individual dispositions to belligerent behaviour, there it is realized in a disposition to honour tradition and a reluctant participation, elsewhere it is realized in a variety of methods of coercion. Pending some ingenious proposal, I claim that there is no general kind of state of individuals that corresponds to the group property.

Social institutions may have important effects on patterns of individual behaviour. However it is realized, a system of racial segregation plainly affects the probabilities with which individuals of different races will encounter different developmental environments. Subjects may occupy a different asocial environment, encounter stimuli in a different order, or may be influenced by an entirely different structure of cultural transmission (see last section). Thus one abstract representation of a social institution may be as a function that assigns to different phenotypic groups different combinations of values of parameters that occur in cultural and biological forces. Seen in this way, I hope that my plea for methodological holism may appear less nebulous. For, to the extent that social institutions can be so represented, they are analogous to such genuine biological factors as population structure.

To understand the origin and persistence of a social institution in a population we need to know *at least* the realization(s) of that institution in the population. So, for example, if the system of segregation originates with the activities of a small subgroup of the entire population, and if it persists despite the prevalence of individual dispositions in favour of racial equality, then we need to know precisely how the attitudes within the original population made it possible for the system to be set up, and how the states of members of subsequent generations contribute to its maintenance. It should be obvious that this approach in terms of tracing individual attitudes need not in any way seek to account for individual dispositions to xenophobia — although there may be populations in which it is precisely such dispositions that maintain systems of segregation. What is less obvious is that the atomistic approach may obscure the operation of important causal variables. Thus it may be a feature of some social institutions that the act of bringing them into being alters the subsequent history of the behaviour of the population, and thus leads to their maintenance.

Imagine two adjacent countries, Jingonia and Chauvinia. Long, long ago, an individual leader (whether a Jingonian or a Chauvinian history does not record) assembled an army to threaten its neighbour. The neighbour responded by also mustering the troops. Ever since, both countries have had standing armies and each has continued to accuse the other of harbouring threats against it. Despite the wishes of almost all the individual citizens of both countries, the expenditure of resources on defence has continued to grow. The Jingonians realize that the Chauvinians are quite like themselves, and, but for the worry that disarmament would leave them defenceless, they would be happy to live in full amity with their neighbours. The correlative attitudes are widespread among the Chauvinians. Why then does the arms race between the two nations continue?

We can easily construct a very simple model. (For more realistic discussion in the context of relations among contemporary nations, see Bueno de Mesquita, 1981). In each generation, the Jingonians have the option of increasing defence or decreasing defence, and the same is true for the Chauvinians. Let $p(n)$ be the (perceived) probability of being overwhelmed by an attack if the increase in defence spending is n, let c be the cost of being overwhelmed by attack, and suppose that $c >> n$. Then the difference to expected utility resulting from an increase in defence spending n is

$$-n-p(n)c$$

Each Jingonian (or each Jingonian who has the power to affect the decision) should choose that value of n that minimizes $n + p(n)c$. Now suppose that, in each generation, the value of $p(n)$ is affected by the actions of the Chauvinians in the preceding generation. So, in generation t, $p_t(n)$ is a monotonically increasing function of $i(t\text{-}1)$ the actual increase in investment of the Chauvinians during the immediately preceding generation. Assume further that the situation is symmetric and that the investment of the Chauvinians is determined so as to minimize $n + q(n)d$, where $q_t(n)$ is a monotonically increasing function of $j(t\text{-}1)$, the increment in defence expenditure among the Jingonians in the preceding generation. Then the entire system is governed by the recurrence relations:

$$j(\mathrm{t}) = \min\{n + p_t(n)\ \mathrm{c}\}$$
$$i(\mathrm{t}) = \min\{n + q_t(n)\ \mathrm{d}\}$$
$$p_t(n) = f(n, i(t-1)) \text{ where } \delta f/\delta n < 0,\ \delta f/\delta i > 0$$
$$q_t(n) = g(n, j(t-1)) \text{ where } \delta g/\delta n < 0,\ \delta g/\delta j > 0$$

It is not hard to see how the recurrence relations may be satisfied by functions that dictate zero investment by either side so long as neither side has previously invested, but lock both sides into an arms race once one side invests.[11]

This example illustrates two important points. First, what may be crucial to the persistence of a social institution may be a set of structural relations, independently of how they are realized at various stages of the history of a population. The story I have told is quite compatible with there being radically different realizations of the social institution at different times, so long as these are compatible with the recurrence relations. Second, the simple mathematics reinforces a point that historians and anthropologists have sometimes emphasized, to wit, that contingent facts in the history of a culture may account for its subsequent inability to escape from a particular kind of trajectory. Moreover, the model teaches us that the widespread discovery of a phenomenon across cultures should not be seen as undermining an explanation in terms of historical contingency. For, if a system will continue to occupy a state so long (and only so long) as a particular parameter is held at the value 0, and if there is a chance, however small, that it will be perturbed in each generation, then, given sufficient time, we may expect to find that the system has departed from the state. Indeed, it will not be improbable that independent disruptions have occurred in an ensemble of similar systems, so that the quirks of history have eliminated the original state in all. Whether this is a plausible account of the origins of human warfare I do not know. It is certainly a possibility — a possibility that has been entertained by humanists from Jonathan Swift to Dr Seuss and that perhaps ought to be considered by social scientists as well.

AN IDEALIZED LIFE CYCLE

I now want to pull together all the points that have been made in preceding sections by offering a picture of the ideal life cycle for an organism with the features I have noted. I shall compare this picture with three other pictures of the micro-evolution of culture.

Let us start with a collection of zygotes. These are the products of the matings of the previous generation and they grow up by passing through a developmental environment. This developmental environment should be thought of as a complex sequence of environmental influences, including many interactions with other developing and developed human beings. The probability that a zygote with a particular genotype will encounter a developmental environment of a particular type is determined by the parental phenotypes, the social institutions present in the previous generation, the distribution of phenotypic properties among members of the previous generation, and the asocial environment (possibly also by the genotype itself if there are genetic dispositions for the developing subject to find itself in particular kinds of situations). As a result of development certain phenotypes are present in the population at the beginning of the adult learning phase. The form of the adult phenotypes is then modified by interactions among the adults and by stimuli from the adult asocial environment (which may have been modified by the activities of the developing subjects). The distribution of adult phenotypes may causally affect the collection of social institutions that will play a role in fixing the probability distribution for the developmental environments for the next generation. The adults mate and produce a new set of zygotes.

The life cycle I have described is represented in Figure 1. Let us now contrast it with the simplest kind of sociobiological picture (Figure 2). According to this representation, we are to think in terms of a fixed asocial environment which, together with the organism's genotype, determines phenotype. Hence, variation in phenotype correlates in a direct and simple fashion with variation in genotype, and it is assumed that particular aspects of the phenotype picked out for study are shaped separately by genetic units.[12] Selection works to

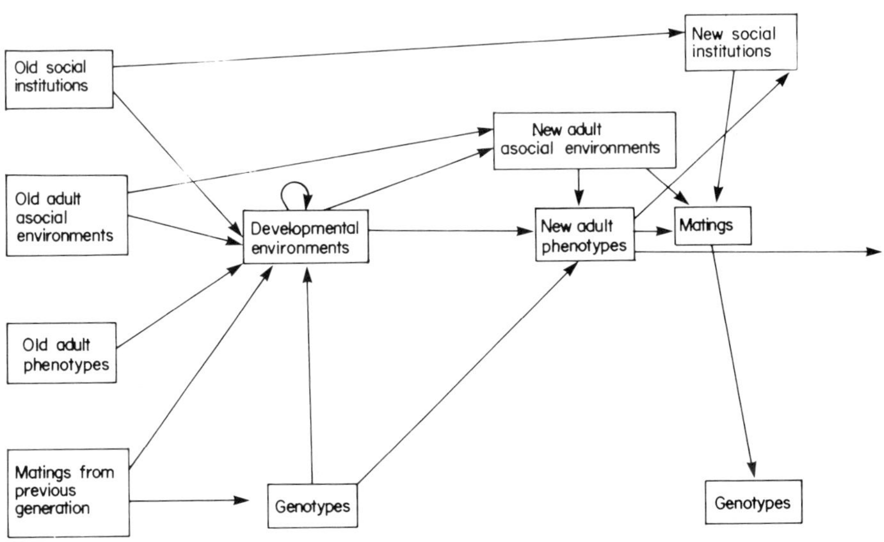

Figure 1

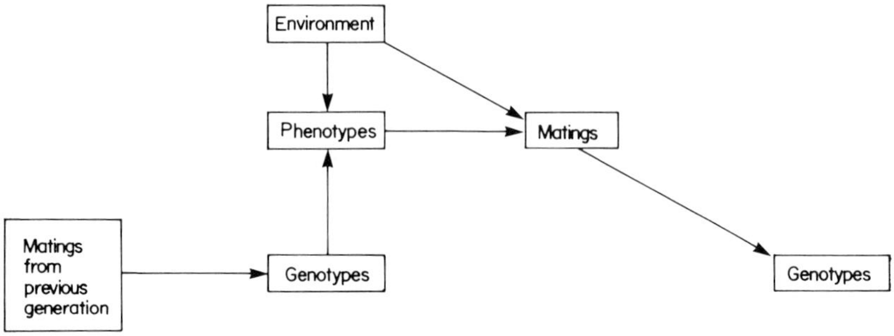

Figure 2

fix those genetic units that code for the fittest available phenotypes. Hence, each pattern of individual behaviour can be understood as the product of selection, and, in so far as social institutions are treated at all, they are simply viewed as the projection on to the entire population of the attitudes and actions of individuals. (The belligerent nation is simply a collection of individually aggressive people).

One way to refine this picture is to introduce some modification of the behavioural phenotype by the surrounding population. Thus, in Figure 3, the behavioural phenotype is no longer simply the product of the underlying genotype and the asocial environment, but may be altered through cultural transmission from the parental generation or through interactions with peers. As the work of Boyd and Richerson (especially their 1985 publication) shows, there are conditions under which the application of this picture gives strikingly different results from the application of the simple sociobiological picture.

However, as the contrast among these three pictures shows quite clearly, a great deal of potentially relevant causal detail is omitted from both Figures 2 and 3. While we continue to think of genetic units as entities that behave like single loci without even heterotic complications, while it is assumed that the phenotype can be atomized in whatever way the investigator chooses, while the developmental phase is left unexplored and the connections among psychological states in the adult learning phase remains unnoticed, while social institutions are simply viewed as projections of patterns of individual behaviour, there is a very real danger that we shall make mistakes about the causal explanations of human behaviour. The problem is that the actual causal history can easily resemble a history of selection. More exactly, given almost any piece of behaviour, it is typically possible to construct a story of selection subject to hypothetical constraints that can be reconciled with everything we know. Figures 2 and 3 can simply be imposed on the situation under study. I suggest that, before we make such impositions, we consider the cases that interest us from the perspective of *something like* Figure 1, asking the kinds of questions about genetics, development, adult psychology and social institutions that that picture brings to the fore. My *caveat* is simply intended to signal the fact that cognitive psychology may enable us to offer a more sophisticated picture.

HYPOTHESES ABOUT INCEST AVOIDANCE

According to our best available information (which may not be very good), people rarely engage in sexual relations with close relatives. Father–daughter incest is very uncommon, brother–sister incest is even less frequent, and mother–son incest is very rare indeed.* What accounts for the fact that people do not copulate with their kin?[13]

One hypothesis, originally proposed by Westermarck and revived by contemporary sociobiologists (van den Berghe, 1983; Shepher, 1983), is that human beings, like other animals, have been subject to selection for incest avoidance. We know that there are often costs of inbreeding; we know that some animals avoid mating with animals with whom they have been reared (Packer, 1979; Pusey, 1980). Allegedly there are data that reveal people as reluctant to mate with those with whom they have been reared (Shepher, 1971, 1983; Wolf and Huang, 1980). Hence, we are invited to conclude that human incest avoidance is the product of a mechanism that directs us not to copulate with those with whom we have been reared, a mechanism that has been shaped by selection because of the costs of inbreeding depression.

There is much to be said about this story. First, as several writers have noted, the models that attribute selective advantage to inbreeding avoidance need to be developed precisely (Bengtsson, 1978; May, 1979). The data from the kibbutz are by no means as clear-cut as sociobiologists like to pretend, both because of the general cultural background in the kibbutzim of the 1960s (a background that stressed purity in intersexual relations) and because of Shepher's failure to compare his hypothesis about the effect of co-rearing with an appropriate null hypothesis (Kaffman, 1977; Hartung, 1985). The discussion by Wolf and Huang is too cautious to support the sociobiological conclusions, and is typically reported by dropping the details about social background that Wolf and Huang rightly emphasize (Kitcher, 1985: 274). Finally, although the studies of animal incest avoidance are rigorous and suggestive, there is no reason to believe that the mechanisms of incest avoidance are the same in humans as in some one of the animal groups (e.g. chimpanzees), or even to think that the phenomena are the same (Spielman, Neel and Li, 1977).

In my judgement, there are three separable aspects of human incest avoidance, each of which needs explaining: (1) the fact (assuming that it is a fact) that people very rarely have sexual relations with close relatives; (2) the fact that

* Here I adopt the sociobiologists' conception of the evidence. Many clinical psychiatrists would disagree.

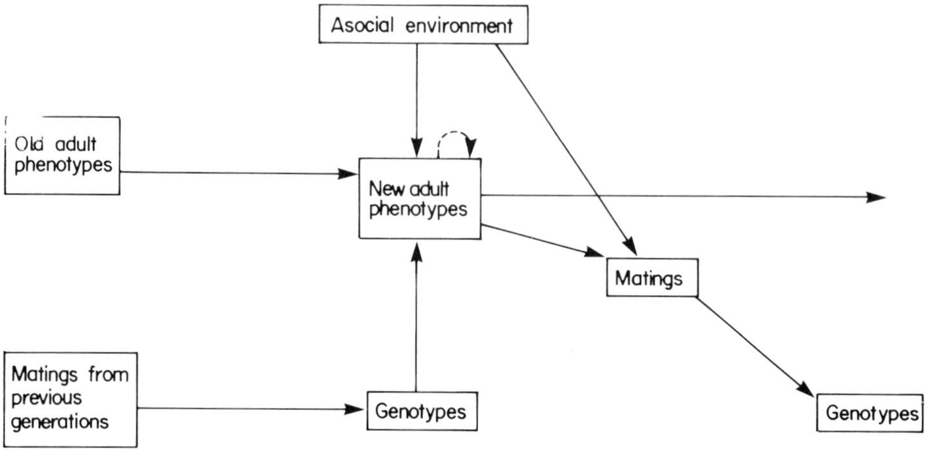

Figure 3

people typically educate their offspring so as not to have sexual relations with relatives; and (3) the fact that virtually all societies have public institutions with sanctions (often very severe) against incest offenders. At best, the sociobiological hypothesis would enable us to understand (1).[14] But it is at least possible that the explanations of (2) and (3) are also connected with the explanation of (1). *One* way to develop the connection is to propose that there has been no natural selection (in humans) for genes that dispose their bearers to avoid incest. Instead, we may propose that families vary with respect to the propensity to advise offspring to avoid incest, that children who are taught incest avoidance avoid incest, that children of families that teach incest avoidance teach incest avoidance to their children in turn, that children of families that do not teach incest avoidance do not teach incest avoidance to their children in turn, and (like the sociobiologists) that lineages in which incest is prevalent suffer the costs of inbreeding (Colwell and King, 1983; Kitcher, 1985).[15] This hypothesis will explain the increasing frequency of incest-avoidance-teaching and the correlative avoidance of incest.

The hypothesis I have just sketched is only one of many possibilities that occur to us once we look at the phenomena of human incest behaviour from the perspective that I have recommended. It would take far too long to map the entire space of possible hypotheses that the framework of Figure 1 allows, but we can appreciate the complexity of that space by using one hypothesis (chosen with relative arbitrariness) to identify some crucial issues.

Suppose, *contra* Shepher, that there is no genetically determined disposition to avoid copulation with those with whom one has been reared.[16] Whether adults (or post-pubertal adolescents) find members of the family (the nuclear family) sexually attractive depends not solely on genotype, but also on the attitudes towards incest taught in the family, the attitudes towards sex taught in the family, the opportunities for sex play with kin at various stages of development, and the attitudes expressed through reward and punishment in the surrounding society. Whether adults (or post-pubertal adolescents) actually express their sexual attitudes towards kin may also depend on their willingness to coerce others, on their fear of sanctions if the frank expression of their feelings should be rejected, and on their tolerance of risk.

It should be easy to see that taboos against incest *and attitudes towards other types of sexual behaviour* may play a large role in lowering the frequency of actual copulations among close kin. Moreover, these taboos and attitudes may have arisen and been maintained through the joint action of a complicated collection of forces. Thus, as anthropologists have frequently pointed out, those who marry outside the family group form a network of alliances that may allow them to get the things they want — and that may enable their group to persist. I claim that we cannot articulate a responsible account of how natural selection has played a role in human behaviour with respect to incest until we know the proximate and developmental determinants of the behaviour and of the attitudes, individual and social, that underlie it.

But let us be really fanciful, since the best way to oppose one fantasy may be by telling another. Suppose that our hominid ancestors passed through a stage in which the species was fragmented into many small populations. Suppose that, in true sociobiological style, each population was deeply suspicious of its neighbours, yielding high costs of migration. Selection for inbreeding tolerance weakened the force of the mechanisms that dispose our primate relatives to avoid copulating with kin. Groups with a relatively large stock of dangerous recessives went extinct, and the habitat became divided among small populations with relatively low costs of inbreeding. At this stage, we can imagine that selection for exogamy went in tandem with selection for the ability to cooperate with members in neighbouring groups. The story of the origin and persistence of marriage rules and incest taboos can thus be constructed as it is on many traditional anthropological accounts (for example, in terms of the benefits of securing alliances with outsiders), with the added complication that, as the effective population size grows, the costs of inbreeding increase.

Now, on this scenario, the adult disposition to avoid copulating with relatives is not genetically determined. Depending on how we develop the story, we can make the development of the disposition sensitive to any number of environmental factors. The fundamental point is that the sociobiological accounts do not have any monopoly on taking biology seriously. There are any number of ways to bring evolution into the explanation of human incest avoidance. We don't know which of them is correct until we understand how the ontogeny of the behaviour actually goes.

My story is just that. It is suggested by a few of the things we do know about incest behaviour. We know that copulation among close kin sometimes occurs, and that one common background condition is the opportunity for the young both to observe sexual relations among adults and to explore one another's bodies (Weinberg, 1955). Moreover, siblings sometimes attest to feelings of sexual desire for one another, and report that they have not expressed those feelings because of fear of rejection (Masters, 1963). Finally, we know that inbreeding coefficients may have been high in some small early human populations (Spielman, Neel and Li, 1977). What we need are far more extensive data on all these issues. When we start to think about constructing a model of incest avoidance according to Figure 1, then we can see how great is our ignorance of numerous potentially important factors. I recommend that we identify our ignorance and strive to eradicate it, rather than beguiling ourselves with simplistic stories.

METHODOLOGICAL CAUTIONS

I want to conclude with a brief response to an obvious objection. When thoughtful students of human social behaviour defend the application of simple evolutionary models, they argue that one must start somewhere and that one must abstract from the complexity of the phenomena. There is a sound idea here. It would be folly to introduce into our explanatory narrative a description of factors whose influence on the outcome is negligible. However, the worry about simple models is that they ignore *important* factors, and that they succeed only in showing that the actual process produces the same outcome as selection would have done.

Consider the three representations of life cycles given by Figures 1–3. Each picture corresponds to a family of models that might be applied to explain some aspect of human social behaviour. Thanks to the work of Boyd and Richerson (1985), we know that the expected trajectories of systems under Figure 3 are different from those under Figure 2. Although I am unable to supply detailed mathematical models, it seems likely that there will be similar differences between the behaviour of Figure 3 models and the behaviour of Figure 1 models. If this is correct, then the transition from Figure 2 models to Figure 3 models, and the transition from Figure 3 models to Figure 1 models is not simply a matter of cluttering up a simple story with parameters that make no real difference. As we take steps towards making the model more realistic, we subvert the conclusions obtained at the previous stage of model building.

Let us say that a model is *fragile* if there is some amendment of it that is closer to the complete description of the actual process which yields conclusions that are significantly different from those of the original model. Then the aim in building simple models is to attain maximum simplicity while avoiding fragility. I claim that neither the routine sociobiological models, nor the more sophisticated models of Boyd and Richerson succeed in avoiding fragility.[17] Because the virtue of simplicity is subservient to the avoidance of fragility, the plea for the use of Figure 2 models or Figure 3 models is unsuccessful.

Now we come to the final twist. There are occasions on which some model corresponding to each conception will yield the same outcome — an outcome moreover which accords with the phenomena as we find them. Instead of chalking up an empirical success for our favourite conception, we should remind ourselves of the purpose of our model building. We are not in the business of *predicting* the form of human behaviour. We are seeking causal explanation, and an adequate causal explanation may assemble all, and only, the causally relevant factors. The fact that Figure 2 models and Figure 3 models are fragile teaches us that the use of these models may ignore causally relevant factors. Hence our urge for simplicity may drive us in quite the wrong directions. We may overlook innumerable possible hypotheses for explaining incest avoidance — including the truth. The Ancients saved the phenomena of astronomy with elaborate hypotheses about crystal spheres. We know better. There are no crystal spheres. To those who claim to have shown how selection fashions human behaviour, we may reply that all they have done is to show how the phenomena may be saved by models of a class, some of whose members are know to be fragile. Human history may only imitate selection.

ACKNOWLEDGEMENTS

I am grateful to the participants in the Liberty Fund Conference for their comments on an earlier version of this chapter. In particular, I would like to thank Patrick Bateson for his advice, encouragement, and the inspiration of his published work on evolution and development of behaviour. Peter Richerson offered me many constructive criticisms, some — but only some — of which I have been able to address in the final version. Finally, I appreciate the kindness of John Dupré and Donald Symons, both of whom have shared with me copies of articles in progress that advance similar ideas to those I present here.

NOTES

1. A state (or property) S is said to supervene on a set of properties (or states) P_1, P_2 . . . just in case two things can be in the same state S and have quite different properties P_i, P_j but it is not possible for two things to have the same property P_k, although one is in S and the other is not.
2. Boyd and Richerson (1986) offer exactly this line of defence of their procedure. I think that this tactic may be unwise for them, since an exactly parallel argument is available to the traditional sociobiologist who wants to ignore culture entirely.
3. The lack of alternatives offers defenders of theories under attack a useful debating point — for they may demand to know how we can do better than the beleaguered theory. Trivers uses this device in response to my (1985) publication. Although the point is illegitimate, since there are occasions on which the rational way to develop a field is to look for something different from any of the alternatives which have so far been proposed, I believe that it may be easiest to perceive the problems of available approaches by outlining something better.

4. Something like a life cycle is offered in Boyd and Richerson (1985), in Lumsden and Wilson (1981) and in Kitcher (1985). I think that each of the proposals is too simple — although I would contend that Boyd, Richerson and I have done substantially better than Lumsden and Wilson.
5. I'd be surprised if this simple picture survives the construction of a serious psychology. I'd be even more surprised if the serious psychology brought aid and comfort to the approaches I want to identify as too simple.
6. There's no need to assume that the conscious is penetrable and the unconscious impenetrable. I am confident that people have a rich system of unconscious beliefs and desires, and that these are sometimes affected by stimuli.
7. The assumptions I've made here may well be false. Perhaps impenetrable states can be induced through penetrable states, and penetrable states might endure throughout the adult-learning phase.
8. 'Determines' is surely too strong. Quantum mechanical indeterminacies may affect what goes on at the macro-level. My suspicion is that there are unlikely to be many natural amplifiers, that quantum mechanics may be the only source of indeterminacy, and that, as a result, the developmental process can be construed as essentially deterministic.
9. The connections among loci may, of course, also be understood in terms of selection. One obvious and important question is whether the power of selection to separate loci is effectively unlimited. I should note that this question is logically independent of issues about the credibility of genic selectionism. In particular, what I have said in this section seems to be compatible with Dawkins's (1982) most sophisticated presentation of that thesis.
10. For example, one can suppose that the actual behaviour is a compromise between greed and natural concern for kin, or, more generally, that there are a number of human interests and motives that are relevant to what is actually done. I should also note that the capacity which we identify as the product of evolution may be quite abstract. See, for example, the lucid discussion of the possibility that human development involves a hierarchy of rules, some of which modify the effects of (lower level) rules, in Bateson (1982).
11. For a simple example, set $f(x,y) = g(x,y) = [y + 1/x - 2c^{-1/2}]$.
12. These do not have to be single loci, but we have to assume that there are no complications due to linkage or epistasis.
13. There should be extensive discussion about the definition of incest. Who are the parties and what must they do? The sociobiological story that I shall consider omits cases in which two people of the same sex engage in sexual activity, cases in which stepsiblings have sexual relations, and cases in which there is sexual behaviour short of intercourse. I shan't worry about these issues for present purposes.
14. Shepher thinks he can go further. Our repugnance at incest causes the taboo. But this is all wrong. Not all groups find incest repugnant — some, for example, think of it as funny. Moreover, there are numerous kinds of behaviour that people do not like — even find horrifying — but which we do not explicitly taboo (often because we think that nobody would be tempted to engage in them). Hence, the emergence of the taboo is not really explained.
15. Jack Hailman and Sidney Fox both raised some interesting questions about exactly how this scenario is supposed to go. I imagine that parents say and do certain things to their children that have the effect of promoting incest avoidance in any case in which incest would be possible. The training behaviour had better not be too specific, for there is an obvious possibility that it will not be heritable in familial lines in which, at some point, all the children are of the same sex. Thus we should think in terms of instructions to respect the privacy (and not to interfere with the

bodies) of other family members, which produce an avoidance of incest behaviour and the disposition to give the instructions is transmitted even in generations in which heterosexual copulation between siblings is not possible.

16. Shepher sometimes formulates his thesis so as to leave it open as to whether the disposition is genetically determined. However, his settled view, presupposed in his explanation of the origins of the taboo, and numerous other remarks (Shepher, 1983: 109) seems to be that there is no environmental input that affects the acquisition of the disposition.
17. But Boyd and Richerson have still performed a valuable service by showing us that the traditional sociobiological models are very fragile. Their detailed models alert us to general problems in applying evolutionary theory in human behaviour, even though, as I believe, they do not provide satisfactory methods for handling all the difficulties.

REFERENCES

Alexander, R. (1979). *Darwinism and Human Affairs*, University of Washington Press, Seattle.

Barash, D. (1979). *The Whisperings Within*, Penguin, London.

Bateson, P.P.G. (1982). Behavioural development and evolutionary processes. In P.P.G. Bateson (ed.), *Current Problems in Sociobiology*, Cambridge University Press, Cambridge, pp. 133–51.

Bengtsson, B. (1978). Avoiding inbreeding at what cost? *J. Theor. Biol.,* **73**, 439–44.

Bock, K. (1980). *Human Nature and History*, Columbia University Press, New York.

Boyd, R., and Richerson, P.J. (1985). *Culture and the Evolutionary Process*, University of Chicago Press, Chicago.

Boyd, R., and Richerson, P.J. (1986). Simple models of a complex world, to appear in Dupré, 1987a.

Bueno de Mesquita, B. (1981). *The War Trap*, Yale University Press, New Haven.

Campbell, J.H. (1982). Autonomy in evolution. In Milkman, 1982, pp 190–201.

Cavalli-Sforza, L., and Feldman, M. (1981). *Cultural Transmission: A Quantitative Approach*, Princeton University Press, Princeton.

Chagnon, N., and Bugos, P. (1979). Kin selection and conflict: an analysis of a Yanomamo ax fight. In Chagnon and Irons, 1979, pp. 213–38.

Chagnon, N., and Irons, W. (1979). *Evolutionary Biology and Human Social Behaviour: An Anthropological Perspective*, Duxbury, North Scituate, Mass.

Colwell, R., and King, M. (1983). Disentangling genetic and cultural influences on human behavior. In D.W. Rajecki (ed.), *Comparing Behavior: Studying Man Studying Animals*, Erlbaum, Hillsdale, NJ, pp. 227–50.

Daly, M., and Wilson, M. (1978). *Sex, Evolution and Behavior*, Duxbury, North Scituate, Mass.

Dawkins, R. (1976). *The Selfish Gene,* Oxford University Press, Oxford.

Dawkins, R. (1982). *The Extended Phenotype*, Freeman, San Francisco.

Dickemann, M. (1979). Female infanticide, reproductive strategies and social stratification: a preliminary model. In Chagnon and Irons, 1979, pp. 321–67.

Dupré, J. (1987a). *The Latest on the Best: Essays on Evolution and Optimality*, Bradford Books/MIT Press, Cambridge, Mass.

Dupré, J. (1987b). The evolution of human behaviour, to appear in Dupré, 1987a.

Durham, W. (1979). Toward a coevolutionary theory of human biology and culture. In Chagnon and Irons, 1979, 39–59.

Gould, S.J., and Lewontin, R.C. (1979). The Spandrels of San Marco and the

Panglossian Paradigm: A critique of the adaptationist programme, *Proc. Roy. Soc. London (B)*, **205**, 581–98. Reprinted in E. Sober (ed.), *Conceptual Issues in Evolutionary Biology*, MIT Press, Cambridge, Mass. 1984.

Hartung (1985). Review of Shepher 1983, *American Journal of Physical Anthropology*, **67**, 169–71.

Hunkapiller, T. *et al* (1982). The impact of modern genetics on evolutionary theory. In Milkman, 1982, pp. 164–89.

Kaffman, M. (1977). Sexual standards and behaviour of the kibbutz adolescent, *Amer. J. Orthopsychiat.*, **47**, 207–17.

Kitcher, P. (1985). *Vaulting Ambition: Sociobiology and the Quest for Human Nature*, MIT Press, Cambridge, Mass.

Kitcher, P. (1986a). Why not the best?, forthcoming in Dupré, 1987a.

Kitcher P. (1986b) Taking culture seriously, review of Boyd and Richerson 1985, *Nature,* **319**, 105–6.

Kurland, J. (1979). Paternity, mother's brother, and human sociality. In Chagnon and Irons, 1979, pp. 145–80.

Lewontin, R.C. (1974). *The Genetic Basis of Evolutionary Change*, Columbia University Press, New York.

Lumsden, C., and Wilson, E.O. (1981). *Genes, Mind, and Culture*, Harvard University Press, Cambridge, Mass.

Lumsden C., and Wilson, E.O. (1983). *Promethean Fire*, Harvard University Press, Cambridge, Mass.

Masters, R.E.L. (1963). *Patterns of Incest*, The Julian Press, New York.

May, R.M. (1979). When to be incestuous, *Nature*, **279**, 192–4.

Milkman, R. (1982). *Perspectives on Evolution*, Sinauer, Sunderland Mass.

Packer, C. (1979). Inter-troop transfer and inbreeding avoidance in *Papio anubis*, *Anim. Behav.*, **27**, 1–36.

Pulliam, R., and Dunford, J. (1980). *Programmed to Learn*, Columbia University Press, New York.

Pusey, A. (1980). Inbreeding avoidance in chimpanzees, *Anim. Behav.,* **28**, 543–52.

Shepher, J. (1971). Mate selection among second-generation kibbutz adolescents and adults: incest avoidance and negative imprinting, *Arch. Sex. Behav.*, **1**, 293–307.

Shepher, J. (1983). *Incest: A Biosocial View*, Academic Press, New York.

Spielman, R.S., Neel, J.V., and Li, F.H. (1977). Inbreeding estimation from population data: models, procedures, and implications, *Genetics*, **85**, 355–71.

Trivers, R.L. (1971). The evolution of reciprocal altruism. In T. Clutton-Brock and P. Harvey (eds), *Readings in Sociobiology*, Freeman, San Francisco, 1979, pp. 189–226.

Trivers, R. (1974). Parent-offspring conflict, ibid, 233–57.

Trivers, R.L. (1985). *Social Evolution*, Benjamin/Cummins, San Francisco.

Van den Berghe, P. (1983). Human inbreeding avoidance: culture in nature.

Weinberg, S.K. (1955). *Incest Behavior*, Aldine, Chicago.

Wilson, E.O. (1975). *Sociobiology*, Harvard Universtiy Press, Cambridge, Mass.

Wilson, E.O. (1978). *On Human Nature*, Harvard University Press, Cambridge, Mass.

Wolf, A., and Huang, C. (1980). *Marriage and Adoption in China. 1845—1945*, Stanford University Press, Stanford.

Author Index

Subject Index

402